Concrete Fracture

A Multiscale Approach

Concrete Fracture

A Multiscale Approach

Jan G.M. van Mier

CRC Press
Taylor & Francis Group
Boca Raton London New York

CRC Press is an imprint of the
Taylor & Francis Group, an **informa** business

CRC Press
Taylor & Francis Group
6000 Broken Sound Parkway NW, Suite 300
Boca Raton, FL 33487-2742

First issued in paperback 2017

Version Date: 20120622

ISBN 13: 978-1-4665-5470-2 (hbk)
ISBN 13: 978-1-138-07352-4 (pbk)

This book contains information obtained from authentic and highly regarded sources. Reasonable efforts have been made to publish reliable data and information, but the author and publisher cannot assume responsibility for the validity of all materials or the consequences of their use. The authors and publishers have attempted to trace the copyright holders of all material reproduced in this publication and apologize to copyright holders if permission to publish in this form has not been obtained. If any copyright material has not been acknowledged please write and let us know so we may rectify in any future reprint.

Library of Congress Cataloging-in-Publication Data

Mier, J. G. M. van
    Concrete fracture : a multiscale approach / Jan G.M. van Mier.
        p. cm.
    Includes bibliographical references and index.
    ISBN 978-1-4665-5470-2
    1. Concrete--Fracture. I. Title.

TA440.M487 2013
620.1'366--dc23   2012017244

'How all things hang together,
when one has the perspective from which to view them …'

JOHN BANVILLE, *THE INFINITIES*

Contents

Preface

This is a book about fracture, about concrete fracture. Since my studies at Eindhoven University of Technology I have developed an interest in how things break; eventually that knowledge helped define the limits in structural design. Perhaps it was by accident that it was all about concrete. At that time, 1978–1979, a large research effort was launched in The Netherlands, named "Concrete Mechanics." The basic idea was to develop numerical models for the analysis of reinforced concrete structures. Experiments would be done as in the past, in structural laboratories, but in addition small-scale tests were necessary to deliver the parameters needed in the new numerical tools. Most of these models were based on the finite element method, which, in those days, attracted more and more users in engineering. Needless to say that the linear version is indeed very useful provided the application goes hand in hand with an excellent intuition for structural behavior. Just using the numerical tools as a black box will not help the design of excellent structures. Every institute in The Netherlands involved in research on structural concrete collaborated in the large research program one way or the other. As a graduate student I became part of it via my thesis supervisor Professor Wim van der Vlugt. The years between 1979 and 1984 span the time of my doctoral research. With hindsight this was a very fruitful period in my research career. Basically I was left alone to do what I felt was needed.

The project was summarized in one line: "Investigate the shear-punch failure of concrete slabs; the preliminary investigation will focus on the multiaxial behavior of concrete." I started along both tracks: doing some punching experiments on slabs, and reading the literature on multiaxial concrete behavior. Finally, the failure of concrete under multiaxial compression seemed to be an interesting area, in particular the softening behavior. Only very limited experimental data were available in this field; the new computer models required full three-dimensional knowledge of plain concrete, and it was a major input parameter for solving the shear-punch problem. I got the opportunity to build a new multiaxial apparatus, which allowed for measuring the complete stress–strain behavior including the softening regime. Ironically, the most interesting result was related to the behavior of concrete under uniaxial compression. I am still grateful today that I could follow my own nose, which has helped to develop an independent position and critical mind.

Ever since my doctoral work, concrete fracture has fascinated me, and at all the places where I have worked for longer or shorter periods of time, at least part of my research focused on the fracture of concrete. Rather than having to work with numerical models based on continuum theory, it was quickly decided to follow new interesting developments from theoretical physics. I became interested when I read an article by Termonia and Meakin (1986), and after I participated in a summer school in Cargèse (France) in 1989 it was absolutely clear: a lattice model would be needed. The inspiration came from work done by, among others, Stéphane Roux and Alex Hansen. I learned a lot in discussions with my good friend, Dusan Krajcinovic, who unfortunately passed away two years ago. Dividing my time between numerical simulations and experiments was not done in those days. In Delft, at the civil engineering department where I worked, there was a strict separation between the two fields: experimentalists performed experiments and "model engineers" built models. Anyhow, the lattice model was built, in the laboratory, in a rather special way, namely as a continuing series of PhD projects, where the coherence between projects always has been my greatest concern. The first version of the lattice model for concrete was written by Erik Schlangen, followed by Adri Vervuurt, Marcel van Vliet, Eduardo Prado, Giovanna Lilliu, and, most recently in Zurich by Hau-Kit Man. Without these excellent researchers the "Delft lattice model" would never have existed, and I am grateful that it has helped considerably in elucidating many of the questions about fracture of concrete. Next to these numerical results, a constant stream of experiments yielded new insight.

I have always tried to organize the doctoral work in such a way that a student hardly could avoid being confronted with experiments. So, most of the aforementioned researchers, plus a number of others helped move the experimental work forward: M. B. Nooru-Mohamed on mixed-mode I and II fracture, Marcel van Vliet worked on size effects in tension, Jeanette Visser on fluid-driven fracture, Jan Bisschop did experiments on drying shrinkage cracking, Ahmed Elkadi on the size effect in compression, Chunxia Shi worked on difficult experiments related to crack stability, Pavel Trtik was active with the microtensile testing, and luckily, because The Netherlands is a small country, I could be involved in the continuation of the multiaxial work that I started at Eindhoven; specifically the collaboration with Rene Vonk, Erik van Geel, Henk Fijneman, and Harry Rutten has been quite memorable. A constant stream of foreign visitors to the lab has helped improve on the results, both the experiments and the simulations. Memorable also is the excellent atmosphere in the Stevin laboratory when my group was shaped between 1988 and 1995, and later on, between 1999 and 2002 when I built a new microlab in collaboration with the computational mechanics group, which in the beginning was led by Rene de Borst, and, after his departure by Johan Blaauwendraad. After my transition to ETH-Zurich in 2002, research continued on the fracture of cement and concrete, but at a slower pace because some of my energy deviated toward the development of new concretes.

The combination of numerical simulations alongside physical experiments is considered particularly fruitful. Unfortunately many believe just doing the modeling will do. It is a recurring theme in this book: just sitting behind a computer will not teach you how a material such as concrete behaves. Hands-on experience is crucial, and seems increasingly more important in times where the entire world is hiding behind a laptop or a tablet or is fumbling with a smart phone. This book presents the state of the art in concrete fracture, and in the last decade we have witnessed a gradual shift of interest toward fracture studies at increasingly smaller size/scales. There are enormous experimental challenges to address in the near future; some of them are mentioned in this book. It is hoped that there will be new initiatives to overcome the difficulties that are encountered when performing fracture experiments on cement at micro-size/scale and smaller, not forgetting that the obtained results should fit in the larger picture of the material science of concrete, particularly the design of new high-performance concrete materials which can be put to good use in the development of efficient and durable structures.

I am grateful to all persons mentioned: they were important in arriving at the results presented in this book. Foremost I am grateful to my wife Ria: she lets me fumble on with this fracture stuff, even when I should actually do something more important.

Jan van Mier
Bergschenhoek, The Netherlands

The Author

Jan G. M. van Mier received his engineering and PhD degrees from Eindhoven University of Technology in 1978 and 1984, respectively. After a post-doc year at the University of Colorado in Boulder he moved to Delft University of Technology. As an associate professor at the Stevin Laboratory a first start was made with the development of the lattice model, and numerous experiments were carried out elucidating the fracture of concrete under a variety of conditions. In 1999 the author was appointed "Antonie van Leeuwenhoek" professor at TU Delft, based on excellence in research, and developed and built the new microlab to immerse in fracture studies at smaller size/scales than before. An offer from ETH Zurich in 2002 was hard to resist, and the work continued there as full professor and director of the Institute for Building Materials. In 2010 the author became president of the International Association for Fracture Mechanics of Concrete and Concrete Structures (IA-FraMCoS), with a mandate till 2013. The author has contributed to numerous publications in journals and conference proceedings and has been active in several national and international research committees and editorial boards. Moreover, he published a textbook on fracture mechanics of concrete in 1997, and is editor of several conference proceedings and special issues of journals. In 1987 the author was awarded the RILEM Robert l'Hermite medal, in 2005 was the corecipient of the 33rd Japanese Cement Association Paper Prize, and is a fellow of RILEM and FraMCoS. The author is married and lives with his family in Bergschenhoek, The Netherlands.

List of Notations and Abbreviations

Abbreviations

AE	acoustic emission
BC	boundary conditions
CBM	crack-band model
CH	calcium hydroxide
CMOD	crack-mouth opening displacement
CMSD	crack-mouth sliding displacement
CSH	calcium silica hydrate
CT	computed tomography
DEN	double-edge notched
DIC	digital image correlation
ESEM	environmental scanning electron microscope
FCM	fictitious crack model
FIB	focused ion beam
FRC	fiber reinforced concrete
hcp	hexagonal close packing; also hardened cement paste
HD-CSH	high-density CSH
ITZ	interfacial transition zone
LD-CSH	low-density CSH
LE	linear elasticity
LEFM	linear elastic fracture mechanics
MIP	mercury intrusion porosimetry
NLFM	nonlinear fracture mechanics
rp	regular packing
RVE	representative volume element
SEM	scanning electron microscope
SEN	single-edge notched
TEM	transmission electron microscope

Roman Symbols

a	semi-crack length
A	area; randomness factor (random lattice)
A/s	randomness (lattice model)
b	beam width
c_b	crack-band width (CBM)
c_f	fracture process zone length
d	particle (aggregate) size; diameter
d_{max}	maximum particle (aggregate) size
D	structure diameter or depth; shear force
E	Young's modulus; modulus of elasticity (LE)
f_c	uniaxial compressive strength
f_t	uniaxial tensile strength
$f(a)$	probability density function of semi-crack length a
$f_{ij}(\theta)$	geometrical function or weight function (LEFM)
F	force
G	shear modulus (LE)
G_c	critical energy release rate (LEFM)
G_f	specific fracture energy (FCM)
h	beam height
H	structure height
I	moment of inertia (LE)
k	spring (normal) stiffness
K	bulk modulus (LE)
K_I	mode I (opening mode) stress intensity factor (LEFM)
K_{II}	mode II (in-plane shear mode) stress intensity factor (LEFM)
K_{III}	mode III (out-of-plane shear mode) stress intensity factor (LEFM)
K_c	critical stress intensity factor
l	beam length
l_{avg}	average beam length (random lattice)
l_b	buckling length
l_c	characteristic length (MFSL)
l_{ch}	characteristic length (FCM)
l_{meas}	measuring length

L structure length

m Weibull modulus

M bending moment

M_t torsion moment

N normal force

p_d capillary pressure

P load

P_f failure probability

P_k particle volume

r radius or interatomic separation distance

r,θ,z polar coordinate system

R gas constant

s cell size (random lattice)

S beam span

t thickness; size of plastic crack-tip zone (NLFM)

T temperature

u displacement (x-direction)

U_E elastic energy (LEFM)

U_s surface energy (LEFM)

v displacement (y-direction)

V volume

V_a aggregate volume fraction

V_{fib} fiber factor

V_m matrix volume fraction

w crack opening (FCM)

w_c crack opening at zero bridging stress

W section modulus (LE); structure width (LEFM)

x,y,z Cartesian coordinate system

Y geometrical function (LEFM)

Matrix Symbols

C   combination matrix

k   force vector

S   stiffness matrix
T   transformation matrix
v   displacement vector

Greek Symbols

α parameter in lattice model; stress ratio σ_2/σ_1

β stress ratio σ_3/σ_1

δ displacement; crack opening

δ_t tensile crack opening

δ_s crack shear displacement

ε strain

ε_i principal strains ($i = 1,2,3$)

γ surface energy

η f_c/f_t

φ rotation (z-direction); or density

σ stress

σ_i principal stresses ($i = 1,2,3$)

σ_c compressive stress

σ_E Euler buckling stress

σ_N nominal stress

σ_s shear stress

σ_t tensile stress

σ_y yield stress

τ shear stress

τ_b frictional stresses at specimen-loading platen interface

μ friction coefficient

ν Poisson's ratio (LE)

Size/Scale Levels for Concrete

n nano-level; 10^{-9} m

μ   micro-level; 10^{-6} m
m   meso-/intermediate-level; 10^{-3} m
M   macro-level/continuum

1

Introduction—Why a New Book on Fracture of Concrete?

Fracture mechanics of concrete remains an important topic of research. The material and structures made of concrete are prone to cracking caused by a variety of reasons. Mechanical loading is one important cause, but other physical loadings such as differential drying, temperature gradients, and chemical attack may also lead to severe cracking and deterioration of structures. The inherent reduction of serviceability time is not acceptable, especially in a world where resources are becoming scarce and global warming threatens the living conditions of all creatures on our home planet. During the production of cement a large emission of greenhouse gases is inevitable; roughly one metric ton of CO_2 is emitted for the production of one metric ton of cement, reason enough to limit the use of cement as a building material, not only through a direct decrease of the amount of cement used, but also by providing longer service time for new and existing structures. Both for the development of new high-performance cement composites, as for the improvement of the durability of concrete structures, fracture mechanics plays a key role.

Concrete cracking is primarily caused by the material's low tensile resistance. For ordinary concrete with a compressive strength of around 40 MPa, we may expect a tensile strength of no more than 10% of the compressive, thus approximately 4 MPa is the maximum. The imbalance between tensile and compressive strength is traditionally taken care of through the use of steel reinforcement placed at those locations where the highest tensile stresses appear in the considered structure. The main reason the joint venture between steel and concrete is successful is their almost identical thermal expansion coefficient. Moreover, the bond between steel and concrete is sufficiently good to allow for short anchorage length of the rebar. In principle the combination is ideal, yet problems arise owing to the relatively high porosity and often good permeability of the concrete cover of the steel reinforcement. Bad workmanship may do the rest, and given sufficient amounts of water and oxygen the rebar may start to corrode once the protective oxide layer is passivated, either through the ingress of chlorides, or through carbonation of the cover concrete. Cracks may facilitate the ingress of corrosive media; therefore there is interest in reducing crack widths to a minimum, or to prevent cracking altogether. The swelling of a corroding rebar may lead to substantial increase of cracking, thereby accelerating the corrosion process, depending on the actual climatic conditions.

In summary, there is an interest in developing high-tensile strength concrete, which resists higher tensile stresses, and thereby reduces the probability for crack nucleation and growth. High-performance concretes are usually much denser than the aforementioned average quality, and as an additional advantage there is reduced permeability and the protective cover may perform much better leading to substantial improvement of the service lifetime of the structure.

Of course there are many influence factors when the durability and service lifetime of concrete structures are considered, but because cracking is the major deterioration mechanism, it seems quite appropriate to give some attention to fracture processes in concrete and concrete structures, investigate causes for cracking, and to develop models that can "predict" the whereabouts of cracks during any stage of the lifetime of a concrete structure. Since the beginning of the 1970s there has been a steadily increasing interest in developing robust and reliable models that can simulate concrete cracking; much of the development is certainly caused by the introduction of the finite element method and other numerical simulation techniques in science and engineering.

One expression that has been heard a lot over the past decades is that improvements in the said computer models may eventually reduce the role of experiments, if not make experiments completely obsolete. Experiments are considered expensive, time consuming and cumbersome, only limited structural conditions can be tested, some loading situations are so complex that testing is considered impossible, and, after all, it is much worldlier to spend one's time behind a computer screen in a neat office, rather than in a dusty concrete laboratory. Unfortunately practice has shown that experiments are still very useful, if not always leading to the development of new and improved models, even though often it is claimed that an ingenious theoretical insight has led to the new model, which was subsequently confirmed by experiment. It is quite clear that the statement is wrong: a theory cannot be confirmed by a single experiment; at best the experiment may falsify the claims made by the modelers. In short, experiment and theory must be considered in unison. Both theory and experiment play their role in the progress of science and engineering. Experiments can be used to falsify existing theoretical points of view, and may subsequently lead to new insights that help to develop improved theories. The interaction is crucial; of extreme importance is that in the highly specialized world we live in, those doing the experiments and those involved in theory development speak the same language, and are prepared to make the necessary sacrifices of rejecting developments that subsequently did not show the expected results, or were somehow in contradiction to "accepted" wisdom.

Studies related to the fracture of materials and structures go back a long time in history. The oldest practice was purely empirical, and many would agree that the earliest theory development dates back to 1920–1921 when Griffith presented his theory of the fracture of glass. Rock and metal fracture have a lengthy history as well, but the fracture of cement and concrete is a relative newcomer. Surprisingly, every time attention shifts toward new materials, new additions or new versions of the classical theories emerge. Seldom are these new developments applicable for the new class of material only, but attempts are made to devise modifications and apply the new insights to a wider class of materials and structures. It remains to be seen whether this is not just a waste of time. With the increased attention to fracture of ductile materials the plastic crack-tip model was developed, which subsequently served as a basis of cohesive crack models widely used nowadays for fracture of cement and concrete, and some advocate using cohesive models for other materials including rocks, ceramics, and polymers as well.

One important change in the cross-over from the plastic crack-tip model to the cohesive crack model used for concrete is that a local fracture criterion suddenly was used to describe global specimen/structure behavior. The size of the plastic zone in metal plasticity, or its equivalent, the cohesive zone in concrete suddenly changed from a very small, local zone, to an area encompassing the entire material-specimen/structure, in particular in common laboratory-scale specimens. The expected result is of course a tremendous dependency of the results on structure/specimen size and boundary conditions. These latter aspects were hitherto successfully incorporated in the linear elastic fracture mechanics theory. Obviously the fracture process in concrete and related materials such as rock and ceramics is more complicated or we simply lack the right insight. Just transferring an existing theory (plastic crack-tip model) to cement and concrete leads to previously unsuspected complications, which need to be resolved before a new truly effective theory becomes available.

Because the cohesive zone in concrete is often considered to be larger than the dimensions of the considered specimen/structure, and the nearby boundaries influence the fracture process, a better approach seems one where everything is considered to be a structure: from the so-called material experiments all the way up to the real structural applications. Size/scale is the only difference. A common material experiment for concrete uses specimens on the order of five times the maximum aggregate size, that is, 100–150 mm. This is considered the minimum specimen size where the material volume can be considered as a representative volume element and continuum ideas can still be used. Forces can be averaged over a surface ($\sigma = F/A$), and strains can be calculated as the relative elongation over the specimen length ($\varepsilon = \Delta \ell/\ell$). As soon as a large crack of a large cohesive zone develops (i.e., larger than half of the specimen size), averaging seems to be less obvious, and perhaps another approach is needed. At the scale of laboratory experiments the scale of the material structure may directly affect the fracture process, as in the case of concrete the particle structure of the material. For larger structures, it may seem that the behavior can be expressed in the usual continuum variables stress and strain, but the final stages of the fracture process rely on the carrying capacity of a compression zone often no more than a few centimeters wide, for example. Thus, the development of a macroscopic theory, based on the aforementioned continuum state variables, or an alternative fracture theory based on energy considerations or stress-intensity factors may benefit from structural analyses at a size or scale just smaller than the size or scale of the real structure. This means that for the usual material size or scale specimen we descend one step on the dimensional ladder and consider the material specimen as a small-scale structure where the aggregate structure of concrete affects the fracture process directly. Thus, the aggregate structure must be included in the model. This may be seen as an enormous complication, yet the widespread use of the finite element method allows analyzing the behavior of structures with complex geometry.

So, this is one of the ideas behind this book. Use so-called meso-level analyses of laboratory-scale specimens for a better understanding of the fracture process and the fracture mechanisms, and apply the obtained knowledge to criticize existing macroscopic fracture models and propose new ways that may lead to improved models. Thus, analyze a laboratory-scale specimen as if it were a full-scale structure and consider the internal material structure as well as the exact boundary conditions of the experiments (restraint, free, or suppressed rotations and translations of supports and loading-point). To help simplify the analysis, the layout of an experiment should at points be improved, for instance, applying a point-load at the exact desired location, or providing a hinge that simply defines the rotation point uniquely with respect to the structure. Trying to apply uniformly distributed displacement or stresses over a larger area, as in standard compression experiments, is not quite simple and at edges and interfaces stress or displacement concentrations may develop leading to premature fracture from those locations.

Thus, one should not underestimate the effects caused by boundary conditions: a different type of failure may occur, sometimes unwanted, but some of the effects may be elucidated by the aforementioned "structural analysis" of material-scale test specimens. For the structural analysis of laboratory-scale experiment relevant details about the material structure must be known, and it must be singled out as to what scale the structure of the material must be. Does it suffice to just incorporate the particle structure (aggregate structure) of the considered concrete, or is it necessary to include certain details of the hydrated cement structure as well? Sensitivity studies may show the most appropriate approach.

In many of the models presented in this book parameters need to be determined. At best these parameters have physical meaning, can be determined in an independent test, and are not just "fitting parameters." The latter type of parameters can be used to fit any theory to any experiment, to a higher or lesser degree, depending on the actual number of such "empirical parameters" (to use a neater terminology). Although superficially sound, the approach obviously lacks in-depth understanding of the physical problem at hand. Here it seems important to use theories and (numerical) model simulations in a slightly different way. The theory, or model, must be considered as an approximation of physical reality. The global curves may be approximated extremely well or extremely poorly, but what is considered as most imminent in the field of fracture is that the fracture mechanism resemble the one observed in the experiment. The sequence in which cracks appear during the entire loading history is crucial: when a model approximates this crack mechanism to a high degree of accuracy, chances are large that the fracture mechanics are right too. It is important to extract the necessary information from the experiments: the global load-deformation behavior is one aspect; the fracture mechanisms and the developing crack patterns must be determined as well. In general we can only see the outer surface of a concrete specimen, therefore ways must be devised to measure interior crack development. Luckily there is a great deal of progress in the field and much of the internal material structure and cracks can be revealed by means of various tomography methods.

To summarize the lengthy text above, this book deals primarily with three aspects: (1) fracture of concrete and concrete structures, (2) models and theories that are capable of approximating the observed fracture behavior to a larger or lesser degree, and (3) the model–experiment interaction. The book is divided into 12 chapters, including this introductory chapter. The goal is to present the subject matter in concise form. There are several reasons to do so. First, several books exist on concrete fracture, most of them focusing in quite some depth on one of the aspects sketched above, thus, either on cohesive modeling, only on higher-order continuum theories for concrete fracture, or just on experiments. Because we have the ambition to show the intricate connection between theory and experiment a more global format has been chosen. As one advantage it is possible to show the relation between theory and experiment and at the same time a useful text becomes available to newcomers in the field. We are at the brink of expansion, namely the application of fracture mechanics for the analysis of durability of concrete and concrete structures under corrosive circumstances. Tackling such processes not only relies on knowing physical deterioration processes in detail (alkali–silica reaction, rebar corrosion, shrinkage and swelling due to moisture or temperature gradients, etc.), but also on the specific fracture model that can best be used under the given circumstances. The present book is considered useful for those starting in this emerging field. It provides a fast insight to existing theoretical approaches, as well as parameter identification and experimental techniques common in studying fracture of cement and concrete materials and structures.

1.1 Contents per Chapter

- Chapter 2 deals with "classical" fracture models, namely linear elastic fracture mechanics (LEFM): the Griffith energy balance, and the Irwin and Orowan solution of the crack-tip stresses and displacements. After the exposure of these linear theories, including the three basic crack modes (tensile opening mode, in-plane shear, and out-of-plane shear modes, referred to as modes I, II, and III, respectively), the plastic crack-tip zone model for fracture in plastic metals, and the cohesive fracture models used for concrete, based on Hillerborg's Fictitious Crack Model (Hillerborg, Modeér, and Peterson 1976) are described. Because the latter type of model also requires rather sophisticated testing (knowledge is needed on the softening behavior of concrete) some attention is given to existing test methods and proposed standard test methods. In view of the aim to provide a concise text, the information is given in compact form and sources for further reading are provided.

- In Chapter 3 the mechanics aspects of lattice modeling are explained. Based on simple framework analysis a discretization of a continuum is obtained. The relevant theoretical background, well known from finite element textbooks, is described in concise form. The equivalence between a shell element and a simple truss as shown by Hrennikoff (1941) is included, demonstrating the limitations of the elastic constants that may hamper such solutions. Yet, moving from truss elements to beam elements eliminates the problem of trusses, and as shown first by statistical physicists lattice beam models are excellently suited for fracture analyses (Roux and Guyon 1985; Herrmann, Hansen, and Roux 1989). Some modifications to the lattice geometry and the inclusion of local material properties are necessary, however, in order to obtain more realistic results. Chapter 3 continues with a comparison of lattice and particle models. Particle models are frequently used for simulating the behavior of soils, rocks, and also concrete. In this book, however, the similarity between the lattice and particle model is presented only briefly. Chapter 3 closes with the description of various fracture laws for lattice-type models, such as the Rankine, a combined tension/bending criterion, a tension/shear criterion, and a possible buckling criterion for compressive failure.

- Incorporating a realistic material structure is possible in a lattice model, but basically in any type of finite element model by simply projecting the material structure on top of the lattice. Relevant properties of the local material phases are then assigned to the respective lattice elements. This simple approach leads to models with geometrically correct phase distributions, which stands in contrast to the usual way of statistical physics, where it is widespread usage to apply statistical distributions, for example, following a Gaussian, Weibull, or any other distribution, in order to include the effects of the material heterogeneity. In Chapter 4 different ways of incorporating heterogeneity in the lattice are discussed, as well as details concerning the overlay of the material structure and the lattice geometry. At the end of Chapter 4 local material properties are given for normal concrete simulations. Of particular interest are the properties of the interfacial transition zone between aggregate and matrix. This topic is presented in a bit more depth because it is so decisive for the behavior of concrete.

- In order to allow the application of lattice models with particle overlay for the simulation of the mechanical behavior of concrete it is essential to show that the elastic properties of the model are correct. This is the topic of Chapter 5: results are shown for the elastic properties of a lattice with particle overlay in 2D and 3D. A comparison is made with the upper and lower bounds for the elastic properties of composites from classical theories such as Hashin–Shtrikman (1963) and the Voigt/Reuss bounds.

- In the following three chapters the fracture behavior of concrete subjected to tensile combined tensile and shear load, and (confined) compressive loading are presented. Results from both experiments and from numerical simulations with lattice-type models are shown. Where necessary additional information about experimental techniques used is given in the appendix, and a comparison with results provided by competing models is made. In particular for compression behavior, such comparisons are provided, because a simple and effective model is still lacking for this important loading regime.

- First, in Chapter 6, the failure of concrete subjected to uniaxial tension is discussed. Classical experiments by Evans and Marathe (1968) are realistically simulated, and details of the fracture process become visible from meso-level analyses. There is a significant influence of the boundary conditions, that is, the free or suppressed rotations of the loading platens. Different fracture mechanisms emerge, which are described in detail. The use of notches is frequently advocated in experiments for determining the softening behavior of concrete. The necessity can be debated, as in Section 6.2. Many researchers feel that uniaxial tensile tests are far too complicated, in particular because of the difficulty of gluing specimens in the test machine, therefore indirect methods are favored. Yet, such indirect experiments are usually difficult to interpret and require back analysis of the true tensile properties using advanced fracture models. The uncertainty of the entire enterprise makes the use of indirect test methods, such as the Brazilian splitting test or the three-point bending test, highly debatable.

- Chapter 7 focuses on the behavior of concrete subjected to combined tension and shear. The combination of loads is very important under many practical circumstances, for example, the shear failure of reinforced concrete beams. True shear failure, as known from metal failure, is hard to imagine for cement and concrete due to the great imbalance between tensile and compressive strength. Tensile failure will under most (unconfined) stress states dominate the failure mode. For a variety of cases, hitherto considered as mode II (in-plane) shear modes can easily be explained from local mode I (tensile opening mode) crack nucleation and growth. The case of torsion is interesting inasmuch as it provides an appropriate transition to compressive failure, the topic of Chapter 8. It appears that from a superficial point of view compressive fracture can be solved in the same way as the fictitious crack model does with tensile fracture. The tensile case was presented in Chapter 2; in Chapter 8 the focus is on experimental results that make the case for compressive localization in the softening regime, models that are capable of describing the localized failure mode, a discussion of surface versus volumetric effects on the localized failure mode, and some alternative approaches that treat softening as the outcome of a crack-growth process. Both results from confined and unconfined compression tests are presented. The last section of Chapter 8 deals with a meso-level model for describing fracture of concrete in compression. A mechanism still appears to be lacking; perhaps the newly presented buckling criterion in Chapter 3 may give the right answer.

- Size effect is a consequence of fracture mechanics. The linear versions of fracture mechanics theory predict that larger structures fail at relatively smaller loading. The traditional way to deal with size effect is the Weibull weakest link theory (Weibull 1939, 1951), which states that larger volumes of material will have a larger probability to contain weak spots, and consequently will fail at lower loads. How rapid the load decrease goes with increasing size is still subject to debate. Two empirical size-effect models were developed in the past two decades, but they are no real improvement to Weibull theory. A major objection to the latest models is that they are purely empirical. Linking the Weibull modulus to the damage distribution may be a more fruitful approach, and from lattice analyses of structures of different size, this is exactly what can be obtained. In Chapter 9 the various theoretical approaches are debated.

- The results presented in Chapters 6 through 8 point toward an alternative approach to dealing with softening, namely softening as the consequence of crack growth phenomena at the meso- and macro-size/scale. In Chapter 10 the consequences of this hypothesis are debated. What emerges is the so-called 4-stage fracture model (Van Mier 2004a, 2008). Softening is treated as the stage of unstable macrocrack growth; nucleation is a more important aspect, but that takes place in the prepeak regime of the stress–strain curve. The softening behavior can be modeled using classical LEFM with a small bridging stress over the crack. The proposed approach opposes the cohesive crack models in the fact that it does not accept the complete tensile softening diagram as a bridging stress profile, nor as a material property. The 4-stage crack model can be applied both for tension and (confined) compression of plain and fiber-reinforced concrete. Because the model is based on classical LEFM it is necessary to determine the geometrical factor, which included specimen size and boundary condition effects, for every newly considered structure.

- The meso-level computations with the lattice model represent a model approach where "up-scaling from a predefined scale level" is performed. In the applications of Chapter 6 through 9 the predefined scale was the aggregate scale of concrete, but at the cost of more lengthy computations it would also have been possible to start from the micro-size/scale, that is, the scale of hydrated cement phases. The number of elements in a given analysis will increase enormously, and advanced solvers, multiprocessor computers, and a substantial amount of wall-clock time will be needed for such detailed analyses. The enormous computational costs can be reduced substantially by means of a new approach based on *multiscale interaction potentials*. Interaction potentials describe the interaction between neighboring particles. The size of the particles determines the shape of the interaction potential acting between them, and corresponds to the actual force-separation behavior at that particular size/scale, from the nano size/scale to the macroscopic size/scale. The so-called multiscale interaction potential approach (Van Mier 2007) is debated in Chapter 11. It is a different way of dealing with the material/ structure entanglement during softening. The outcome is surprising: softening must be used as a structural property, and solving the kinematic and equilibrium equations in structural analysis can be done without relying on a separate constitutive law.

- Finally in Chapter 12 a number of conclusions are drawn. Many questions are posed in the book; some are resolved, and some are still open to debate. The role of models is that they support experiments. They should provide a workable hypothesis for carrying out new experiments, and provide new knowledge and insight on fracture of cement and concrete. A model can only become a theory or law if it contains physical-based parameters that can be determined from independent experiments. Many of our engineering models, probably close to 99% of them, are empirical in nature, and an exact fit to experimental data is quite meaningless if the model parameters are not well defined. Softening goes to the heart of a long-standing problem in structural engineering: where does the structure become a structure, and at what point does the material stop being a material. One hopes that the present excursion along a variety of fracture problems is of help to the reader in adding some new ideas to the existing kaleidoscope.

- At the end of the book a list of notations and a key-word index have been included. A number of topics have been transferred to the Appendix, such as some remarks about computational efficiency of the lattice model (Appendix 1), two simple results from LEFM (Appendix 2), testing related topics such as test stability (Appendix 3), which is of eminent importance when dealing with softening, and an overview of crack detection techniques (Appendix 4), which are of great use when determining fracture mechanisms. Finally in Appendix 5 the effects of internal and external confinement on the behavior of concrete under compressive loads are discussed. Each of these appendices contains examples that are used at several places in the main text.

2

Classical Fracture Mechanics Approaches

2.1 Stress Concentrations

Linear elastic fracture mechanics (LEFM) dates back to 1920–1921 when Griffith proposed his energy approach for the brittle fracture of glass. Any material, including a very smooth homogeneous material such as glass, contains imperfections. These imperfections are the source of stress concentrations, which may lead to failure of the material well below its theoretical strength. Based on a sinusoidal approximation of the atomic bond potential,

$$\sigma = \sigma_{max} \cdot \sin\left[\frac{\pi}{r}(x - r_0)\right] \tag{2.1}$$

where σ_{max} is the peak stress in the atomic bond stress-spacing diagram and r is the increase of the original lattice spacing r_0 of the atoms, it is possible to calculate the theoretical strength of crystalline solids, which leads to (Kelly and MacMillan 1986):

$$\sigma_{max} = \frac{E}{\pi} \tag{2.2}$$

The Young's modulus E relates stress with strain following $\sigma = E.\varepsilon = E.x/r_0$. For example, for alkali-resistant glass fiber, with a Young's modulus $E = 70$ GPa (Gupta 2002), the predicted theoretical strength according to Equation (2.2) would be $\sigma_{max} = 23$ GPa, whereas in reality about 70 MPa is measured on a single fiber. The strength of the fiber is very much affected by its diameter. Surface defects result in premature failure at stress levels quite below the maximum attainable value. In the glass rod of Figure 2.1a, the crack seems to have nucleated from the small white line at the bottom of the mirror area. The rod, which was a simple off-the-shelf product, was highly polluted on the outside as can be seen in Figure 2.1b. The crack nucleated from an imperfection, and appeared to have started symmetrically in the beginning. After the rather smooth mirror-zone, surface roughness gradually increased into the mist- and hackle-zones.

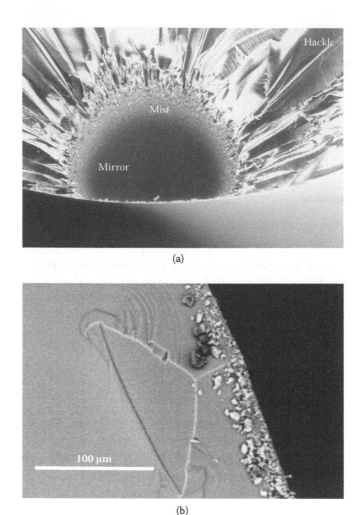

FIGURE 2.1
Mirror-mist-hackle zones after loading a glass rod to failure (a), and close-up of the point where the crack nucleated. (After Van Mier. 2000. *De Kunst van Breken and Scheuren [The Art of Fracturing].* Inaugural Lecture, Delft University of Technology [in Dutch]. With permission.)

Stress concentrations in real materials are, for instance, caused by pores, inclusions, interfaces between distinct material phases, and the like. A simple example is a circular hole with radius $r = a$ in a flat plate, as shown in Figure 2.2a. When the plate is stretched at infinity, the stresses in the neighborhood of the hole are no longer uniformly distributed, with the highest stress concentrations appearing at the sides of the hole (points A and B). At the edge along the hole the radial and shear stress components (σ_{rr} and $\tau_{r\theta}$) are equal to zero, whereas the tangential stress $\sigma_{\theta\theta}$ is equal to

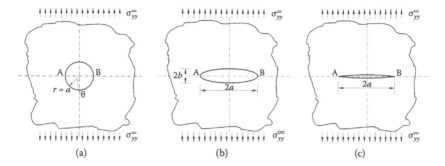

FIGURE 2.2
Plate with perfectly circular hole (a), elliptical hole (b), and slit (c).

$$\sigma_{\theta\theta} = \sigma_{yy}^{\infty} - 2\sigma_{yy}^{\infty} \cos 2\theta \qquad (2.3)$$

which is derived directly from linear elasticity theory; see Timoshenko and Goodier (1970) for the full solution. The term σ_{yy}^{∞} is the externally applied stress. It is easy to see that along the sides of the hole ($\theta = \pi/2$ and $3\pi/2$) stresses are three times the external stress, and above and below the hole ($\theta = 0$ and π) compressive stresses equal to $-\sigma_{yy}^{\infty}$ are found. Thus, if the plate material fails in tension, the actual measured stress is three times lower than the material could withstand without the circular hole.

When the shape of the hole is changed to elliptical, the stress concentrations become more severe (Figure 2.2b). When the semi-axes of the ellipse are equal to a and b, the tangential stresses at A and B increase to

$$\sigma_{\theta\theta} = \sigma_{yy}^{\infty} \left(1 + \frac{2a}{b}\right) \qquad (2.4)$$

The "flatter" the ellipse, the higher the stress concentrations, and the lower the measured failure stress of the plate will be. For a slit, that is, when $b \to 0$ (see Figure 2.2c), the tangential stress at the tip will become infinitely large ($\sigma_{\theta\theta} \to \infty$). At this point fracture mechanics becomes of interest. The displacement jump caused by the slit should be approached differently than through the application of classical elasticity.

2.2 Linear Elastic Fracture Mechanics (LEFM)

The above situation, farfield tension, leads to one of three fracture modes that are distinguished in classical fracture mechanics. In Figure 2.3 the three

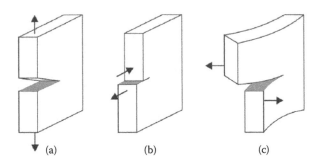

(a)   (b)   (c)

FIGURE 2.3
Three fracture modes: (a) opening mode I, (b) in-plane shear mode II, and (c) out-of-plane shear mode III.

common modes are depicted: the opening mode or mode I, in-plane shear or mode II, and out-of-plane shear or mode III. Combinations of the three modes are referred to as "mixed-modes," for example, tension and shear can either be mode I+II or mode I+III.

In engineering practice most interest centers on mode I. The other modes appear less frequently but are nevertheless of great importance. The crack-tip stresses were derived by Irwin (1958), bringing in a new material constant in the process, namely the stress intensity factor K. It is obvious that somehow the singularity has to be dealt with and changing the failure criterion to one based on stress intensity rather than on stress appears to be a viable approach. A recurring theme in this book is the separation between material and structural aspects of fracture, which comes in quite naturally in the equations derived by Irwin (1958). So, the problem here is a slitlike rack in an infinite plate subjected to farfield tension or shear. In Figure 2.4 the situation is sketched along with the definition of near-tip stresses in a Cartesian coordinate system. The expressions for the three

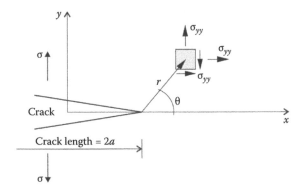

FIGURE 2.4
Tip of a slitlike crack with length $2a$ in an infinite plate subjected to farfield tension. Of interest are the local stresses σ_{xx}, σ_{yy}, and τ_{xy} in point P defined by polar coordinates (r,θ).

stress components are nowadays quite familiar, and can be found in many textbooks; see, for instance, Broek (1983), Lawn (1993), and Suresh (1991). Because we need the expressions in Chapter 10, the main equations are listed here and their characteristics are summarized.

For mode I (tensile or opening mode) the three stresses σ_{ij} and displacements u_i at P are:

$$
\left\{
\begin{array}{c}
\sigma_{xx} \\
\sigma_{yy} \\
\sigma_{xy}
\end{array}
\right\}
=
\frac{K_I}{\sqrt{2\pi r}}
\left\{
\begin{array}{c}
\cos\dfrac{\theta}{2}\left[1-\sin\dfrac{\theta}{2}\sin\dfrac{3\theta}{2}\right] \\[3mm]
\cos\dfrac{\theta}{2}\left[1+\sin\dfrac{\theta}{2}\sin\dfrac{3\theta}{2}\right] \\[3mm]
\sin\dfrac{\theta}{2}\cos\dfrac{\theta}{2}\cos\dfrac{3\theta}{2}
\end{array}
\right\}
\tag{2.5}
$$

$\sigma_{zz} = 0$ for plane stress and $\sigma_{zz} = \nu(\sigma_{xx} + \sigma_{yy})$ for plane strain, $\sigma_{xz} = \sigma_{yz} = 0$, and

$$
\left\{
\begin{array}{c}
u_x \\
u_y
\end{array}
\right\}
=
\frac{K_I}{2E}\sqrt{\frac{r}{2\pi}}
\left\{
\begin{array}{c}
(1+\nu)\left[(2\kappa-1)\cos\dfrac{\theta}{2}-\cos\dfrac{3\theta}{2}\right] \\[3mm]
(1+\nu)\left[(2\kappa+1)\sin\dfrac{\theta}{2}-\sin\dfrac{3\theta}{2}\right]
\end{array}
\right\}
\tag{2.6}
$$

$u_z = -\dfrac{\nu z}{E}(\sigma_{xx} + \sigma_{yy})$ for plane stress and $u_z = 0$ for plane strain.

In Equation (2.6) $\kappa = (3 - \nu)/(1 + \nu)$ for plane stress, and $\kappa = 3 - 4\nu$ for plane strain; K_I is the mode I stress-intensity factor, defined as $K_I = \sigma_{yy}^{\infty}\sqrt{\pi a}$ with dimension MPa$\sqrt{\text{m}}$. For mode II (in-plane shear mode) the expressions are more or less identical:

$$
\left\{
\begin{array}{c}
\sigma_{xx} \\
\sigma_{yy} \\
\sigma_{xy}
\end{array}
\right\}
=
\frac{K_{II}}{\sqrt{2\pi r}}
\left\{
\begin{array}{c}
-\sin\dfrac{\theta}{2}\left[2+\cos\dfrac{\theta}{2}\cos\dfrac{3\theta}{2}\right] \\[3mm]
\sin\dfrac{\theta}{2}\cos\dfrac{\theta}{2}\cos\dfrac{3\theta}{2} \\[3mm]
\cos\dfrac{\theta}{2}\left[1-\sin\dfrac{\theta}{2}\sin\dfrac{3\theta}{2}\right]
\end{array}
\right\}
\tag{2.7}
$$

$\sigma_{zz} = 0$ for plane stress and $\sigma_{zz} = \nu(\sigma_{xx} + \sigma_{yy})$ for plane strain, $\sigma_{xz} = \sigma_{yz} = 0$, and

$$
\left\{ \begin{array}{c} u_x \\ u_y \end{array} \right\} = \frac{K_{II}}{2E}\sqrt{\frac{r}{2\pi}} \left\{ \begin{array}{c} (1+\nu)\left[(2\kappa+3)\sin\dfrac{\theta}{2}+\sin\dfrac{3\theta}{2}\right] \\[2ex] -(1+\nu)\left[(2\kappa-3)\cos\dfrac{\theta}{2}+\cos\dfrac{3\theta}{2}\right] \end{array} \right\} \tag{2.8}
$$

$$
u_z = -\frac{\nu z}{E}(\sigma_{xx}+\sigma_{yy}) \text{ for plane stress and } u_z = 0 \text{ for plane strain.}
$$

The definition for κ is the same as in Equation (2.6), and K_{II}, the mode II stress intensity factor defined as $K_{II} = \sigma_{xy}^{\infty}\sqrt{\pi a}$. And, finally, for mode III (out-of-plane shear mode),

$$
\sigma_{xx} = \sigma_{yy} = \sigma_{zz} = 0
$$

$$
\sigma_{xy} = 0
$$

$$
\left\{ \begin{array}{c} \sigma_{xz} \\ \sigma_{yz} \end{array} \right\} = \frac{K_{III}}{\sqrt{2\pi r}} \left\{ \begin{array}{c} -\sin\dfrac{\theta}{2} \\[2ex] \cos\dfrac{\theta}{2} \end{array} \right\} \tag{2.9}
$$

$$
u_x = u_y = 0
$$

$$
u_z = \frac{4K_{III}}{E}\sqrt{\frac{r}{2\pi}}\left[(1+\nu)\sin\frac{\theta}{2}\right] \tag{2.10}
$$

As with the other modes, K_{III} is the mode III stress intensity factor; that is, $K_{III} = \sigma_{yz}^{\infty}\sqrt{\pi a}$.

The expressions Equations (2.5)–(2.10) are valid only in the immediate vicinity of the crack-tip. The expansion with goniometric functions has been cut off after the first two terms; higher-order terms can in general be neglected, except of course for very special circumstances, which we do not discuss in the framework of this compact book. The equations can be cast in the same generic form as follows:

$$
\sigma_{ij} = \frac{K}{\sqrt{2\pi r}}f_{ij}(\theta) \tag{2.11a}
$$

$$u_i = \frac{K}{2E} \sqrt{\frac{r}{2\pi}} f_i(\theta) \tag{2.11b}$$

The stresses, as may be clear from the above equations, all show a singularity when $r \to 0$, which makes it impossible to use a stress criterion for crack propagation. The stress intensity factor K takes over and is now the quantity to compare to a critical value K_c in order to judge whether the crack grows. This is the point where continuum mechanics breaks down. The appearance of a singularity prevents further application of continuum mechanics in the presence of slitlike cracks, unless of course, as done in many applications, not calculating a stress at a point, but smeared over a larger area, as we see, for instance, in the plastic crack-tip model. There the notion of stress returns. From a physical point of view it is hard to imagine that a singularity would actually occur. For instance, in a crystalline solid, atomic bonds at the crack-tip must be broken one at a time for a crack to propagate. If the atomic potentials are correct, the subsequent fracturing of bonds occurs at a finite stress (or rather force inasmuch as it is a point of contact). Just from this fact it is good practice to start doubting continuum mechanics for the present application.

The stresses at the crack-tip, Equation (2.11a), are in part dependent on a so-called geometrical function, or weight function, $f_{ij}(\theta)$. This function depends on the actual specimen/structure geometry and the boundary conditions, that is, the circumstances under which the crack-tip is loaded. To make matters clearer we can write the stress intensity factor as follows:

$$K = \sigma \sqrt{\pi a} f(\xi) \tag{2.12}$$

where $\xi = a/W$ is used to describe the influence of the crack length and specimen dimensions (denoted by W, i.e., the width of the specimen in the direction of the crack growth). For the center crack in an infinite plate, as sketched in Figure 2.2c, $f(\xi) = 1$, and K reduces to $\sigma\sqrt{\pi a}$. When the specimen has finite width W, the geometrical factor is approximated by:

$$f(\xi) = f\left(\frac{a}{W}\right) = \left[\frac{W}{\pi a} \tan\left(\frac{\pi a}{W}\right)\right]^{1/2} \tag{2.13}$$

Several solutions, with varying accuracy, can be found in handbooks, such as the one by Tada, Paris, and Irwin (1973), as well as in the aforementioned textbooks on fracture mechanics.

For a single-edge notched tensile specimen, loaded between hinged boundaries, Tada et al. (1973) give the following expression for the geometrical function $f(\xi)$:

$$f(\xi) = f\left(\frac{a}{W}\right) = \sqrt{\pi}\left(\frac{2W}{\pi a}\tan\frac{\pi a}{2W}\right)^{0.5}\left[\frac{0.752 + 2.02\dfrac{a}{W} + 0.37\left(1 - \sin\dfrac{\pi a}{2W}\right)^3}{\cos\dfrac{\pi a}{2W}}\right] \quad (2.14)$$

which has an accuracy better than 0.5% for all a/W, whereas for the same single-edge notched specimen, but now loaded between fixed boundaries (i.e., the rotations of the loaded ends of the specimen are prevented), a more complicated formula results (Marchand, Parks, and Pelloux 1986):

$$f(\xi) = \frac{F_1(\xi)\left(1 - \dfrac{F_2(\xi)C_{12}(\xi)}{F_1(\xi)[12L/W + C_{22}(\xi)]}\right)}{\left(1 + \dfrac{W}{L}\left[C_{11}(\xi) - \dfrac{C_{12}(\xi)^2}{(12L/W + C_{22}(\xi))}\right]\right)} \quad (2.15)$$

where F_1 and F_2 are the geometrical functions for a normal load (Eq. (2.14)) and pure bending (see Eq. (2.16) below), respectively, and C_{ij} are dimensionless crack compliances containing all information regarding the relations between normal load N, bending moment M, displacement u, and rotations φ. The crack compliances can only be computed numerically. It is beyond the scope of this book to include the full details of the solution, and the interested reader is referred to Marchand et al. (1986).

The function F_2 in Equation (2.15) can be found in Tada et al. (1973) as follows:

$$F_2 = f\left(\frac{a}{W}\right) = \sqrt{\frac{2W}{\pi a}\tan\frac{\pi a}{2W}}\left(\frac{0.923 + 0.199\left(1 - \sin\dfrac{\pi a}{2W}\right)^4}{\cos\dfrac{\pi a}{2W}}\right) \quad (2.16)$$

with an accuracy better than 0.5% for all values of a/W.

Using Equations (2.14)–(2.16) in a simple expression that shows the effect of relative crack length on residual stress carrying capacity of a cracked specimen (Equation (A1.4) in Appendix 1), it is possible to show the effect of hinged versus fixed boundaries in an uniaxial tensile test on a single-edge notched plate. The result is shown in Appendix 2, Figure A2.3, and is in agreement with experimental observations in concrete fracture, except for the tail part of the softening curve; see Figure 2.11 and related results in Van Mier, Schlangen, and Vervuurt (1995). A similar analysis as for the boundary rotation effect in uniaxial tension tests can be carried out showing the size effect emanating from linear elastic fracture mechanics. LEFM predicts that larger structures carry smaller loads than smaller geometrically similar structures; see Appendix 2. We return to the boundary rotation effect and the size effect in Chapters 6 and 9, respectively.

The above-mentioned solution of Irwin succeeds the original energy approach by Griffith by several decades. The elastic energy U_E released in an area of radius $2a$ (the total crack length) is in balance with the total surface energy U_S needed to create the new crack area $2a.1$ (assuming unit thickness of the plate containing the slitlike crack). The elastic energy can be written in the familiar form

$$U_E = \pi a^2 \frac{\sigma^2}{E} \tag{2.17}$$

The surface energy is the product of the crack area and the specific surface energy γ:

$$U_S = 2a \cdot \gamma \tag{2.18}$$

Crack growth occurs when the rate of energy release dU_E/da exceeds the increase in surface energy dU_S/da following:

$$\frac{dU_E}{da} \geq \frac{dU_S}{da} = 2\gamma \tag{2.19}$$

which leads to:

$$\sigma = \left[\frac{E\gamma}{\pi a} \right]^{1/2} \tag{2.20}$$

where E is the Young's modulus in plane stress, which must be replaced by $E' = E/(1 - v^2)$ in the case of plane strain. Thus, as the externally applied stress exceeds the stress given by Equation (2.20), the crack will propagate. Note that crack propagation is unstable, unless specific boundary conditions restrain the crack. The difficulty lies in establishing accurate values of the surface energy γ. In the literature γ is often replaced by the critical energy release rate G_{Ic}, where the subscript I refers to mode I. It is equal to the amount of energy consumed in the fracture process. The critical energy release rate is related to Irwin's critical stress intensity factor, as one can easily derive from Equation (2.12) for the infinite plate ($f(\xi) = 1$) and Equation (2.20) with G_{Ic} instead of γ. This leads to the well-known relation:

$$K_{Ic} = \sqrt{EG_{Ic}} \tag{2.21}$$

2.3 Plastic Crack-Tip Model

The singularity from the Irwin analysis leads to problems. The applicability of continuum mechanics seems restricted in the case of a slitlike crack in an elastic plate. For crystalline solids such as we mentioned at the beginning of this chapter, it seems to make sense to include the atomic potential as a realistic way of dealing with crack-tip processes. The sinusoidal atomic potential of Equation (2.1) is depicted in Figure 2.5. The area under the curve is the amount of energy needed to separate two atoms, that is, to break the bond. This energy is in fact twice the specific surface energy for the solid under consideration. Based on the cohesion caused by this potential, Barenblatt (1962) devised a cohesive crack model that served as a basis for subsequent developments in concrete fracture mechanics. The cohesive stresses caused by the atomic bonds act along the plane of the crack near its tip.

Barenblatt posed an important hypothesis, namely that the cohesive zone at the crack-tip should be small compared to the size of the whole crack. As we show later on, in their analogy of the model applied to concrete fracture, Hillerborg, Modeér, and Peterson (1976) violated this assumption. Dugdale (1960) proposed a similar model for plastic metals, where the interatomic cohesive forces used by Barenblatt were replaced by the yield stress of the metal. In Figures 2.6a,b the principle of the plastic crack-tip model is shown. The size of the plastic zone is t, advancing in front of the stress-free crack with length $2a$. The LEFM analysis would lead to a stress singularity at the

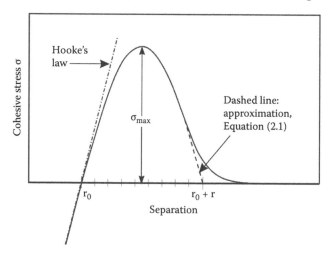

FIGURE 2.5
Atomic potential or cohesive stress–separation curve describing the bond between two atomic planes in a crystalline solid. The real behavior is shown as a solid line; the sinusoidal approximation of Equation (2.1) is shown as a dashed line. At the point where the curve intersects the *x*-axis, the slope of the curve is equal to the Young's modulus as commonly used in Hooke's law.

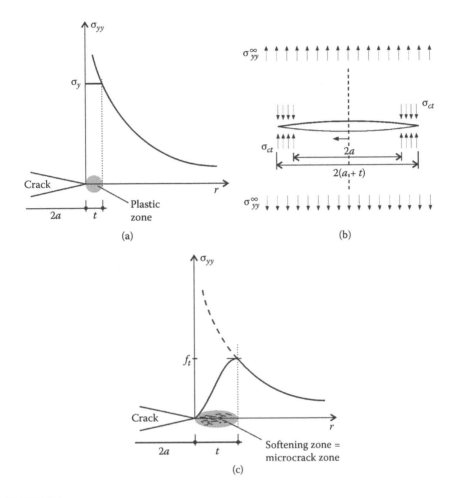

FIGURE 2.6
Principle of the plastic crack-tip model (a), closing pressure at crack-tips (b), and similarity of the Fictitious Crack Model (c).

tip of the stress-free crack, but through the assumption of the cohesive zone, also referred to as the process zone, the maximum stress is limited to the yield stress σ_y. In the subsequent analysis it is assumed that the total crack length is $2(a + t)$, and over the tip region of this longer crack, that is, over the segments of length t, a closing pressure σ_{ct} acts equal to σ_y. Now two stress intensities are calculated: one from the crack-tip stresses σ_y (which are assumed to be constant over t), and the other from the farfield tensile stress σ_{yy}^∞; see Broek (1983):

$$K_{\sigma_{yy}^\infty} = \sigma\sqrt{\pi(a+t)} \qquad (2.22a)$$

$$K_{\sigma y} = 2\sigma_y \sqrt{\frac{a+t}{\pi}} \arccos \frac{a}{a+t} \qquad (2.22b)$$

From the condition that the two stress intensities should cancel when super-imposed, it follows that the size of the plastic crack-tip zone t is:

$$t = \frac{\pi^2 (\sigma_{yy}^\infty)^2 a}{8\sigma_y^2} \qquad (2.23)$$

Much can be said about the actual stress distribution over the crack-tip cohesive zone, and much can be said about the shape of the plastic zone. Because this is not extremely important for the discussion to follow, the interested reader is referred to Lawn (1993) and Broek (1983). What is of importance though, is the stress intensity due to an arbitrary wedge-stress distribution in the process zone, which leads to:

$$K_{\sigma y} = 2\sqrt{\frac{a+t}{\pi}} \cdot \int_a^{a+t} \frac{\sigma_{ct}(x)dx}{\sqrt{(a+t)^2 - x^2}} \qquad (2.24)$$

where $\sigma_{ct}(x)$ denotes the stress distribution in the crack-tip plastic zone, and the x-coordinate runs from the origin at the center of the crack toward the tip as indicated in Figure 2.6b. An example of a useful function for the crack-tip wedging stress is the interatomic potential of Figure 2.5; see also Equation (2.1).

2.4 Fictitious Crack Model (FCM)

In 1976 Hillerborg, Modéer, and Petersson proposed an extension of the aforementioned plastic crack-tip model for concrete fractures. The so-called "Fictitious Crack Model" included a process zone similar to the plastic crack-tip zone, although the stress distribution would not be uniform as in the Dugdale model, and the maximum "yield stress" was much smaller, namely equal to the uniaxial tensile strength of concrete; see Figure 2.6c, which is similar to Figure 2.6a except for the changes in the crack-tip process zone. Microcracks would be present in the process zone, but stresses could still be transmitted. In Hillerborg et al. (1976) and Petersson (1980) the model is described in detail. The closing pressure over the crack-tip zone was thought to resemble the shape of the softening curve of a concrete test specimen loaded in uniaxial tension. There are some drawbacks to the model, which, however, are frequently ignored. So, let us summarize the most important aspects of the model

In doing so, it is interesting to repeat some of the original assumptions and starting points here; see also Van Mier (2004a). The first one is:

> The fracture zone is formally represented as a crack with the ability to transfer stresses. Such stress transferring crack is called a fictitious crack, as distinguished from a real crack, which cannot transfer tensile stresses perpendicular to itself. (Hillerborg 1992, p. 488)

And, reflecting about the nature of the process zone:

> At the same time the assumptions of Figure 2.6 [i.e., the correspondence between the closing pressure in the Dugdale/Barenblatt model and the Hillerborg model] may be looked upon as a reality. Stresses may be present in a microcracked zone as long as the corresponding deformation is small. This has been clearly demonstrated in tension tests, using very rigid testing equipment, e.g., by Evans and Marathe (1968). (see Hillerborg et al. 1976)

The experiments of Evans and Marathe (1968; see Figure 2.10) appear to have played a crucial role in the development of the Fictitious Crack Model. These experiments were among the first to demonstrate that significant load transfer was possible after the tensile strength of the material was exceeded. We discuss these results, as well as some further consequences later on in this chapter, and in Chapter 6 in more depth. The hint toward microcracks in the process zone is quite essential, because it is exactly this point that has been used by many others in similar model developments (e.g., the crack-band model by Bažant and Oh 1983). Petersson (1980, p. 80) refers to the microcrack zone as follows:

> It is supposed that when the tensile strength is reached, a micro-cracked zone starts developing. The micro-cracked zone continues to grow when the load increases and finally some of the microcracks join together and a real crack opens.

This then continues:

> The micro-cracked zone is able to transfer stress. The greater the number of microcracks, i.e., the wider the micro-cracked zone. (Petersson 1980, p. 80)

Thus, it is clear that the process zone is assumed to consist of many closely spaced microcracks; the larger the deformations are (i.e., the virtual crack opening of the fictitious crack), the larger the number of microcracks.

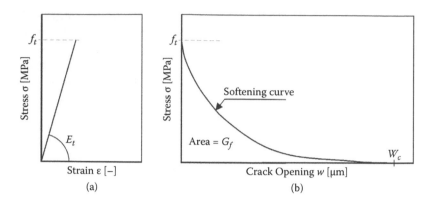

FIGURE 2.7
Fictitious Crack Model: separation in prepeak stress–strain curve (a), and postpeak stress–crack opening diagram (b).

These statements have led to a significant research effort to determine the exact size of the "micro-cracked zone in front of the stress-free macrocrack." In Mindess (1991) the outcomes of many such studies have been compared. The main conclusions were that the extent of the microcrack zone seemed to depend on the resolution of the crack detection technique used, as well as on the structural conditions and the size of the specimen. In Chapter 6 we describe the true nature of the fracture process in tension in greater detail. Suffice it to say here that the above citations from the 1976–1992 papers of Hillerborg et al. and Petersson were a wild guess and real proof of the true nature of the process zone was lacking. The only hunch was that when the frontal zone at the crack-tip was sprayed with water, it would absorb the water more quickly than other (uncracked) parts of the concrete specimen. The matter is of importance because many experimentalists were driven in the wrong direction, and actually started to measure the extent of the process zone.

The Fictitious Crack Model is shown schematically in Figure 2.7. In the prepeak regime stress and strain are state variables. The uncracked material is often assumed to exhibit perfect linear elastic behavior, as shown in Figure 2.7a. The most important parameters in this stage are the Young's modulus E_t and the tensile strength f_t. The material cracks as soon as the tensile strength is reached, where the material can then be characterized by the softening diagram of Figure 2.7b. Because the crack-opening displacement cannot be smeared over a specific part of the specimen's volume, Hillerborg et al. (1976) proposed that the state variables should change to stress and displacement. Obviously this is a complication: at the peak of the stress–strain diagram a "phase change" occurs, from the uncracked to the (micro) cracked material. The stress that can be carried by the cracked concrete can be described by the softening curve. The shape of this curve is therefore essential information, as is the maximum crack opening w_c at which the crack becomes stress-free.

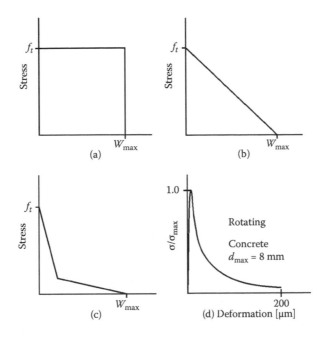

FIGURE 2.8
Schematic stress-crack opening relations for concrete: (a) perfectly plastic, (b), linear softening, and (c) bilinear softening. These models are approximations of the real curve depicted in (d). The test result is obtained from a tensile test between freely rotating loading platens. (After Van Mier. 1997. *Fracture Processes of Concrete: Assessment of Material Parameters for Fracture Models*. With permission.)

The shape of the softening curve is often approximated by means of simple linear or bilinear functions, as shown in Figure 2.8. Deviations from truly observed softening curves can be very large, as can be seen from a comparison of the approximate curves with the measurement result of Figure 2.8d. Therefore many researchers have invested time in developing improved representations such as exponential or power law functions. Directly based on the Dugdale/Barenblatt plastic crack-tip model, Reinhardt (1984) proposed a simple softening power law:

$$\frac{\sigma_t}{f_t} = 1 - \left(\frac{w}{w_c}\right)^k \tag{2.25}$$

with $k = 0.31$ and $w_c = 175$ μm giving the best comparison to experimental data published in the same paper. A slightly different form was proposed by Foote, May, and Cotterell (1986):

$$\frac{\sigma_t}{f_t} = \left[1 - \frac{w}{w_c}\right]^n \tag{2.26}$$

In both equations f_t is the maximum tensile strength of the concrete and w_c is the maximum crack opening at which the crack becomes stress-free.

Hordijk (1991) made an exponential best-fit to test data by Cornelissen, Hordijk, and Reinhardt (1986a,b) using

$$\frac{\sigma(w)}{f_t} = \left[1 + \left(c_1 \frac{w}{w_c}\right)^3\right]\exp\left(-c_2 \frac{w}{w_c}\right) - \frac{w}{w_c}(1 + c_1^3)\exp(-c_2) \qquad (2.27)$$

Four parameters are needed, namely $f_t = 3.2$ MPa, $c_1 = 3$, $c_2 = 6.93$, and $w_c = 160$ μm. With these values for the parameters the fracture energy $G_f = 99.7$ J/m² (see Equation (2.28)). Two purely empirical parameters c_1 and c_2 appear; they are a consequence of the chosen function and have no physical meaning. The comparison of Equation (2.27) derived by Hordijk with the experimental data by Cornelissen et al. is included in Figure 2.9. Linear and bilinear softening relations are also shown. In all three cases the same fracture energy has been assumed, namely $G_f = 99.7$ J/m². This leads to the values for the maximum crack opening w_c shown along the *x*-axis. Equations (2.25)–(2.27) are sometimes used in finite element calculations. The success of the model is demonstrated in many papers; see, for example, Elfgren (1989), Van Mier (1987), and CUR (2003).

It appears that the slope of the softening curve directly beyond peak stress is of utmost importance: the bilinear and nonlinear curves will likely

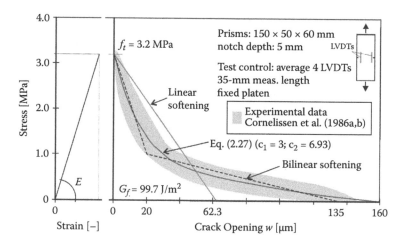

FIGURE 2.9

The prepeak stress–strain curve (left) is characterized by two parameters E and f_t, whereas the postpeak or softening curve (right) has a nonlinear shape. Fitting data from Cornelissen et al. (1986a,b) to Equation (2.27) derived by Hordijk (1991) requires four parameters, two of which (c_1 and c_2) are purely empirical and relate to the choice of the function only. For comparison, linear and bilinear softening relations have been included, both leading to the same value for the fracture energy $G_f = 99.7$ J/m².

yield satisfactory results, whereas linear softening may fail to do so (see Van Mier 1986b). Often simulations are not based directly on the fictitious crack model, which actually requires that the crack develops between element boundaries, but on the crack-band model (CBM; Bažant and Oh 1983), which is a modification of FCM. In CBM the crack-opening displacement in the process zone is smeared out over a band of width $c_b = n.d_{max}$, where d_{max} is the maximum aggregate size of the concrete, and $n = 3$. The factor n is also a fitting factor, in spite of the fact that it was named a material property. A unique relationship between the actual process zone width and the maximum aggregate size of concrete has never been satisfactorily established. In recent years many more complicated models have been developed, such as higher-order continua. In the context of this book it suffices to say that over 90% of the finite element models appear to be based on the simple earliest models mentioned above.

One remark should be made about the representation of the experimental data in Figure 2.9. A collection of data points from tests by Cornelissen et al. (1986a,b) is shown. These points derive from many experiments, and do not reveal the shape of the actually measured curves, which means that some effects important to our understanding may simply be missed, for instance, the effect of the rotational stiffness of the loading platen (see Figure 2.11). Using a test apparatus with loading platens fixed against rotation leads to a specific "bump" in the softening curve, which is simply averaged out in the data representation of Figure 2.9. We return to the effect of the rotational stiffness of the loading platen later in this chapter.

There are two important quantities that can be derived from the Fictitious Crack Model: the specific fracture energy G_f (already mentioned above) and the characteristic length l_{ch}. These parameters are used to try to quantify the brittleness of concrete. The specific fracture energy G_f is defined as the amount of energy needed to create one unit crack area, and is equal to the area under the softening curve of Figure 2.7b. The softening curve, and thus also the fracture energy can only be determined in a stable displacement-controlled uniaxial tension test. It is important that the deformations are corrected for the elastic deformations in the total measuring length; see also Section 2.5 and Appendix 3. Thus, the fracture energy can be written as

$$G_f = \int_0^{w_c} \sigma(w)dw \qquad (2.28)$$

Higher specific fracture energy would mean a more ductile material. Yet, the shape of the softening curve is in this context very important. When a steep stress-drop occurs, for example, directly after maximum stress, the overall behavior may still be judged as brittle.

The second parameter, the characteristic length l_{ch} of the material is defined as

$$l_{ch} = \frac{EG_f}{f_t^2} \qquad (2.29)$$

Hillerborg and coworkers proposed that the characteristic length is a measure of the brittleness of the concrete. A 16-mm concrete has a higher specific fracture energy than a 2-mm mortar. With constant E, the value of l_{ch} would depend on the ratio between G_f and f_t. Differences in tensile strength may occur due to a different w/c-ratio to achieve similar workability, the cement content, age, and other (environmental) factors. A tendency for more heterogeneous concretes to have larger characteristic lengths has been observed. There is a catch, however. The brittleness of a material may differ when the structural conditions change. This was recognized by Elfgren (1989, p. 399) who proposed a brittleness number as the quotient of the elastic energy stored in the structure and the fracture energy. The elastic energy U_E is controlled by the volume of the structure (i.e., the third power of a characteristic length of the structure L), whereas fracture energy U_s is a surface measure and depends on L^2 only. The result is:

$$\frac{U_E}{U_s} = \frac{L^3 f_t^2 / E}{L^2 G_f} = \frac{L f_t^2}{E G_f} = \frac{L}{l_{ch}} \qquad (2.30)$$

and a measure for the size of the structure comes into play. A structure is brittle when the stored elastic energy is higher than the energy needed to create the critical fracture plane; that is, $U_E \gg U_s$. When the balance is reversed, $U_E \ll U_s$, ductile behavior is observed. This result is quite interesting, and again hints toward the influence of the structural environment on the fracture properties of a material. The interdependence of structural and material aspects of fracture is at the core of coming to a true understanding of fracture mechanics. It is the main theme of this book.

2.5 Determination of FCM Parameters

The softening diagram of any material and structure can only be measured using advanced testing techniques. Nowadays a fast servohydraulic or electromechanical test machine is used; in the past systems with bars arranged in parallel to the test specimens allowed for a first insight in the postpeak behavior of concrete (and rock). In Appendix 3 the stability of fracture experiments is addressed. For the determination of the FCM-parameters it was proposed to use a displacement-controlled uniaxial tension test. The loading platens should be fixed against rotation, thereby imposing uniformly

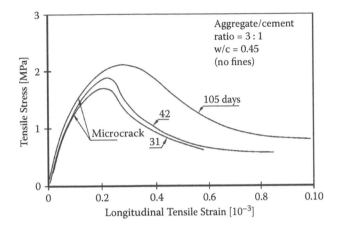

FIGURE 2.10
Stress–strain behavior of concrete subjected to uniaxial tension. (From Evans and Marathe. 1968. *Mater. Struct. (RILEM)*, 1(1): 61–64. With permission.)

distributed deformations at the specimen's ends. The FCM was developed with the now classical tensile test results of Evans and Marathe (1968) in mind. In Figure 2.10 some of their results are shown. For a concrete with a w/c-ratio of 0.45 and an aggregate-to-cement ratio of 3:1 the stress–strain curves were measured at varying ages. Using the parallel-bar setup shown in Appendix 3 (Figure A3.1a) stable softening curves were measured. These tests, together with those of Hughes and Chapman (1966) are probably the earliest known examples showing the softening behavior of concrete in tension.

The curves were very smooth and showed that after reaching a maximum (f_t), the tensile stress would decrease again, very slowly, with increasing longitudinal strain. The notion of using displacements in the postpeak regime was not introduced until 1976 with the FCM. Interestingly, Evans and Marathe marked the onset of microcracking in the diagrams, which was well in the prepeak regime, and not postpeak as assumed in the FCM (see Section 2.4). As said, Evans and Marathe seemed not to be aware of the localization of deformations in a single crack during the postpeak regime, at least not so much as to propose a change to a different model, that is, a model with different state variables for the postpeak behavior. Therefore by simply dividing the total deformations by the specimen length the longitudinal strains were determined. Nowadays it is common either to mention the measuring length used, or to show softening curves in terms of stress and displacement. Note that extensometers or LVDTs are more useful measuring devices when cracks appear; strain gauges might break and can in general not be used in fracture experiments other than showing the growth of a crack: when the strain-gauge breaks we know a crack is passing.

The exact boundary conditions in the experiments of Evans and Marathe are not completely clear. Hillerborg et al. specified that a test between fixed

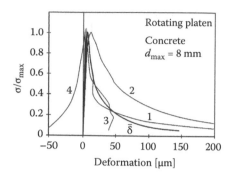
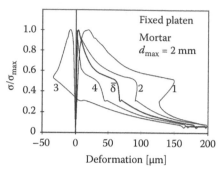

FIGURE 2.11

Effect of rotational stiffness of the loading platen on the softening behavior in uniaxial tension. (After Van Mier, Schlangen, and Vervuurt. 1995. Lattice type fracture models for concrete. In H.-B. Mühlhaus (Ed.), *Continuum Models for Materials with Microstructure*. With permission.) Cylindrical specimen of 100-mm diameter and 200-mm length were loaded either between fixed platen (a) or between freely rotating platens (b). The curves numbered 1–4 show the deformations measured at four sides of a specimen; the bold curve denoted $\bar{\delta}$ shows the average behavior.

boundaries is needed, or at least between plates with very high rotational stiffness compared to the bending stiffness of the specimen itself. In the late 1980s it was found that this led to a more or less pronounced bump in the postpeak regime, depending on the size of the specimen; see Van Mier (1986b), Van Mier and Nooru-Mohamed (1990), Hordijk (1991), and many others. In Figure 2.11 results by Van Mier, Schlangen, and Vervuurt (1995) are presented, which show the bump very clearly when fixed platens are used, whereas a smooth curve is obtained when the loading platens are free to rotate. Obviously there is more to the uniaxial tension test than previously thought. The problem raised, of course, is whether softening can be used as a material parameter. Figure 2.11 is quite clear: there is a profound influence of the structural conditions in the experiment on the test result. Either the selected test is not correct, and other loading arrangements must be established, or softening is simply a structural property, and as such cannot be used unrestrictedly in finite element analyses as proposed with the development of the FCM and the crack-band model.

There have been attempts to minimize the effect of boundaries, for example, by using very short specimens. Hordijk (1991) showed that the postpeak behavior was smoother when the specimen size was reduced to $50 \times 60 \times 50$ mm^3 (with two 5-mm deep notches at half height), yet there appears to be an effect on the maximum stress, which is considerably lower than that of 250-mm long prisms, and the peak is more "rounded." Likely boundary effects still play a role but the effects from bending are reduced when the specimen slenderness decreases. It all adds to the conclusion that softening is a structural property; the effect of boundaries and specimen size becomes notable around peak stress.

Uniaxial tension tests are not so easy to conduct; at least that is the general opinion. There are several practical problems that need to be solved. An exhaustive overview of important test-related factors has been given in Van Mier (1997) and Van Mier and Shi (2002). Most researchers tend to follow the proposal by Hillerborg and use nonrotating boundaries; see, for example, Hordijk (1991) and Mechtcherine (2007). The argument is that uniform displacement is applied, but that is only true when the total specimen length is considered. Flexure of the specimen cannot be avoided because the heterogeneity of the material will always cause fracturing from one side, in most cases a corner (if prismatic specimens are used). This asymmetric failure is not dependent on centric placement or alignment of the specimen in the loading chain, but is simply the consequence of testing a highly heterogeneous material. Thus, as shown in Figure A.2.1 in Appendix 2, crack growth will start from one side and continue to grow toward the other side, but at some moment a second crack will nucleate from the opposite side when boundaries are prevented from rotating. As a result the total crack area exceeds the specimen cross-section, and for a proper evaluation of the specific fracture energy, Equation (2.28), it is necessary to measure the exact crack area. Things become easier, as far as determining the specific fracture energy is concerned, when freely rotating platens are used. In that case only a single crack develops perpendicular to the loading direction, and it has been shown that the fracture energy is lower compared to the result from experiments between fixed platens, and likely matching the "true" specific fracture energy of the tested material; see Van Mier et al. (1995).

The second major problem in performing uniaxial tension tests is gluing the specimen between the loading platens. Gripping the specimens is not really possible, and gluing is considered the best solution for applying tensile loading. Many commercial two-component epoxies are available, with tensile bond strength far exceeding the tensile strength of plain concrete (for fiber concrete, especially high-performance fiber concrete, the bond strength of the epoxy may not be high enough), but often the concrete has to be completely dry in order to guarantee a proper adhesion to the loading platens. Essential steps in gluing are providing a rough concrete surface, taking precautions that a uniform glue layer thickness is applied, and that the glue surfaces are not polluted by some greasy substance. It may take some practice before the best combination is found, but once everything is set there are no longer any appreciable difficulties.

Notches are often used to crack the specimen at a known location. The major reason is to have a fixed place where the transducers for test control can be positioned, but an additional reason is that the stresses at the glue contact to the loading platens are somewhat lower than the stress in the notched area. In Section 6.1.4.2 we return to the use of notches and the shape of the tensile specimen in general.

Because of the aforementioned difficulties encountered in tensile testing many researchers prefer to measure the parameters for FCM via indirect

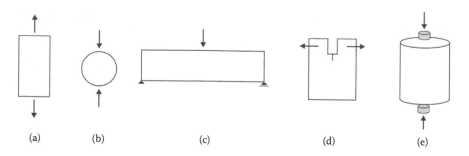

FIGURE 2.12
Direct and indirect tests for determining the tensile strength of concrete: (a) uniaxial tension test, (b) Brazilian splitting test, (c) 3-point bending test, (d) wedge-splitting test, and (e) double punch test.

test methods, including the 3-point or 4-point bending test (Figure 2.12c), the wedge-splitting test (Figure 2.12d; Tschegg and Linsbauer 1986 and Linsbauer and Tschegg 1986), the Brazilian splitting test (Figure 2.12b, proposed by Akazawa in 1943 and Carneiro in 1949; see Nilsson 1961) or the double-punch test (Figure 2.12e; Chen 1970). These tests are favored because in general they can be performed in a standard compression test rig and no complicated gluing procedure is needed. The Brazilian test, using either cubes or cylinders, will provide an estimate of the uniaxial tensile strength at best. The outcome depends very much on details such as the width of the strips used to load the specimen. These matters are usually prescribed in model codes, and it is quite essential to perform tests as laid down in codes in order to allow for comparison with results from others. Bending tests or wedge-splitting tests are not useful for determining the tensile strength of concrete; at best the specific fracture energy can be determined. Guinea, Planas, and Elices (1992), Planas, Elices, and Guinea (1992), and Elices, Guinea, and Planas (1992) have published in-depth analyses of problems met in the 3-point bending tests, and indicate how the fracture energy should be derived, eliminating the known error sources as much as possible.

For the determination of the FCM parameters the bending test (or wedge-splitting test) must be combined with the Brazilian splitting test: the bending test (or wedge-splitting test) gives the specific fracture energy; the Brazilian test provides an estimate for the uniaxial tensile strength (see Planas et al. 2007). Using inverse parameter identification techniques, the softening diagram as needed in the FCM (usually a bilinear diagram is fitted) may be derived, but uncertainty about uniqueness of the obtained solution will always remain; see, for example, Wittmann et al. (1987). In Section 6.2 we return to these indirect methods. There a meso-level analysis provides insight to the actual behavior in such experiments, and clarifies why some of the test methods should not be used.

Finally, because in this book we consider each experiment that leads to complete failure of the specimen as a structural experiment, it is not considered useful to address one of these tests as leading to "the best" or "pure" material properties. This point of view is abandoned completely. It can always be shown that specimen size and boundary conditions affect the outcome of any fracture experiment, and thus one always determines structural properties of the complete specimen/machine system. Thus, instead, all experiments are analyzed as being a structure using meso-level models that require simple input parameters only, such as local tensile strength and the Young's modulus of the various material phases inside the concrete (matrix, aggregate, interfacial transition zone). Before analyzing the various loading cases we next explain the meso-level approach.

3

Mechanics Aspects of Lattice Models

3.1 Short Introduction to Framework Analysis

A continuum can be discretized in a lattice of truss or beam elements. This has been known for a long time: in 1941 Hrennikoff showed how the element parameters should be set to obtain the same elastic properties in a truss lattice and a shell element loaded in plane stress. A lattice can consist of different types of elements. Linear elements are always used, but the connectivity may vary. Truss elements (Figure 3.1a) have two degrees of freedom (dof) in each node (displacements u_i and v_i); beam elements also include rotations (φ_i) and have 3dof per node (Figure 3.1b). Thus for the beam element of Figure 3.1b the displacement vector is

$$\bar{\mathbf{v}}^T = \left| \begin{array}{cccccc} \bar{u}_i & \bar{v}_i & \bar{\varphi}_i & \bar{u}_j & \bar{v}_j & \bar{\varphi}_j \end{array} \right| \tag{3.1}$$

where the subscript i and j refer to the two nodes. For the simple truss element of Figure 3.1a the two rotations φ_i and φ_j are missing. The hyphen in $\bar{\mathbf{v}}$ indicates that displacements are considered in the local coordinate system. This is the situation for simple linear elements deforming in a two-dimensional plane. We start by briefly describing the set of equations needed in a 2D truss or beam-lattice. Generalization to three dimensions is quite elaborate but straightforward.

Three sets of equations must be solved to arrive at a solution in structural mechanics: the kinematic equations describe the relation between displacements and strains, the constitutive equations relate stress and strain, and equilibrium equations describe the relationship between internal stresses and external forces. For an elastic material the relation between strains ε and displacements $\bar{\mathbf{v}}$ is given through the combination matrix $\mathbf{C}$ following

$$\varepsilon = \mathbf{C}\bar{\mathbf{v}} \tag{3.2}$$

For the simple truss element of Figure 3.1a, three degrees of freedom are needed to describe the rigid body motion, which leaves one dof to describe

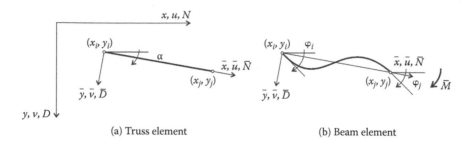

FIGURE 3.1
Positive directions for forces and displacements for truss element (a) and beam element (b).

the deformation of the element; in this case only the axial deformation is needed: $\varepsilon = \bar{u}_j - \bar{u}_i$. Similarly, for the beam element of Figure 3.1b the rigid body motion is described by 3 dof, and thus 3 dof are left for describing the relation between displacements and strains, following:

$$\varepsilon_1 = \bar{u}_j - \bar{u}_i \tag{3.3a}$$

$$\varepsilon_2 = \bar{\varphi}_i - \frac{\bar{v}_j - \bar{v}_i}{l} \tag{3.3b}$$

and

$$\varepsilon_3 = \bar{\varphi}_j - \frac{\bar{v}_j - \bar{v}_i}{l} \tag{3.3c}$$

The constitutive equations are defined through the stiffness matrix **S**,

$$\sigma = \bar{\mathbf{S}}\varepsilon \tag{3.4}$$

For the simple truss element the relation between stress and strain is quite straightforward and is written as

$$\bar{\mathbf{S}} = \frac{EA}{l} \tag{3.5}$$

where E is the Young's modulus of the (elastic) material and A is the cross-sectional area of the truss element. Finally, we need the relation between stresses and external forces, which can be derived using the combination matrix **C** following

$$\mathbf{C}^T\boldsymbol{\sigma} = \bar{\mathbf{k}} \qquad (3.6)$$

where $\bar{\mathbf{k}}$ is the vector containing the nodal forces N_i, D_i, N_j, and D_j as indicated in Figure 3.1a. Positive directions are as shown in this figure.

The above is written in terms of the local coordinate system of the considered truss (or beam) element. In the global lattice elements may have different orientations and it is necessary to transform everything to the same global coordinate system. Using the transformation matrix **T** local forces and displacements are transformed following

$$\bar{\mathbf{v}} = \mathbf{T}\mathbf{v} \qquad (3.7)$$

and

$$\mathbf{k} = \mathbf{T}^T\bar{\mathbf{k}} \qquad (3.8)$$

The transformation matrix **T** takes the following form,

$$\mathbf{T} = \begin{vmatrix} \mathbf{T}_i & 0 \\ 0 & \mathbf{T}_j \end{vmatrix} \qquad (3.9)$$

where $\mathbf{T}_i$ and $\mathbf{T}_j$ are identical if both element ends have the same orientation:

$$\mathbf{T}_i = \mathbf{T}_j = \begin{vmatrix} \cos\alpha & \sin\alpha \\ -\sin\alpha & \cos\alpha \end{vmatrix} \qquad (3.10)$$

and α is the angle between the truss element and the positive x-axis as shown in Figure 3.1.

Finally the complete relation between external forces and displacements can be formulated as

$$\mathbf{k} = \mathbf{T}^T\mathbf{C}^T\bar{\mathbf{S}}\mathbf{C}\mathbf{T}\mathbf{v} = \mathbf{S}\mathbf{v} \qquad (3.11)$$

S is the stiffness matrix for a single element in the global coordinate system. The entire matrix for the global lattice, containing many elements, is then constructed placing exactly those stiffness terms at their respective places. Where two or more truss elements meet in the same node, the stiffness terms are added. In the end a symmetric diagonal matrix emerges, which is quite suitable for numerical processing. The entire procedure is described in detail in various textbooks, for example, Nijenhuis (1973). We do not go into further detail in this chapter where the goal is to describe a numerical model globally based on the finite element method only, and indicating its potential use for the analysis of particle composites. Before doing that, however, it is interesting to consider the elastic properties of the lattice. For that purpose we write in full detail the equations for a simple truss element. First we compute $\mathbf{CT} = \mathbf{D}$.

$$\mathbf{CT} = \begin{vmatrix} -1 & 0 & 1 & 0 \end{vmatrix} \cdot \begin{vmatrix} \cos\alpha & \sin\alpha & 0 & 0 \\ -\sin\alpha & \cos\alpha & 0 & 0 \\ 0 & 0 & \cos\alpha & \sin\alpha \\ 0 & 0 & -\sin\alpha & \cos\alpha \end{vmatrix} \quad (3.12)$$

$$= \mathbf{D} = \begin{vmatrix} -\cos\alpha & -\sin\alpha & \cos\alpha & \sin\alpha \end{vmatrix}$$

Next we determine the stiffness matrix from

$$\mathbf{D}^T\overline{\mathbf{S}}\mathbf{D} = \mathbf{T}^T\mathbf{C}^T\overline{\mathbf{S}}\mathbf{C}\mathbf{T} = \begin{vmatrix} -\cos\alpha \\ -\sin\alpha \\ \cos\alpha \\ \sin\alpha \end{vmatrix} \cdot \frac{EA}{l} \cdot \begin{vmatrix} -\cos\alpha & -\sin\alpha & \cos\alpha & \sin\alpha \end{vmatrix} =$$

$$= \mathbf{S} = \frac{EA}{l} \begin{vmatrix} \cos^2\alpha & \cos\alpha\sin\alpha & -\cos^2\alpha & -\cos\alpha\sin\alpha \\ \cos\alpha\sin\alpha & \sin^2\alpha & -\cos\alpha\sin\alpha & -\sin^2\alpha \\ -\cos^2\alpha & -\cos\alpha\sin\alpha & \cos^2\alpha & \cos\alpha\sin\alpha \\ -\cos\alpha\sin\alpha & -\sin^2\alpha & \cos\alpha\sin\alpha & \sin^2\alpha \end{vmatrix}$$

$$(3.13)$$

which can be written in compact form as

$$\mathbf{S} = \begin{vmatrix} \mathbf{S}_{11} & \mathbf{S}_{12} \\ \mathbf{S}_{21} & \mathbf{S}_{22} \end{vmatrix} \quad (3.14)$$

This compact form is the usual representation in truss and frame analysis. Close scrutiny of Equation (3.13) reveals

$$\mathbf{S}_{11} = -\mathbf{S}_{12} = -\mathbf{S}_{21} = \mathbf{S}_{22} = \frac{EA}{l} \begin{vmatrix} \cos^2\alpha & \cos\alpha\sin\alpha \\ \cos\alpha\sin\alpha & \sin^2\alpha \end{vmatrix} \quad (3.15)$$

When the local and global coordinate systems coincide, $\alpha = 0$ and all the elements in the 2 × 2 matrix are zero except for the first diagonal element, which is equal to 1.

3.2 Equivalence between a Shell Element and a Simple Truss (Hrennikoff)

Let us now consider a simple truss that is supposed to replace a shell element loaded in plane stress. In Figure 3.2 the two cases are shown side by side in a global Cartesian coordinate system (x,y). This situation was first considered by Hrennikoff (1941), but the solution can be found in many textbooks. Here we follow the solution provided in Nijenhuis (1973). The shell element is replaced by a simple truss containing six elements. The main question is how the sectional area of the various truss elements must be selected to obtain the same elastic deformations as the shell element. In Figure 3.2 the sectional areas are indicated as A_1, A_2, and A_3 for the horizontal, vertical, and diagonal elements, respectively. The length of the horizontal and vertical ele-

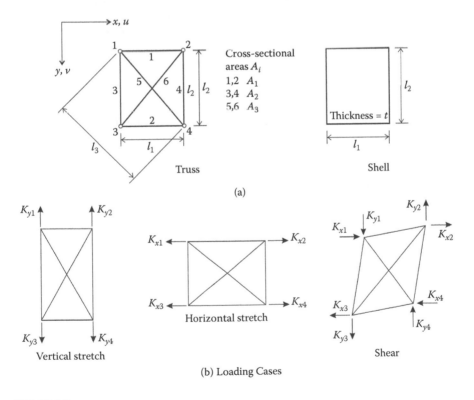

FIGURE 3.2
Shell element and equivalent truss lattice under generalized plane stress (a). For determining the geometrical properties of the lattice and the shell element three loading cases must be considered (b): vertical stretch, horizontal stretch and shear. (Adapted from Nijenhuis. 1973. *De Verplaatsingsmethode: Toegepast voor de Berekening van (staaf) Constructies.* With permission from Elsevier.)

ments is equal to $l_1 = l$ and $l_2 = cl$; consequently the diagonals have length $l_3 = l(1 + c)^{1/2}$. The thickness is denoted by the symbol t ($t < l$).

The deformation of the shell element under plane stress is well known:

$$\varepsilon_x = \frac{1}{E}\sigma_x - \frac{v}{E}\sigma_y, \quad \varepsilon_y = \frac{v}{E}\sigma_x - \frac{1}{E}\sigma_y, \quad \text{and} \quad \gamma_{xy} = \frac{2(1+v)}{E}\tau_{xy} \qquad (3.16)$$

The 6-element truss which is compared to the shell element has four nodes, and the relation between the nodal forces and displacements is described completely by

$$\begin{vmatrix} K_{x1} \\ K_{y1} \\ K_{x2} \\ K_{y2} \\ K_{x3} \\ K_{y3} \\ K_{x4} \\ K_{y4} \end{vmatrix} = \begin{vmatrix} S(1)+S(3)+S(5) & -S(1) & -S(3) & -S(5) \\ -S(1) & S(1)+S(4)+S(6) & -S(6) & -S(4) \\ -S(3) & -S(6) & S(2)+S(3)+S(6) & -S(2) \\ -S(5) & -S(4) & -S(2) & S(2)+S(4)+S(5) \end{vmatrix} \cdot \begin{vmatrix} u_1 \\ v_1 \\ u_2 \\ v_2 \\ u_3 \\ v_3 \\ u_4 \\ v_4 \end{vmatrix}$$

$$(3.17)$$

where the $S(i)$ are all 2×2 matrices as shown in Equation (3.15) and are the individual stiffness components for the elements $i = 1, \dots, 6$. In each submatrix the appropriate value of α for each truss element must be substituted. After specifying the external nodal forces on the truss-work to resemble the stresses on the shell element, and after some mathematical manipulation (Nijenhuis 1973), two sets of equations are derived that must be fulfilled simultaneously.

First set:

$$A_1 = lt\frac{c^2 - v}{4v(1+v)c} \quad \text{and} \quad A_2 = lt\frac{2+3v-c^2}{4(1+v)} \qquad (3.18)$$

Second set:

$$A_1 = lt\frac{2c^2 - 1 + 3vc^2}{4(1+v)c} \quad \text{and} \quad A_2 = lt\frac{1 - vc^2}{4v(1+v)} \qquad (3.19)$$

It is quite straightforward to show that Equations (3.18) and (3.19) can only be fulfilled simultaneously when $v = 1/3$, which leads to the final result

$$A_1 = \frac{3}{16}\frac{3c^2-1}{c}lt, \quad A_2 = \frac{3}{16}(3-c^2)lt, \quad \text{and} \quad A_3 = \frac{3}{16}\frac{(1+c^2)^{3/2}}{c}lt \quad (3.20)$$

For a square plate, $c = 1$, and the three cross-sectional areas are $A_1 = A_2 = 3lt/8$ and $A_3 = 3\sqrt{2}lt/8$.

In the above example a regular rectangular lattice with crossing diagonals was used. The geometry of the lattice may vary from other regular schemes to completely random geometries. The analysis to match the stiffness of shell and truss-work becomes increasingly more tedious. In general, for more complicated lattices one must revert to numerical analysis. The same approach can be followed for 3D lattices, but in that case an analytical solution as given above is almost unthinkable.

In this book we are interested in developing a model based on a lattice for simulating the fracture of heterogeneous materials. Fracturing poses a problem because nonlinearity must be introduced in these computational schemes, which are suitable for elastic analysis. The problem with fracturing is that the stiffness of a fractured element may eventually become zero, also on the main diagonal of the stiffness matrix, which then becomes singular. One way around this problem is to remove the element when it is fractured, and reconstruct the stiffness matrix with one element less. Even then, when the lattice is built from truss-elements (normal force lattice), unstable situations may arise depending on the connectivity of the remaining elements. This may be one, among several other arguments, to revert to a beam lattice. Instabilities in the lattice after removal of even a large number of beams are hardly possible because of the other degrees of freedom in such elements, in particular the bending stiffness.

So, let us consider the element stiffness matrix for a beam. The C-matrix for a beam with 6 dof (2D beam element) is based on Equation (3.3):

$$\varepsilon = \mathbf{C}\bar{\mathbf{v}} = \begin{vmatrix} -1 & 0 & 0 & 1 & 0 & 0 \\ 0 & 1/l & 1 & 0 & -1/l & 0 \\ 0 & 1/l & 0 & 0 & -1/l & 1 \end{vmatrix} \cdot \begin{vmatrix} \bar{u}_i \\ \bar{v}_i \\ \bar{\varphi}_i \\ \bar{u}_j \\ \bar{v}_j \\ \bar{\varphi}_j \end{vmatrix} \quad (3.21)$$

The stiffness relations now include bending, and the various components can be derived giving unit displacements or rotations to the beam element. The resulting **S**-matrix for a single beam element is equal to

$$\bar{S} = \begin{vmatrix} \dfrac{EA}{l} & 0 & 0 \\[2ex] 0 & \dfrac{4EI}{l} & \dfrac{2EI}{l} \\[2ex] 0 & \dfrac{2EI}{l} & \dfrac{4EI}{l} \end{vmatrix} \tag{3.22}$$

where E is the Young's modulus of the (elastic) material, A is the cross-section of the beam ($A = bh$ for rectangular beams, and $A = \pi d^2 / 4$ for beams with a circular cross-section, which is more suitable for 3D analyses), and I is the moment of inertia ($I = bh^3/12$ for rectangular beams).

The relation between nodal forces and nodal displacements in the global coordinate system with $\alpha = 0$ ($\mathbf{T} = \mathbf{T}^T = \mathbf{I}$) can now be obtained by going through all the transformations done before for the truss element in Equations (3.12) and (3.14):

$$\mathbf{S} = \mathbf{T}^T\mathbf{C}^T\bar{\mathbf{S}}\mathbf{C}\mathbf{T} = \mathbf{C}^T\bar{\mathbf{S}}\mathbf{C} =$$

$$= \begin{vmatrix} \dfrac{EA}{l} & 0 & 0 & -\dfrac{EA}{l} & 0 & 0 \\[2ex] 0 & \dfrac{12EI}{l^3} & \dfrac{6EI}{l^2} & 0 & -\dfrac{12EI}{l^3} & \dfrac{6EI}{l^2} \\[2ex] 0 & \dfrac{6EI}{l^2} & \dfrac{4EI}{l} & 0 & -\dfrac{6EI}{l^2} & \dfrac{2EI}{l} \\[2ex] \dfrac{EA}{l} & 0 & 0 & \dfrac{EA}{l} & 0 & 0 \\[2ex] 0 & -\dfrac{12EI}{l^3} & -\dfrac{6EI}{l^2} & 0 & \dfrac{12EI}{l^3} & -\dfrac{6EI}{l^2} \\[2ex] 0 & \dfrac{6EI}{l^2} & \dfrac{2EI}{l} & 0 & -\dfrac{6EI}{l^2} & \dfrac{4EI}{l} \end{vmatrix} = \tag{3.23}$$

$$= \begin{vmatrix} S_1 & 0 & 0 & -S_1 & 0 & 0 \\ 0 & S_2 & S_3 & 0 & -S_2 & S_3 \\ 0 & S_3 & 2S_4 & 0 & -S_3 & S_4 \\ -S_1 & 0 & 0 & S_1 & 0 & 0 \\ 0 & -S_2 & -S_3 & 0 & S_2 & -S_3 \\ 0 & S_3 & S_4 & 0 & -S_3 & 2S_4 \end{vmatrix}$$

The shorthand forms

$$S_1 = \frac{EA}{l}, \quad S_2 = \frac{12EI}{l^3}, \quad S_3 = \frac{6EI}{l^2}, \quad \text{and} \quad S_4 = \frac{2EI}{l}$$

are useful when it comes to setting up the equations for a small framework. In compact form the above stiffness matrix can be written as

$$S = \begin{vmatrix} S_{11} & S_{12} \\ S_{21} & S_{22} \end{vmatrix} \qquad (3.24)$$

but for these more complicated beam elements it is not possible to simplify matters any further as we did for the truss element in Equations (3.14) and (3.15).

3.3 Effective Elastic Properties of Beam Lattices

It is interesting to see what changes in the lattice of Figure 3.2 when the truss elements are replaced by beams. How should the dimensions of the beams be selected to obtain the same elastic response as the simple shell element loaded in plane stress? This problem was resolved by Schlangen and Garboczi (1997) for regular triangular lattice configurations based on evaluation of the elastic energy stored in a single unit cell subjected to uniform strain. Only the final result is given here:

Bulk modulus

$$K = \frac{\sqrt{3}}{2} \frac{EA}{l} \qquad (3.25a)$$

Shear modulus

$$G = \frac{\sqrt{3}}{4} \frac{EA}{l} \left(1 + \frac{12I}{Al^2}\right) \qquad (3.25b)$$

Poisson's ratio

$$\nu = \frac{K - G}{K + G} = \frac{1 - \dfrac{12I}{Al^2}}{3 + \dfrac{12I}{Al^2}} \qquad (3.25c)$$

With $I = (1/12)bh^3$ and with $A = bh$ the expression for the Poisson's ratio can be rewritten as

$$\nu = \frac{\left(1-\left(\dfrac{h}{l}\right)^2\right)}{\left(3+\left(\dfrac{h}{l}\right)^2\right)} \tag{3.26}$$

Figure 3.3 shows the relation between ν and h/l in graphic form. The Poisson ratio is independent for the choice of b and depends on the ratio h/l of the beam elements only. Thus by choosing the appropriate h/l ratio the Poisson ratio of the beam lattice can be set to resemble actual values of the material to be analyzed. There are some restrictions, however, inasmuch as the highest value for $\nu = 1/3$ when $h/l = 0$ and the lowest value appears for $h/l \to \infty$, namely $\nu = -1$.

A regular triangular lattice has some disadvantages when analyzing cracking, in particular when heterogeneous materials are considered. Because in 2D, in regular triangular lattices (assuming that all beams have equal length) three symmetry axes appear, preferential crack paths are defined by the lattice geometry. Mesh sensitivity can thus be expected, but this can be avoided by bringing some randomness into the lattice through a variation of the elastic properties or fracture thresholds of the lattice. Alternatively it is possible to revert to triangular lattices with random beam lengths, for example, the random lattice developed by Mourkazel and Herrmann (1992). The construction of the random lattice is explained in Section 4.5. Here we discuss the elastic properties. For lattices with random beam lengths Vervuurt (1997) analyzed

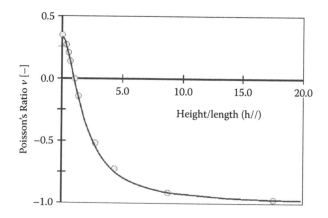

FIGURE 3.3
Relation between beam size h/l and Poisson's ratio ν in a regular triangular lattice. (After Schlangen and Van Mier. 1994. *Computer Methods and Advances in Geomechanics.*)

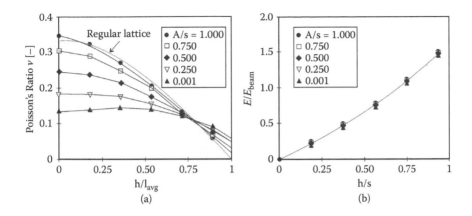

FIGURE 3.4

(a) Relation between effective Poisson ratio v and h/l_{avg} ratio for a lattice with random beam lengths. Also included is Equation (3.25c), the Poisson ratio for a regular triangular lattice. The factor A/s is the randomness parameter, which may vary between zero (regular triangular lattice) and 1.0 (full randomness); see Section 4.5. (b) Relation between Young's modulus E/E_{beam} and h/s for random lattices with $0 < A < 1$. (After Vervuurt. 1997. *Interface Fracture in Concrete*. With kind permission of Dr. Adri Vervuurt.)

the effective Poisson ratio and Young's modulus as a function of the height h over average length l_{avg} of the lattice beams (h/l_{avg}) and as a function of the randomness parameter A/s (see Figure 4.11b in Section 4.5). Several random lattices were generated, each measuring 50×50 nodes. For each randomness (A/s = 1.0, 0.75, 0.5, 0.25, and 0.001) several lattices were generated and the results presented in Figure 3.4 are the average values of all these analyses. In addition Equation (3.25c), that is, the Poisson ratio for a regular triangular lattice, has been included in Figure 3.4a. For the regular lattice the Poisson ratio will be smaller than zero if h/l exceeds 1.0, and this part of the curve has been omitted. The effect of randomness on the Poisson ratio is considerable. For small A/s the random lattice geometry almost resembles a regular triangular lattice, and the Poisson ratios are almost the same (compare a regular lattice with A/s = 0.001). When the randomness increases, the value of v decreases compared to that of a regular lattice, at least as long as $h/l_{avg} <$ 0.75; the trend reverses for larger values of h/l_{avg}. For materials such as mortar and concrete the experimentally determined Poisson ratio lies between 0.15 and 0.20. Considering the empirical relations in Figure 3.4a, it is obvious that larger h/l_{avg} must be selected to achieve such a value if the randomness of the lattice is large. If v = 0.2, then h/l = 0.58 for a regular triangular lattice according to Equation (3.25c). In order to obtain v = 0.2 for a random lattice with A/s = 0.5, 0.75, and 1.0, h/l_{avg} = 0.48, 0.57, and 0.61, respectively. This means that rather "stubby" beams (height is almost equal to length) must be used, which makes the application of Bernoulli beams rather debatable. For that reason several authors have suggested replacing the Bernoulli beams by Timoshenko beams.

Figure 3.4b shows that when the height of the beams is determined, the local Young's modulus of the beams E_{beam} can be determined from the global Young's modulus of the lattice, denoted by E. The influence of randomness A/s has almost disappeared. Note that along the horizontal axis h/s has been plotted instead of h/l_{avg}. Linearity causes the E/E_{beam} to be constant when h/s is fixed.

The geometry of the lattice with random beam lengths is isotropic. The angles between adjacent beams vary. On average six beams connect in each node. When a uniform strain is applied to such a lattice, nodes are not at a center of symmetry and have to move to achieve equilibrium. This line of reasoning was followed by Schlangen and Garboczi (1996), and by varying the values for A (area) and I (moment of inertia) for each individual beam in the random lattice they achieved a uniform elastic lattice. The drawback was, however, that some of the beams had to be assigned a negative cross-sectional area A in order to obtain uniform elastic behavior. One of the other difficulties encountered is of course that A and I are not independent of each other, therefore a unique solution cannot be obtained. For a random lattice homogenized in this way the following expressions were found for bulk modulus K, shear modulus G, and Poisson ratio ν, respectively:

$$K = \frac{E_{beam}}{4A_{tot}} \sum_{i=1}^{N} A_i l_i \tag{3.27a}$$

$$G = \frac{E_{beam}}{4A_{tot}} \sum_{i=1}^{N} \left[\frac{12I_i}{l_i} + \frac{4\alpha_i}{l_i^3}\left(A_i - \frac{12I_i}{l_i^2} \right) \right] \tag{3.27b}$$

$$\nu = \frac{K-G}{K+G}, \quad -1 < \nu < \frac{1}{3} \tag{3.27c}$$

In these equations E_{beam} is the Young's modulus of a beam, A_{tot} is the total area of the lattice, and $\alpha_i = (x_1 - x_2)^2(y_1 - y_2)^2$ for the unrestrained endpoints (x_1, y_1), (x_2, y_2) of the ith beam. Note that these equations are valid only for the homogenized lattice obtained using the procedure described by Schlangen and Garboczi. If the method is changed, or the lattice is newly generated, the equations must be derived again. It is quite questionable if this procedure is helpful. One of the goals of using a lattice with random beam lengths is to eliminate mesh dependency in the fracture patterns. The randomness is a convenient way to incorporate material heterogeneity, as was the original intention for materials such as concrete and rock. Consequently the strain field will become nonuniform, and the best approach seems to keep A and I the same for all beams in the lattice. The simplicity derived from this, at least at the moment, would be preferable instead of the added complexity for obtaining an homogenized random lattice as proposed by Schlangen and Garboczi. The homogenized random lattice leads to nice mesh-independent crack patterns, but the same is true for the direct application of the (unhomogenized) random lattice; see Schlangen and Garboczi (1996) and Vervuurt (1997), respectively.

3.4 Similarity between Beam Lattice Model and Particle Model

For simulating the mechanical behavior of particulate materials such as sand often so-called particle models are used, for example, the models developed by Cundall and Strack (1979), Thornton and Antony (2000), Luding (2004), and many others. Basically the material is considered as an assemblage of spheres which is a close representation for soils and other granular media, but may be a bit far-sought for concrete and rock, except perhaps for sandstone. Many geological materials, or rather geomaterials when concrete is included, show quite similar mechanical behavior at the macroscopic level, which may lead to the idea to model all these materials with the same type of particle models. In general equal-sized particles are used, with spherical shape, although in some cases certain particle variations (in size and shape) were considered. The particles are generally considered as rigid, and all deformation is localized in the contacts between the spheres, which may transmit normal and shear forces; shearing not only includes sliding friction but also rolling friction (see, e.g., Iwashita and Oda 2000). The contact force-displacement relations for normal and shear forces are of particular interest because they define the mechanical response, and various proposals have been made over the years. We return to these matters in Chapter 11. Similarities between the particle models and a beam lattice have been demonstrated by Beranek and Hobbelman (1992, 1994). They considered a particle model where a thin layer of "matrix-material" appeared between two neighboring spheres. The spheres were all assumed to be undeformable, and consequently all deformations occurred in the contact layer. The properties of the contact layer are the Young's modulus E^*, the Poisson ratio v^*, and shear modulus G^*, where the Young's modulus and shear modulus are related following

$$G^* = \psi^* E^* \tag{3.28}$$

Beranek and Hobbelman considered a hexagonal close-packing (hcp), which reduces to either a regular triangular lattice or a regular diamond-shaped lattice for various 2D intersections of a 3D hcp stack. In the example presented here only the 2D equivalence between an assemblage of disks (the

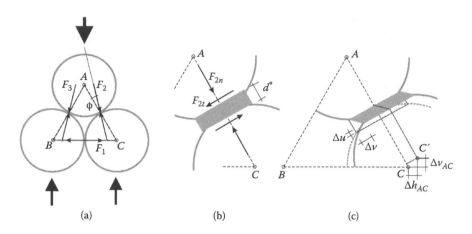

FIGURE 3.5
Assemblage of three rigid disks A, B, and C subjected to vertical compression (a), close-up of the contact layer with thickness d^* and area A^* (b), and deformed contact layer under normal and shear loading (c). (From Beranek and Hobbelman. 1992. *Glossary of the Mechanical Behaviour of Masonry*. With permission from CURNET.)

distance between the centers of two neighboring disks is twice the radius and equal to l, which is the length of the elements in the beam lattice) and the regular triangular beam lattice is shown. Figure 3.5a–c shows the particle stack considered; all relevant parameters are indicated. Of particular interest are the displacements Δu and Δv of the contact layer, which has thickness d^* and area A^*.

The contact layer is subjected to a normal force N and shear force D (see Figure 3.6), which results in displacements Δu and Δv. The displacements can be written as

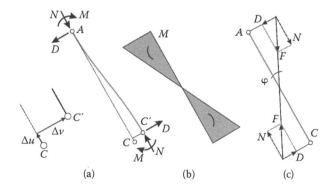

FIGURE 3.6
Beam model: definitions for an equivalent beam: (a) external forces and displacements for a single beam, (b) moment distribution, and (c) compressive load. (From Beranek and Hobbelman. 1992. *Glossary of the Mechanical Behaviour of Masonry*. With permission from CURNET.)

$$\Delta u = \varepsilon_{nn} d^* = \frac{d^*}{E^*} \sigma_{nn} = \frac{d^*}{E^* A^*} N \tag{3.29a}$$

$$\Delta v = 2\varepsilon_{nt} d^* = \frac{d^*}{G^*} \sigma_{nt} = \frac{d^*}{\psi^* E^* A^*} D \tag{3.29b}$$

The ratio between the normal and shear displacement is then,

$$\frac{\Delta v}{\Delta u} \frac{1}{\psi^*} \frac{D}{N} = \frac{1}{\psi^*} \tan \varphi \tag{3.30}$$

The angle φ is shown in Figure 3.5a, and derives directly from equilibrium considerations:

$$\tan \varphi = \sqrt{3} \frac{\psi^*}{2 + \psi^*} \tag{3.31}$$

For a comparison with a beam lattice consider the situation sketched in Figure 3.6a–c. The assemblage of disks is replaced by a regular triangular beam lattice. The dashed lines in Figure 3.5a are the lattice beams connecting the particle centers.

The demand is that the lattice beams deform in such a way that the nodal displacements of the centers of the particles are identical to those in the particle model where all deformations are lumped in the contact zone. As debated by Beranek and Hobbelman, the bending moment is zero at the location of the contact zone. Only normal and shear forces must be considered and thus the normal stiffness EA and bending stiffness EI must be selected in such a way that the deformations are the same. For the beam the displacements are as follows:

$$\Delta u = \frac{Nl}{EA} \tag{3.32a}$$

$$\Delta v = \frac{Dl^3}{3EI} - \frac{Ml^2}{2EI} = \frac{Dl^3}{12EI} \tag{3.32b}$$

where l is the length of the lattice beam, or the distance between two particle centroids. Similar to Equation (3.30) we can now calculate the displacement ratio

$$\frac{\Delta v}{\Delta u} = \frac{Al^2}{12I} \frac{D}{N} = \frac{Al^2}{12I} \tan \varphi \tag{3.33}$$

Equating to Equation (3.29) leads to,

$$I = \frac{\psi^*}{12} Al^2 \tag{3.34}$$

which should be fulfilled to guarantee equal displacements in the particle and the beam lattice model. For a circular cross-section of the beam

$$A = \pi r^2, \quad I = \frac{\pi}{4} r^4$$

and thus follows for the radius of the beam's cross-section $r = 0.378l$ when it is assumed that $v^* = 1/6$. Likewise, for a square cross-section with $A = b^2$, $I = b^4/12$ it follows that $b = 0.655l$. This is a rather stubby beam with a large height compared to its length, and it is debatable whether simple Bernoulli beam theory still applies. Most likely Timoshenko beams would be more appropriate as argued by several authors.

The relation between the effective Poisson ratio and the h/l ratio of the lattice beam of Equation (3.26) can also be used to determine the effective properties of the contact zone in the particle model, given of course the various assumptions made. One advantage of the particle model and the idea that all deformations are lumped in the contact zone is that a Mohr–Coulomb type failure criterion can easily be formulated. Such a failure criterion is considered quite appropriate for geomaterials, as we debate in the next section.

3.5 Fracture Criteria

Part of the lattice model is of course the constitutive equation, in particular the fracture criterion used. The simplest approach is removing lattice elements for which the loading has exceeded a certain critical threshold. The "gap" between the remaining elements is considered to be a crack. When the element is removed, the lattice contains one element less, and the excess load is redistributed over neighboring elements upon reloading. Thus, in each step the external load on the lattice is increased and the element closest to the critical threshold is removed. Because the critical element is simply removed, the load carried by that particular part of the lattice is relaxed instantaneously; that is, no softening is assumed at the local level of the lattice beams. In Chapter 2 we already suggested that softening should be considered a structural property, that is, depending on the structural boundary conditions in the immediate surroundings of the element. By simply assuming a steep load drop after the critical (strength) threshold has been reached is no more than saying, "We don't know what the actual softening behavior at the meso-level looks like," or, "We cannot untangle the structural influences from the lattice-beam element at this stage of the debate."

The (strength) threshold is obviously an important element in a lattice analysis. What should best be used depends to some extent on the problem to be solved and the lattice layout used (truss or frame-lattice, 2D or 3D lattice). As we have seen from the foregoing discussion on the elastic properties of a lattice, beam lattices can be fitted over a wider range to match the elastic constants of (uncracked) concrete. Truss lattices have some serious limitations, as shown in Section 3.2. There is another argument to use beam lattices in fracture simulations, namely, the lattice may become unstable when too many beams are removed, as in the case of very dense crack patterns. Beam lattices are just slightly more robust under that circumstance.

The simplest possible fracture law is where the effective tensile stress in a truss or beam lattice element is tested against the (local) uniaxial tensile strength of the material. Thus,

$$\sigma_{eff} = \sigma_{tens} = \frac{N}{A} \geq f_t \tag{3.35}$$

where N is the normal force of the lattice element and A is its cross-sectional area. The above criterion can be used in 2D or 3D lattices. For 3D simulations this simple Rankine-like criterion is considered the best choice, mainly because of its simplicity; see Lilliu (2007).

In the case of a 2D beam lattice, one might not just look to the tensile stress caused by the normal force N, but include the effects of bending, following,

$$\sigma_{eff} = \frac{N}{A} \pm \alpha \frac{(|M_i|, |M_j|)_{max}}{W} \geq f_t \tag{3.36}$$

where M_i, M_j are the bending moments in the nodes i and j, respectively, and W is the section modulus (see Figure 3.7a). The coefficient α (between 0 and 1) can be added to limit the effect of bending in the fracture law. For further debate on this criterion see Schlangen (1993). Here it is just mentioned that the criterion, Equation (3.36), is well suited for simulating the fracture of concrete subjected to tensile or combined tensile and shear loading. For compressive loading the model seems less ideal; see Chapter 8. In three dimensions it is required to include bending in the x- and y-direction (the z-axis is oriented along the beam axis), and likely also torsion; see Lilliu and Van Mier (2003).

Fracture of concrete subjected to compression is accompanied by a considerable amount of tensile microcracking, and it is tempting to assume that the above criteria would suffice for compression as well. Yet failure seems

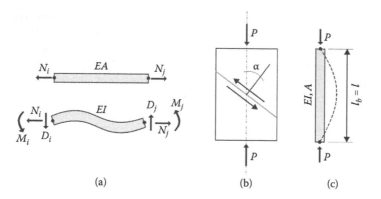

(a)   (b)   (c)

FIGURE 3.7
Fracture laws for lattice beam elements: (a) normal force or combined normal force and bending criterion, (b) prismatic element subjected to axial compression with inclined failure plane, and (c) column between pinned supports loaded in compression (buckling instability).

to be caused by additional mechanisms. The microcracking is important, but should best be seen as a means to weaken the material, before compressive failure may proceed. Following the classical work of Coulomb (see, for instance, Timoshenko 1983) the failure of solids in compression may be the result of shear failure and sliding along an inclined plane to the loading direction. If the normal to the shear plane makes an angle α to the axis of the compression load P (as shown in Figure 3.7b), and failure along the inclined plane is governed by overcoming the tensile resistance f_t of the material only, the maximum compression load P_{max} derived from

$$P \cdot \sin \alpha = P_{max} = \frac{f_t A}{\cos \alpha} \tag{3.37}$$

where A is the cross-sectional area of the prism. When $\alpha = 45°$, the maximum failure load is achieved, $P_{max} = 2A \cdot f_t$, that is, just two times the maximum failure load carried under uniaxial tension. For concrete, brick, and several types of rock the ratio of compressive to tensile failure stress is in the range of 5–15, and thus the factor 2 is far too low. Therefore Coulomb suggested that friction in the inclined plane plays a major role. If the friction coefficient is denoted by μ, the maximum compressive load becomes:

$$P_{max} = \frac{f_t A}{\cos \alpha (\sin \alpha - (\cos \alpha)/\mu)} \tag{3.38}$$

The friction coefficient μ for the shear plane is difficult to measure, but could lie between 1.0 and 2.0. The angle α of the shear plane can more easily be determined from experiment, and for concrete may range between 60 and

80° (i.e., shear-band angle between 10 and 30°), which not only depends on the composition of the material, but also on the frictional restraint between the loading platen and the specimen as shown in Section 8.2. Coulomb calculated that the maximum load would develop when $\tan\alpha = 2$. Depending on the friction coefficient between 1.0 and 2.0, this leads to maximum compression loads between 5 and 3.33 times the tensile failure load ($f_t \cdot A$). When $\mu = 0.5$, the goniometric function in the denominator becomes indeterminate.

Of course the above calculation is rather crude. Friction undeniably plays an important role in compressive fracture; consider, for example, the behavior of confined concrete (see Section 8.3). Yet an additional mechanism may include buckling of slender parts of the material that have been split off initially through tensile microcracking. Buckling instability is governed by the buckling length of the element. Considering Euler buckling, the maximum load a column between pinned ends of length l (see Figure 3.7c) can carry is equal to

$$P_E = \frac{\pi^2}{l_b^2} \cdot EI \tag{3.39}$$

In this case the buckling length l_b corresponds to the actual length l of the column. The effect of pinned, fixed, or free supports leads to a variation of the buckling length l_b between 0.5 and 2 times the original column length l.

The buckling stress σ_E may be calculated by dividing P_E by the area of the column A, leading to

$$\sigma_E = \frac{\pi^2 EI}{l_b^2 A} \tag{3.40}$$

Now assuming the column (lattice element) has a square cross-section with side d leads to $I = d^4/12$ and $A = d^2$, and thus

$$\sigma_E = \sigma_{compr} = \frac{\pi^2}{12} \cdot E \cdot \left(\frac{d}{l_b}\right)^2 \tag{3.41}$$

For this criterion to work the ratio η between tensile and compressive failure stress must be in the neighborhood of 10. The discussion focuses on the actual buckling length, which, fortunately, depends on the structural boundary conditions of the considered lattice element, that is, the rigidity of the fixations caused by the flexural stiffness of neighboring lattice elements connected to the same node. Thus, by using a buckling criterion for compressive failure the structural boundary conditions automatically become part of the solution, which is exactly what we are after. We return to these matters in

Chapter 8, where we show that matters are just a bit more complicated than suggested here.

In conclusion, there are several ways of including compressive failure in the lattice model, yet the above criteria have not been tested exhaustively today and require further research. It should be mentioned that in all of the above-mentioned criteria, failure is immediate and catastrophic as soon as the critical load (or stress) is reached. This means that no softening is considered at the local level, that is, the meso-level where individual material phases such as the aggregate, matrix, and interfacial transition zone (ITZ) are distinguished. All three phases are assumed to behave perfectly elastic and purely brittle. Softening fracture laws, based on the fictitious crack model or the crack band model can of course be applied in lattice models (see, e.g., Ince, Arslan, and Karihaloo 2003), but given the uncertainties with softening as mentioned in Chapter 2 there is a clear preference to assume perfectly elastic, purely brittle behavior.

4

Lattice Geometry and the Structure of Cement and Concrete

4.1 Size/Scale Levels for Cement and Concrete

In Chapters 2 and 3 we limited the discussion to situations where the material is assumed to be homogeneous and isotropic. If such a material is replaced by a lattice, the geometry of the lattice structure cannot be ignored because anisotropy may be introduced. The main question is now how can we simulate the behavior of real materials using a lattice model. Geomaterials such as concrete, rock, ice, clay, and so on all have a rather heterogeneous material structure. Other materials, such as metals and glass, are also heterogeneous, but at a much smaller size/scale. In normal continuum-based approaches the structure of the material is ignored and the average properties of the material are considered. For determining average properties it is important that a representative material volume is considered; that is, the volume has to be large enough. For small material volumes the scatter may become disproportionally large if the structure of the material is coarse. As an extreme example a dam concrete may be mentioned containing aggregates of up to 150-mm diameter. Obviously a sample volume of 100 mm^3 would not suffice. Every time a new sample is cut from a larger volume of material and the Young's modulus is determined, the value may (in the extreme case) vary between the Young's modulus of the matrix and that of the aggregate material, which may be a variation by a factor of 6 to 7 in the case of normal gravel concrete. For a better understanding of the situation it is necessary to discuss to some extent the material structure of concrete.

Figure 4.1 shows the structure of concrete at different size/scale levels. Along the top part examples of atomic structures are shown: a crystal, a quasi-crystal, and an amorphous structure. These structures can of course be modeled directly by means of a lattice model, provided that the geometry is dealt with in an appropriate way. The figures in the middle row, Figure 4.1b, show cement and concrete at three different size/scale levels. At the far right the aggregates in concrete are visible at the so-called meso-level (or intermediate level). The size of these aggregates may vary substantially, up to 150 mm for the aforementioned dam concrete. In the other direction

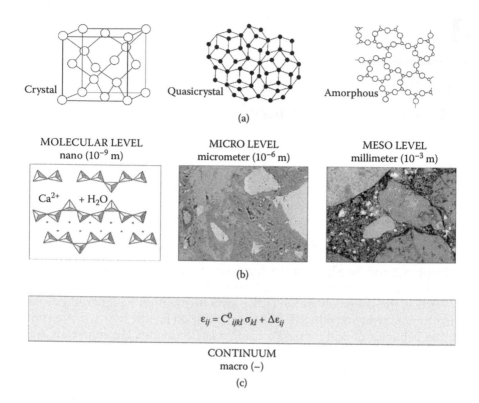

FIGURE 4.1
(a) Three examples of atomistic material structures (crystal, quasi-crystal, and amorphous);
(b) structure of cement and concrete at nano-, micro- and meso-level (from left to right); and
(c) continuum interpretation applicable at all size/scale levels as long as requirements of the
RVE are met. (From Van Mier. 2007. *Int. J. Fract.*, 143(1): 41–78. With permission from Springer.)

(i.e., at smaller scales) submicron particles are used to reduce pore space. The
role of aggregates in concrete is to reduce costs. Cement is the most expensive component, and reducing the amount leads to a cost reduction of concrete as a whole. Moreover, making a composite helps to improve some of
the mechanical properties of the material, as shown later (e.g., in Chapter 6).

The right image in Figure 4.1b shows details as small as a tenth of a milli-
meter. The large sand grains are easily recognized as large gray patches with
hardly any internal structure except for porosity which is visible as small
black spots. Whether the sand grains have internal structure depends on
the type of rock. Between the sand grains the cement matrix is visible. The
matrix may contain small sand grains or other fillers smaller than μm-size.
Furthermore there is hydrated cement (light gray), unhydrated cement
(white), and quite a lot of porosity (black) in the matrix. A blown-up image
of the matrix can be seen in the middle of Figure 4.1b. This image shows
hydrated Portland cement without any additions. The hydrated cement
appears in two gray shades, which reflect the commonly made distinction

between high-density CSH (calcium silicate hydrates) and low-density CSH. Also in this image unhydrated cement is visible as whitish/light-gray.

Finally, on the far left in Figure 4.1b the molecular structure of CSH is shown. At the nano-level not all details of the structure have been determined today owing to the incredible difficulties met in sample preparation, for example, for observation in TEM (transmission electron microscopy). Cement and concrete are very prone to drying, and actually water forms an important part of the structure of these materials. Water contributes to the strength of cement and concrete at micro- and nanoscales; we return to this important aspect of cement behavior in Chapter 11.

At all size/scale levels it can be proposed to represent the material by means of a continuum. This view is depicted in Figure 4.1c. The material structure effects are incorporated in (nonlinear) constitutive equations. As mentioned, but this cannot be emphasized too often, it is important that the considered sample used for tuning the constitutive equations should be representative, meaning that the sample volume be large enough to allow for averaging. From an experimentalist point of view it may be argued that the RVE is defined when the scatter in experiments drops to a minimum value; at least this argument may be used when considering the elastic properties of the material under consideration. For example, for the concrete of Figure 4.1b (right image), with a maximum particle size d_{max} = 2 mm the minimum sample size representative for a continuum formulation is on the order of 16–20 mm, that is, 8–10 d_{max}. For fracture the arguments may follow another line of reasoning. Because cracks may eventually become as large as the test specimen itself, a representative volume can never be defined.

It is hoped that the short summary on the structure of cement and concrete clarifies that these materials are complex. They are far away from ideal crystals or other regular material structures. Randomness is an important aspect and one of the problems to resolve is how to incorporate the random material structure into a numerical simulation model. The lattice and particle models mentioned in the previous sections commonly have a regular geometry and clearly some further steps are needed to transform these regular structures into a random structure resembling cement or concrete. The various ways to incorporate effects from a random material structure are clarified in this chapter. Note that the emphasis is on lattice models; particle models are discussed briefly in Chapter 11 (multiscale interaction potentials).

4.2 Disorder from Statistical Distributions of Local Properties

An easy and quite straightforward manner to incorporate heterogeneity in a lattice model is to assign random values for the Young's modulus and/or for the failure threshold of the individual lattice elements. So, every element will have different material properties following a certain assumed probability distribution. For example, one may select a Gaussian or Weibull distribution. The choice for a given distribution has direct consequences, for example, on scaling of structural strength, brittleness, or ductility, and so on. Matching of global behavior is the key to justifying a certain choice for a probability distribution. One argument to revert to statistics may be that the computational effort should be reduced as much as possible. This is a shifting boundary because computational capacity increases constantly. In the future very likely structures of very large size (10–100 m) can be analyzed at the micro-level. For the case of a regular triangular lattice in the past we analyzed two different distributions and compared the results with the so-called "particle-overlay methods" that are discussed in Section 4.5. The common approach in statistical physics is to base all considerations on a certain probability distribution, which can be fixed from the beginning of the analysis or even vary while an analysis progresses. In the example shown here the properties were fixed at the beginning of the analysis, and could not change while the fracture process (which was the emphasis) took place. The Young's moduli of all the lattice elements were kept at the same constant value, but the breaking thresholds of the lattice beams were varied following a Gaussian or a Weibull distribution. For the Gaussian distribution,

$$f(x) = \frac{1}{\sqrt{2\pi}\sigma} e^{-(x-\mu)^2/2\sigma^2} \qquad \text{for } -\infty < x < \infty \qquad (4.1)$$

the mean value μ and the standard deviation σ are the two important parameters to be defined.

In Figure 4.2a for $\mu = 6$ MPa the shape of the Gauss-bell is shown for varying σ. For a material such as concrete a very wide distribution must be selected in view of the large range of properties: the strength of aggregates may be substantially higher than that of the cement matrix and, even more important, higher than the strength of the interfacial transition zone (ITZ) or bond zone between aggregate and matrix. We return to these matters in Section 4.6.

For the Weibull distribution the probability density function is

$$f(x) = \frac{\beta}{\delta}\left(\frac{x}{\delta}\right)^{\beta-1} e^{-(x/\delta)^\beta} \qquad \text{for } x > 0 \qquad (4.2)$$

where the two main parameters are the scale parameter δ and the shape parameter β. In Figure 4.2b the function has been plotted for $\delta = 4$ and varying β. Whereas the Gauss distribution is symmetric around the mean value, the Weibull distribution shows a larger probability for smaller strength values. The number of "weak" elements increases, which might actually represent the case for concrete.

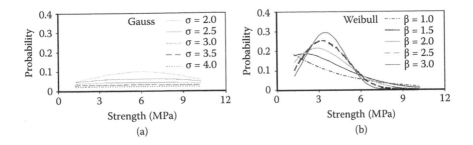

FIGURE 4.2
(a) Gaussian probability density function for selected values of σ and μ = 6 MPa; and (b) Weibull functions for varying β at δ = 4 MPa. (Reprinted from Van Mier, Van Vliet, and Wang. 2002. *Mech. Mater.*, 34: 705–724. With permission from Elsevier.)

The choice of a certain distribution has a clear effect on the fracture behavior. Before explaining the entire procedure in detail, we show two examples here of analyses of a uniaxial tension test in 2D (Figure 4.3). The sample is pulled in deformation control in a vertical direction. The analysis consists of a number of consecutive linear analyses, therefore the typical zigzag pattern develops in the force deformation diagrams. The distinction between the two analyses is that for the Gauss distribution microcracks appear to be more widespread over the entire specimen (the crack pattern with 63 beams removed is at peak load). In the analysis with a Weibull distribution major cracks start to develop at a relatively early stage (see, e.g., the situation at peak load with 50 beams removed), the peak load is lower, and a particular phenomenon called "bridging" appears at later stages of the fracture process (beyond peak, 400 beams removed).

The problem encountered in lattice analyses based on a random distribution of element properties is that it remains difficult to relate the distribution of local material properties to a certain probability density function. Moreover, as can be seen in Figure 4.1b (right image; meso-level), aggregates form discrete clumps, and the use of a statistical distribution will very likely not lead to distributions related to the spatial arrangement of particles in a real concrete. In connection to the spatial arrangement of aggregates in the real material the ITZ also will appear at certain locations, which is not just a matter of chance.

4.3 Computer-Generated Material Structures

As an alternative to the statistical distributions discussed in the previous sections one may revert to generating a particle structure of concrete, using realistic particle size distributions. Quite common for a concrete material is a

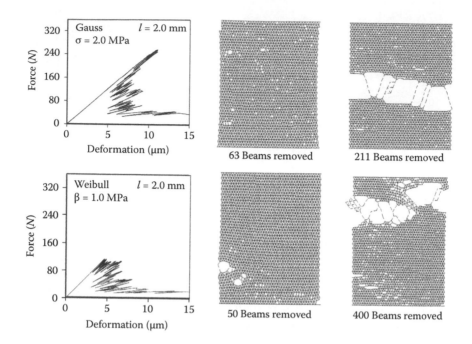

FIGURE 4.3

Axial force deformation diagrams and crack patterns for regular triangular lattices with beam length 2 mm. In the top example a Gaussian distribution of strength thresholds is used with $\sigma = 2.0$ (see Figure 4.2a); in the bottom example a Weibull distribution following Figure 4.2b is used with $\beta = 1.0$. (Reprinted from Van Mier, Van Vliet, and Wang. 2002. *Mech. Mater.*, 34: 705–724. With permission from Elsevier.)

so-called "Fuller distribution." The Fuller distribution leads to a dense packing of spherical particles; the size distribution follows the equation

$$p = 100\sqrt{\frac{d}{d_{max}}} \tag{4.3}$$

where d = the particle diameter and d_{max} is the size of the largest particle. A typical particle distribution for concrete with maximum particle size of 32 mm leads to the following: 100% of all particles are smaller than 32 mm, 71% are smaller than 16 mm, 50% are smaller than 8 mm, 35% are smaller than 4 mm, 25% smaller than 2 mm, 18% for 1 mm, 13% for 0.5 mm, and 9% smaller than 0.25 mm, and 6% for 0.125 mm. If the aggregate volume is known, the aggregate particle distribution may be discretized and for each particle size the number of actual particles may be determined. If for a certain volume of concrete the number of particles is known, they should be placed in the considered volume. The procedures may vary from random placement to ballistic deposition (see Figure 4.4). Random placement is the

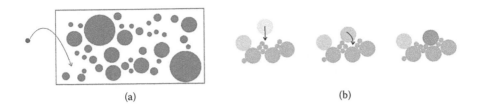

(a)   (b)

FIGURE 4.4
(a) Rand random-placement method; and (b) ballistic deposition of particles.

simplest method, but generally takes quite some time when many small particles have to be included. Basically one starts with the largest particles and places them one by one at a randomly selected position (center coordinate x_i, y_i, z_i for the ith particle). If a newly placed particle overlaps with a previously placed particle a new random coordinate must be selected, and the particle placed again. If there are no new overlaps, the location is accepted, and the next particle is placed. The process continues until all particles have been assigned a specific location. When placing a particle the decision to accept a location may depend on certain additional rules, such as, for example, the minimum distance from a neighboring particle. In the past often a rule prescribing a minimum interparticle distance $D_{min} = 1.1 \ (d_i + d_j)/2$, with d_i and d_j the diameters of two neighboring particles. The effect obtained by using the minimum distance rule is that a thin ribbon of cement matrix is always present between neighboring aggregate particles. This corresponds to one of the basic assumptions in concrete technology: just enough cement must be included to cover the total surface of all aggregates with a minimum thickness layer of cement. It will be obvious that it will take increasingly longer to place the next particle while the placement process progresses.

Therefore for problems involving a very dense distribution (high aggregate volume P_k), it may be better to revert to alternative methods such as ballistic deposition or a method where particles are initially assumed to be very small, and subsequently are inflated to fill the space (software package Space; Stroeven 1999). Ballistic deposition is less time consuming. In this case, again, particles are placed one by one. Now particles are randomly drawn from the entire population, and dropped from a randomly selected position along the top surface of the volume that must be filled (left image of Figure 4.4b). If there are no particles in the box, the dropped particle simply hits the bottom and is placed directly under the drop point. When some particles have already been placed, and the next dropped particle hits one in the box, certain rules should be obeyed, such as rolling along the surface, and so on (middle image of Figure 4.4b). The porosity of the sample will depend on the drop rules and may vary substantially. When a stable position is reached (right image of Figure 4.4c) the next particle can be added.

The computer-generated material structure must be constructed a bit differently when a 2D analysis is made. In that case the spherical particles

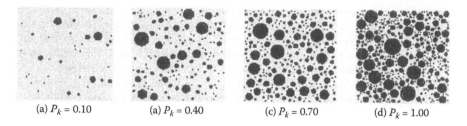

(a) $P_k = 0.10$ (a) $P_k = 0.40$ (c) $P_k = 0.70$ (d) $P_k = 1.00$

FIGURE 4.5
Computer-generated particle structure representing concrete with aggregates of size $1 \leq d \leq 16$ mm (Fuller distribution) for different particle content P_k. (After Van Vliet. 2000. *Size Effect in Tensile Fracture of Concrete and Rock*. Reprinted with kind permission of Dr. Marcel van Vliet.)

reduce to cylinders with height equal to the thickness of the analyzed structure. Using a method developed for aggregate interlock models in concrete (see, e.g., Pruijssers 1988) the distribution of intersection circles in a random planar section of a 3D particle distribution can be described following

$$P(D < D_0) = P_k \cdot \{1.065 D_0^{0.5} d_{max}^{-0.5} - 0.053 D_0^4 d_{max}^{-4} - 0.012 D_0^6 d_{max}^{-6} +$$
$$- 0.0045 D_0^8 d_{max}^{-8} + 0.0025 D_0^{10} d_{max}^{-10}\} \tag{4.4}$$

This equation gives the probability that an arbitrary point in the considered volume, lying in an intersection plane, is located in an intersection circle of diameter $D < D_0$. P_k is the aggregate volume as percentage of the total volume. This equation is based on the aforementioned Fuller distribution; see, for example, Schlangen (1993).

In Figure 4.5 four different examples of computer-generated particle structures using Equation (4.4) are shown. The amount of aggregates varies between $P_k = 0.10$ and 1.00. Note that for $P_k = 1.0$ the entire volume does not consist of aggregate. This is related to the minimum separation distance between particles, and moreover, the images of Figure 4.5 have already been overlaid with a regular triangular lattice (see Section 4.5). As a consequence some of the aggregate area is lost and incorporated in the ITZ.

4.4 Material Structure from Direct Observation

From two-dimensional sections of the considered materials, or by means of CT scans in three dimensions, it is possible to include the real structure of concrete or cement in a micromechanical analysis. The technique is, of course, not limited to cement or concrete but applies to any material. Mimicking the real structure of materials actually leads to a model approach referred to as "reality modeling," where the simulations must be considered as virtual experiments. The consequence is that, as in physical experiments, each computation has to be repeated several times, and the scatter of the results must be analyzed.

An early example of numerical simulations using a realistic material geometry can be found in Roelfstra, Sadouki, and Wittmann (1985). Using the so-called numerical concrete model, which is based on the finite element model (continuum elements combined with interface springs for the ITZ), a 2D section of concrete containing Rhone Valley gravel was analyzed. The real morphology of the grains was incorporated in the finite element mesh. The real structure of concrete is quite complicated, as was shown in Figure 4.1, and one of the most important aspects is to make the contrast between the various phases that should be distinguished in the analysis as large as possible. The simplest way is to use digital images. Thresholding the grayscale may be an easy way of getting a good contrast between the material structural features that should be included in the mechanical model, which implies that one should have some idea about the most important phases and their interactions in advance.

Nowadays, it has become relatively straightforward to perform micromechanical analyses in three dimensions; see, for example, Lilliu (2007). In order to obtain results that can be compared to experimental data, the three-dimensional character of many processes must be included. An example in concrete is drying shrinkage, which is a surface phenomenon; it can be shown quite easily that the damage caused by drying shrinkage in concrete has a significant effect on the size effect on tensile fracture strength (see Figure 9.6). Three-dimensional analysis should become the standard approach rather than the exception when analyzing the behavior of materials. The material structure should therefore be modeled in three dimensions as well. Computed tomography (CT) is an excellent technique to measure the material structure of concrete (and other materials) in three dimensions.

The technique will of course work best when a balance between the size of microstructural features and the resolution of the tomograph are in balance. CT scans are based on density differences, and in order to help matters a bit, at least for concrete, aggregates that have a much higher density than the surrounding cement matrix could be used, such as basalt or granite. Typically, one would select aggregates with a higher Young's modulus than the surrounding cement. For basalt and granite the Young's modulus is around 70 GPa, whereas for the surrounding cement something like 10–15 GPa is the average value. The other extreme would be to leave out the aggregates completely. In that case large voids are present in the cement and we are actually dealing with a highly porous material. Foamed cement is an example of such a material, and Figure 4.6 shows a three-dimensional CT scan of a small cylinder (height/diameter = 6.0/6.7 mm) made of cement containing small potato-shaped voids. The pores were created by mixing the cement with protein foam. The image was made using a μCT-40 tomograph from

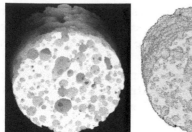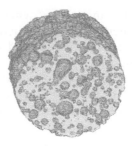

FIGURE 4.6

CT scan of a small foamed-cement cylinder. The sample can be cut and sliced in any direction, and as such reveal the internal structure. The two images to the right are discretized lattice structures, one with a lattice beam length of 0.1 mm, and at the far right with 0.05 mm. The amount of detail in the material structure is best preserved when the lattice beam length decreases. The price for more resolution is of course that the number of elements rapidly increases: the two models contain 683,940 and 6,311,581 elements, respectively. (After Meyer et al., 2009.)

SCANCO MEDICAL AG in Switzerland. The device has a spatial resolution of 6 μm, and can also be used for crack detection. Foamed concrete is a model for real concrete, where the voids are simply places where the aggregates have been left out. An additional advantage is that the interfacial transition zone has been removed from the analysis, and therefore complications arising from the complex ITZ structure are temporarily excluded (see Sections 4.5 and 4.6). This does not mean that the ITZ is ignored, but for studying other aspects of concrete behavior it may be helpful to temporarily look to a simpler model.

For other concretes containing solid aggregates (and pores, which are partly due to the hardening process of cement, and in part caused by mixing the concrete) the same procedure can be followed. In Figure 4.7 two reconstructed particle structures are shown: from a concrete prism containing crushed basalt particles in Figure 4.7a, and from oval-shaped marble particles in Figure 4.7b; see Man and Van Mier (2008b). In these scans it helped to make the matrix very porous by selecting a w/c ratio of 0.3 and using CEM I 42.5. Finer sands were omitted. The density of the cement matrix, the basalt, and marble aggregates was 2,200, 3,000, and 2,700 kg/m³, respectively. The concrete prisms were scanned in the University Hospital in Zurich in a Siemens SOMATOM Definition CT-scanner. For these coarse aggregate mixtures (size range between 8 and 15 mm) the resolution of the scanner was certainly appropriate; for cracks it would not have sufficed.

The same technique can also be applied at smaller scales. For example, Figure 4.8 shows the reconstruction of unhydrated cement grains and pores in a cylindrical sample of Portland cement with height/diameter of 250/130 μm. The extreme small sample was tomographed in the synchrotron at the PSI in Villigen, Switzerland, with a spatial resolution of about 0.7 μm. The

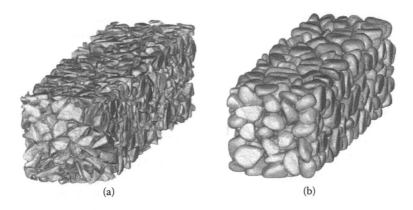

(a)　　　　　　　　　　　　　(b)

FIGURE 4.7
Reconstructed three-dimensional particle structures from CT scans: (a) 45% crushed basalt particles are present; (b) 45% oval-shaped marble aggregates; size range 8–15 mm. These models are overlaid with a lattice structure similar to the example of Figure 4.6. (From Man and Van Mier. 2008b. *Int. J. Fract.*, 154(1–2), 61–72. With permission from Springer.)

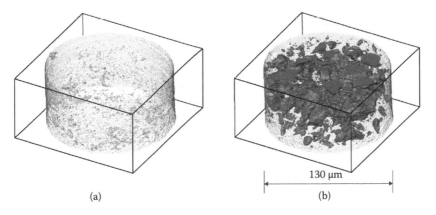

130 μm

(a)　　　　　　　　　　　　　(b)

FIGURE 4.8
(a) Reconstruction of voids and (b) unhydrated cement particles in a small cylindrical sample (diameter 130 μm) of partially hydrated Portland cement. (After Trtik et al. 2007. *Proc. 6th Int'l. Conf. on Fracture Mechanics of Concrete and Concrete Structures (FraMCoS-VI).*)

last images do not reveal the cement structure completely. There are two CSH phases: low- and high-density CSH, which can be seen in the middle image of Figure 4.1b. Thus, a model for hardened cement paste would actually need at least three material phases: two types of hydrates and the unhydrated cement particles, which are usually surrounded by the hydrated CSH phases. Pores may appear at many different locations: in CSH, in the unhydrated particles, but also between two neighboring (hydrated) cement grains. We do not dwell further on these details at this point, but return to these matters in Chapter 11.

In Figures 4.5 and 4.6 the lattice overlay was already made. In the following section the technique to incorporate the material structure into the mechanical model is elucidated.

4.5 Lattice Geometry and Material Structure Overlay

The lattice model is particularly attractive in the way the mechanics model and the material structure are combined. The lattice is the mechanical model; the material structure, in whatever form, is simply projected on top of the lattice and various properties are assigned to the lattice elements depending on their specific location in the projected material structure. The lattice and material structure are thus two independent features of the model. An alternative to the overlay method is the finite element model where the mesh must be constructed in such a way that the edges of the finite elements fall exactly along specific boundaries in the material structure. Especially when it comes to random material structures in three dimensions, the latter method is quite elaborate and the simple projection scheme is the easiest approach, even if this would mean that some of the elements appear in two material phases. The drawback of the overlay method is that some details of the material structure may simply disappear depending on the size of the lattice elements used. We return to these matters later in this section.

The lattice geometry is an important aspect of the mechanical model. The connectivity of the lattice elements is decided by the spatial arrangement of the individual beams. The spatial arrangement may take a regular or random configuration. One form was already shown in Figure 3.2: the Hrennikoff lattice has a rectangular configuration with crossing diagonals. The two-dimensional beam lattice may have twofold or threefold symmetry, or, as referred to before, it may have a completely random structure. In Figure 4.9 the regular square, regular triangular, and random triangular

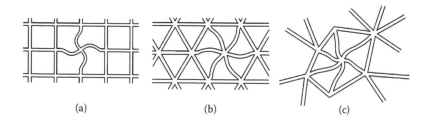

(a)   (b)   (c)

FIGURE 4.9

Examples of two-dimensional lattices: (a) regular-square lattice, (b) regular-triangular lattice, and (c) lattice with random beam-lengths.

lattice are shown in two dimensions. Generation of a regular lattice is quite straightforward: nodes are placed on the regular grid and connectivity is specified. For the random lattice a bit more effort is needed.

In three dimensions, a regular lattice can be based on the hexagonal close packing (Figure 4.10a), the face-centered cubic packing (Figure 4.10b), or a 3D variant of the lattice with random beam length (Figure 4.10c). The hcp- and fcc-lattices derive from packing equally sized spheres. Man (2009) showed that the number of elements differs when a hcp- or fcc-lattice is used.

Two examples of constructing lattices with random beam lengths are the following. In the first method a particle distribution is generated as described in Section 4.3. Next, the centers of neighboring particles are connected. Nodes in a circular area with radius $4d_{max}$ are checked for possible

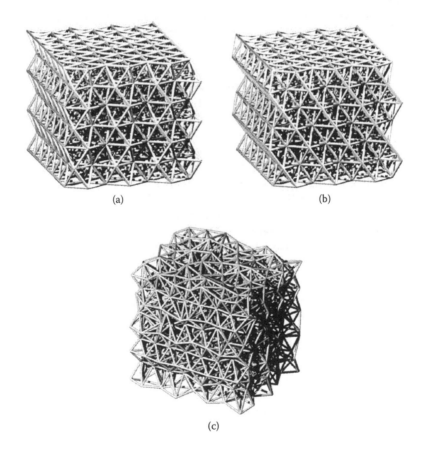

(a)   (b)

(c)

FIGURE 4.10
Examples of three-dimensional lattice geometries: (a) hexagonal-closed packing or hcp-lattice, (b) face-centered-cubic packing of fcc-lattice, and (c) lattice with random beam lengths (randomness $A/s = 0.5$; see below). The examples in (a) and (b) are based on a close-packing of equally sized spheres. (After Man. 2009. *Analysis of 3D Scale and Size Effects in Numerical Concrete*. Reprinted with kind permission from Dr. Hau-Kit Man.)

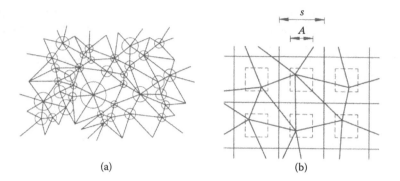

(a)   (b)

FIGURE 4.11
Two methods for generating a random lattice: (a) based on predefined aggregate structure according to Section 4.3, and (b) based on a regular square grid. (After Vervuurt et al. 1995. *Proc. 2nd Int'l. Conf. on Fracture Mechanics of Concrete and Concrete Structures (FraMCoS-2)*.)

connectivities. Given a set of nodes, always the three nodes that are closest to each other are connected by beam elements. The resulting lattice can be considered as the backbone of a particle stack as shown in Figure 4.11a. Such models are frequently used, for example, the one by Beranek and Hobbelman (1994) referred to in Chapter 3, albeit in their particular case a regular structure is used inasmuch as all particles have the same radius.

The second method is the random lattice developed by Mourkazel and Herrmann (1992). This specific form has been used in several analyses in the past. In Figure 4.11b the construction of the lattice with random beam length is clarified. The starting point is a square grid of size s. In each cell of the grid a point is selected at random. The same procedure (i.e., the Voronoi/Delaunay tessellation) is used to connect the points. Connecting lines are then the beams in the lattice, which all have a different length. One problem that may be encountered in this structure is when by accident two nodes appear at a very small distance, almost zero, along the same edge of two neighboring cells. In that case the short beam length may pose a problem in the stiffness matrix, because these elements in the matrix may have a much larger value than all the other matrix elements (see Equation (3.23)). Solving the set of equations may cause numerical problems, especially the inversion of the stiffness matrix. As a remedy, the area where the random grid point is selected may be reduced as sketched in Figure 4.11b. The size A of the subcells may be as small as zero, in which case no diagonals can be generated. By choosing a very small $A/s = 0.001$, this problem is circumvented. In the case $A = s$, we have returned to the initially described random lattice.

In between these extremes the "degree of randomness" of the random lattice may be varied by selecting a different subcell size. In Figure 4.12 the average beam length l_{avg}/s has been plotted for various degrees of randomness (between $A/s = 0.001$ and 1.0). These results were obtained by generating 175 meshes for each degree of randomness (A/s). The white dots in the

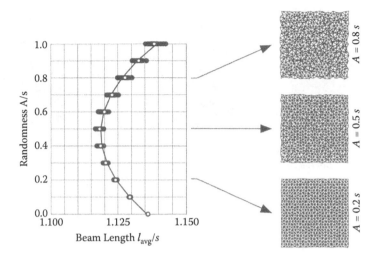

FIGURE 4.12
Relation between randomness A/s and average beam length l_{avg}/s. (From Chiaia, Vervuurt, and van Mier. 1997. *Eng. Fract. Mech.*, 57(2/3): 301–318. With permission from Elsevier.)

figure show the average beam length of all these 175 meshes whereas each individual result has been plotted by a black dot. Next to the graph appear three random meshes, with A/s = 0.8, 0.5, and 0.2. For the lowest A/s-value, the appearance is already quite regular.

For 3D lattices a similar approach was developed by Lilliu (2007). In the two-dimensional case nodes are checked for connectivity in an area of 6 × 6 cells, or 36 potential neighbors. In the three-dimensional case the number of potential neighbors increased to 178. In the three-dimensional model the connectivity is established between one selected node and its three nearest neighbors that belong to a sphere. In most of the analyses performed by Lilliu, the smallest possible randomness was selected; that is, A/s = 0.001.

The next step is combining the material structure and the mechanical model. The procedure is simple and straightforward. The lattice mesh is generated and the previously constructed material structure is projected on top of the lattice. Next the location of the lattice nodes is determined in the respective material phases that are distinguished. If both nodes of a lattice element fall within a single phase, the properties of the element will receive those of that particular material phase. In Figure 4.13 the procedure is shown for a computer-generated particle structure on a regular triangular lattice.

Figure 4.13a shows the computer-generated particle structure; Figures 4.13b,c show the projection of a lattice with different beam lengths (l_{beam} = 0.5 and 2.0 mm, respectively), and in Figure 4.13d the procedure for assigning aggregate, matrix, and ITZ-properties is elucidated. For aggregate and matrix the decision is based on the location of the two element nodes in the same phase. If one node is located in matrix and the other in the aggregate phase, the

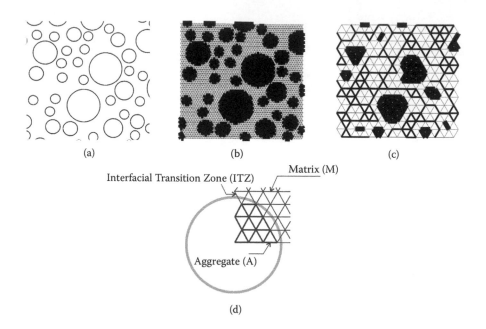

FIGURE 4.13
Projection of a computer-generated particle structure on top of a regular triangular lattice structure: properties are assigned depending on the location of the lattice nodes in the projected material structure.

element is said to be an interface element of bond element. For concrete it has long been known that the interface between aggregate and matrix is very weak, and plays a prominent role in the fracture process. Zimbelmann (1985) measured tensile bond strengths between 0.1 and 1.2 MPa depending on the type of aggregate material. No or just a little reactivity appears between most natural aggregates and the cement matrix, except perhaps when calcite is used. The roughness of the aggregate surface may have a positive effect as this may promote mechanical interlock. The bond stress is thus mostly based on adhesion between matrix and aggregates: physical forces are more important than chemical bonds. At a more microscopic level the ITZ appears as a highly porous zone (35% porosity compared to just 10% for the bulk cement matrix), as was measured by Scrivener (1989).

Before discussing in a bit more detail the identification of the parameters of the various material phases some further remarks should be made regarding the lattice overlay shown in Figure 4.13. The original particle distribution is quite well captured when a fine lattice mesh is used, as shown in Figure 4.13b. When a coarser mesh is used (Figure 4.13c) the grains lose their perfect round shape, and smaller particles in the distribution may get lost completely when they are smaller than twice the length of an individual lattice element. Moreover, quite a large part of the aggregate area (in 2D; aggregate volume in 3D) is lost to the ITZ when the lattice element length increases. The latter loss of aggregate area is caused by the initial assumption that the ITZ thickness equals the length of a single lattice element. In Figures 4.13b and c the element lengths are 0.5 and 2.0 mm, respectively, which is more than a factor of 1,000 larger than the thickness of the "real ITZ," which was determined at about 40 μm by Scrivener (1989). The obvious solution would be to take a more realistic ITZ thickness, but when the simplicity of the approach is to be preserved and all lattice elements have the same length, the number of elements would increase vastly, and likely the computation would explode. Of course one may decide to envelop each aggregate particle in a zone of very short elements, but not only will this affect the stiffness matrix and the numerical solution, the generation of the lattice mesh will be more complicated, and the computational effort would still increase substantially.

One of the most important aspects appears thus to come to a balanced approach at all levels. It does not pay to increase accuracy at one point if not all other aspects in a lattice analysis are improved at the same time. What is meant by "other aspects" are actually all details of a lattice analysis that play a role: lattice mesh size, fracture law, quality of properties of the various material phases, realistic aggregate shapes, aggregate size range, including the internal structure in the cement matrix (which would lead to a scale decrease below the aforementioned ITZ thickness; see also Section 4.6), computational effort needed, available computational infrastructure, and so on. Thus, for example, one might wish to increase the accuracy of representing real aggregate shapes (Man and Van Mier 2008b). The price to pay is a longer computation because the lattice element length must be reduced. At the same time, the uncertainty of many material parameters has not been improved, such as the interpretation of the ITZ behavior. Ultimately, the drive to make the resemblance between "computational aggregate geometry" and "real aggregates" more perfect would end in severe computational problems, without adding to a better understanding of the fracture process. At the same level one should understand efforts to improve the fracture law in a lattice analysis. It has been proposed that a softening law would be needed (e.g., Ince, Arslan, and Karihaloo 2003) at the level of the lattice elements. As a consequence of introducing a nonlinear (softening) fracture law an iterative procedure would be needed to solve the problem. This again leads to a vast increase of computational effort. Inasmuch as a phenomenological softening law would not improve any fundamental understanding of fracture, the entire exercise is considered futile and a waste of time, which is confirmed by the analyses in Chapter 6.

In Figure 4.14 the dependence of the phase fractions (aggregate, matrix, bond) on the lattice beam length is shown for a two-dimensional model containing aggregate particles with diameter $2 \leq d \leq 8$ mm, $P_k = 0.75$ (Equation (4.4)). When $l_{beam} = 0.5$ mm, the smallest aggregates will have three beams over the diameter. However, with increasing beam length the number of beams over the diameter will decrease until at $l_{beam} > 1$ mm the smallest

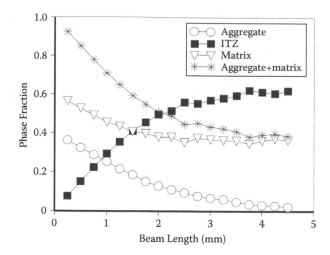

FIGURE 4.14

Effect of lattice-beam length l_{beam} on the phase fractions (aggregate, matrix, ITZ) in a 2D model with intended Fuller particle size distribution $2 \le d \le 8$ mm, $P_k = 0.75$. (After Van Vliet. 2000. *Size Effect in Tensile Fracture of Concrete and Rock.* With kind permission of Dr. Marcel van Vliet.)

2-mm particles will disappear from the material structure. In that case the aggregate particle is transformed completely into ITZ material. The trend increases with increasing lattice-beam length. Thus, Figure 4.14 shows that with increasing lattice-beam length the aggregate fraction (and also the matrix phase) decreases and the ITZ fraction increases. The way the material phases are distributed over the specimen's area (in 2D) or volume (in 3D) affects the mechanical properties of the composite. We return to these matters in Chapter 5 (elastic properties) and Chapter 6 (tensile fracture).

Let us now return to the identification of the material properties of the various phases in a typical lattice problem of concrete. The usual meso-level approach would distinguish between three material phases, namely aggregate, cement matrix, and ITZ. The determination of the elastic properties, the fracture strength, and fracture energy of the bulk cement matrix and aggregate is rather straightforward. Small representative volumes of these materials are simply loaded in displacement control, in tension, to failure, and the Young's modulus; the tensile fracture strength, and the fracture energy can be derived from the stress-crack opening diagram. Note, however, that establishing "softening parameters" may cause some problems due to the ill-posed definition of this parameter, a recurring theme in this book; see also Chapters 2, 10, and 11. The matrix material should represent the composition of the matrix in the lattice analyses: fine-grained sand that is not explicitly represented in the material structure must be considered as part of the cement matrix. For example, it is quite normal not to include aggregate particles of diameter smaller than 1 mm in the lattice model. Obviously, the reason is to reduce the computational effort needed. The matrix in the model of Figure 4.13 would therefore be a cement–sand mixture with sand grains of size $d < 1$ mm, and with the specified water/cement ratio.

4.6 Local Material Properties

Several parameters must be known before a lattice analysis can be performed. In the simplest meso-level representation, three material phases are distinguished, namely the cement matrix, the aggregates, and the interfacial transition zone. A linear-elastic beam analysis requires values for the Young's modulus and the Poisson ratio of these three phases. When fracture is simulated using the sequential beam-removal approach, the threshold tensile strengths for testing Equations (3.35)–(3.36) are needed, again for each of the three material phases. For the cement matrix and the aggregate material one can determine these parameters from small samples of the base material. If the concrete contains river gravel, one is confronted with a mixture of different aggregate types. It is normal practice to assume that all aggregates have the same quality, which is quite appropriate inasmuch as the weaker cement matrix and the ITZ largely determine how the concrete will perform.

Determination of the ITZ properties is most problematic because producing a representative interface is not easy. When concrete is cast, all phases are mixed with water in a container, and cement particles may stick to the wet aggregate surfaces at an early stage of hydration. The wetting of all particles and the water movement during hydration are decisive for the properties of the developing ITZ. This is not easily reconstructed in a simple test geometry where, for example, a small block of cement is cast against an aggregate surface. Although such a test may at least give a rough indication (see e.g., the experiments by Zimbelmann 1985), doubt will remain about the validity of using properties obtained from such tests for real concrete simulations. Also, sawing specimens from larger blocks of real concrete is not a solution because microcracks may be introduced due to the sawing and grinding process. Inverse analysis or back-analysis of the ITZ stiffness and strength may be an option, where, however, the uniqueness of the obtained result remains an important issue.

In addition, sensitivity studies, such as the one demonstrated in Figure 4.15 may help to narrow down the regime in which specific values of interface strength may lie. Figure 4.15 shows the results from two simulations of a single particle embedded in a rectangular plate of cement. The particle has the same diameter through the thickness of the plate. The aggregate may be a low-strength porous aggregate as may be found in lightweight concrete or the aggregate particles may be made from dense strong granite, resembling the material found in commonly used river gravel. The effect of changing the aggregate may affect the ratio between aggregate/matrix/ITZ strength,

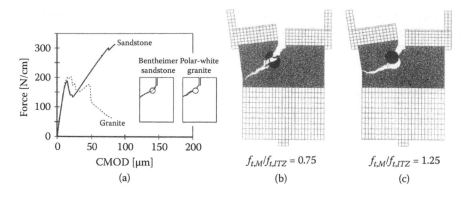

FIGURE 4.15

(a) Fracture behavior observed in single particle configurations in a plate subjected to horizontal splitting forces. (After Vervuurt and Van Mier. 1995. *Proc. 2nd Int'l. Conf. on Fracture Mechanics of Concrete and Concrete Structures (FraMCoS-2)*.) Results for a Bentheimer sandstone particle and a granite aggregate are shown in the same diagram. Results of lattice simulations by Vervuurt (1997) are shown in (b)–(c) for a sandstone model and granite model, respectively. (From Vervuurt. 1997. *Interface Fracture in Concrete*. With kind permission of Dr. Adri Vevuurt.)

and with that the mechanical/fracture response of the concrete as a whole. Relative to the aggregate strength the interface proved to be very weak for the smooth granite particles, whereas the ratio changed for the more porous sandstone where the ITZ proved to be relatively stronger than the aggregate particle. The corresponding experimentally observed fracture mechanisms are shown in the inset of Figure 4.15a.

For the granite, after starting from the tip of the saw-cut the main crack grows toward the aggregate particle. During crack growth the load decreases (after the first peak). Upon reaching the aggregate particle the crack may either continue to grow along the upper or lower ITZ, or it may extend through the aggregate particle. In the particular example shown the crack extended along the upper ITZ zone after the load was increased a little. The second load drop in Figure 4.15a is caused by crack growth along the interface. The situation is quite different when the strength contrast among aggregate, matrix, and ITZ is changed, as will be the case for a porous sandstone particle. The first crack growth toward the aggregate grain is identical as in the previous example. After the drop the load must now increase substantially until the crack continues to grow through the aggregate particle. The aggregate particle is thus the weakest element and no longer the ITZ.

As mentioned, the exact values of ITZ strength and stiffness are hard to measure from such experiments. Instead, a sensitivity study, for example, on E_M/E_{ITZ} and on $f_{t,M}/f_{t,ITZ}$ may help to narrow down the range of relevant values for a given concrete mixture. In general it appears that the strength ratio $f_{t,M}/f_{t,ITZ}$ is more important than differences in the stiffness between matrix and interface phases; see Vervuurt (1997).

TABLE 4.1

Suggested Model Parameters for Normal Concrete

	Tensile Strength [MPa]	Young's Modulus [GPa]
Aggregate (A)	10	70
Matrix (M)	5	25
Interfacial Transition Zone (ITZ)	1.25	25

Suggested model parameters for normal strength concrete containing hard dense aggregates with a larger stiffness than the surrounding cement matrix are given in Table 4.1. Note that the lattice model can be used for any type of material, and that the same procedure of material structure overlay can be adopted. Examples are known where lattice models have been applied for masonry, wood, ceramics, and various types of rock, among others.

5

Elastic Properties of Lattice with Particle Overlay

5.1 Upper and Lower Bounds for the Young's Modulus of Composites

The simplest approach for calculating the Young's modulus of a composite is to revert to analytical models. Quite basic are the parallel and series models. In the parallel model two layers of material are loaded in the direction of the layers. Assuming that the two layers have different Young's moduli (i.e., E_a for the aggregate layer and E_m for the matrix layer), the Young's modulus of the composite can be computed with

$$E = E_m V_m + E_a V_a \tag{5.1}$$

where V_m and V_a refer to the matrix and aggregate volume fractions, respectively. In the series model the two layers are loaded perpendicular to the main layer direction, and we obtain

$$\frac{1}{E} = \frac{V_m}{E_m} + \frac{V_a}{E_a} \tag{5.2}$$

In Figure 5.1 the parallel model is a straight line connecting E_y for 0% and 100% aggregate fraction; the series model comes near the lowest curvilinear line, which is the lower Hashin bound. The upper curved line is the upper Hashin bound that is explained below. Experimental results, for example, those obtained by Wittmann, Sadouki, and Steiger (1993) on mortar, are well in between these extremes, which are considered as the absolute upper and lower bounds for the Young's modulus of a 2-phase composite. Deviations appear when the aggregate volume fraction exceeds 50%. We return to these deviations in Section 5.2.

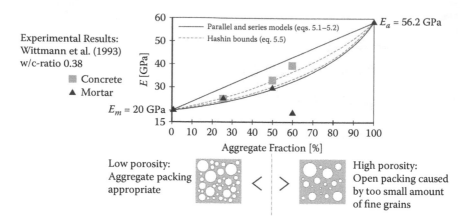

FIGURE 5.1

Bounds on Young's modulus for a two-phase concrete consisting of spherical aggregate particles embedded in a continuous cement-matrix with w/c-ratio = 0.38. (The experimental results shown were obtained by Wittmann, Sadouki, and Steiger. 1993. *Micromechanics of Concrete and Cementitious Composites*.) The upper-bound (straight line) is the parallel model; the two lower curves are the Hashin bounds, which are a refinement of Equations (5.1) and (5.2).

Of course there are many refinements possible. Newman (1968) mentions several variations on the parallel and series models. Well known, and also extensively used are the bounds developed by Hashin and coworkers: see Hashin (1965), Hashin and Shtrikman (1963), and Hashin (1983).

For a 2D-model, the lattice approach resembles a transverse section of a fiber composite with circular cross-section having diameters extending from infinite to finite size. Each cylinder can be thought to consist of a circular fiber enveloped by a concentric matrix shell. Given a wide distribution of fiber sizes, a plane can be assumed to be completely filled. The bounds on the transverse properties of such an assemblage of fibers was determined by Hashin (1965), namely the transverse plane–strain modulus bounds $K^{(-)}$ and $K^{(+)}$, and the upper and lower bounds for the transverse shear modulus. For analyzing the elastic modulus of the 2D-particle composite, where the particles are actually cylinders of different diameters (see also Chapter 4), the Hashin bounds can be used as a comparison to the outcome of numerical simulations. Only the bulk modulus needs to be considered.

Hashin (1965) gives the following upper and lower bound for the plane–strain bulk modulus K as

$$K^{(-)} = K_m + \frac{V_a}{\dfrac{1}{K_a - K_m} + \dfrac{V_m}{K_m + G_m}} \tag{5.3a}$$

and

$$K^{(+)} = K_a + \frac{V_m}{\dfrac{1}{K_m - K_a} + \dfrac{V_a}{K_a + G_a}} \tag{5.3b}$$

K and G are the bulk and shear modulus of the aggregate and matrix phases, respectively, and are equal to

$$K = \frac{E}{3(1 - 2v)} \quad \text{and} \quad G = \frac{E}{2(1 + v)} \tag{5.4}$$

with, of course, the respective indices (*a* and *m*) for aggregate and matrix phases. The V_i stand for volume fractions. The Poisson ratio v has been set equal to 0.2 in all the numerical simulations for both the aggregate and matrix material, which simplifies the equations. Using Equation (5.4) the upper and lower bounds of the Young's modulus of the composites can be expressed in terms of the Young's moduli of the composite phases and their volume fractions:

$$E^{(-)} = E_m \left\{ \frac{(5 - 4v)E_m + [2V_m(1 + v) + V_a(5 - 4v)] \cdot (E_a - E_m)}{E_m(5 - 4v) + 2V_m(1 + v)(E_a - E_m)} \right\} \tag{5.5a}$$

and

$$E^{(+)} = E_a \left\{ \frac{(5 - 4v)E_a + [2V_a(1 + v) + V_m(5 - 4v)] \cdot (E_m - E_a)}{E_a(5 - 4v) + 2V_a(1 + v)(E_m - E_a)} \right\} \tag{5.5b}$$

For the three-dimensional case the aggregates are spherical, and not the cylinders that appear in a 2D simulation. The 2D simulation can therefore at best be a rough approximation of the real behavior of a 3D particle composite. For a composite of spherical inclusions "*a*" embedded in a matrix phase "*m*," Hashin and Shtrikman (1963) derived the bounds for the bulk modulus as follows:

$$K^{(-)} = K_m + \frac{V_a}{\dfrac{1}{K_a - K_m} + \dfrac{3V_m}{3K_m + 4G_m}} \tag{5.6a}$$

and

$$K^{(+)} = K_a + \frac{V_m}{\dfrac{1}{K_m - K_a} + \dfrac{3V_a}{3K_a + 4G_a}} \tag{5.6b}$$

In Hashin and Shtrikman (1963) the bounds for the shear modulus are also given, expressed in terms of the volume fractions and the elastic properties of the two phases. Here we refrain from giving those because only limited analyses for E and K are available, which are summarized in the next section.

5.2 Effective Young's Modulus of a Two-Phase Aggregate-Matrix Composite

As mentioned, the elastic properties of a lattice should best be determined by means of a simple numerical linear elastic framework analysis. Any finite element code supporting beam elements with 3 dof per node can be used. The numerical simulations are particularly useful when a particle composite such as concrete is analyzed. The effect of the material structure on the elastic properties can be determined, like the effect of particle volume on effective elastic modulus and the Poisson ratio, as was done by Van Vliet (2000) and Lilliu (2007) for 2D and 3D lattices, respectively. One important prerequisite is that the analyzed structure is larger than the representative volume of the material. For a coarse-grained material such as concrete and several types of rock the RVE can be quite large and often the characteristic dimension of the specimen to be analyzed exceeds 100 mm. For a single elastic analysis this is not really problematic, even when the number of elements would be over 10^5 or larger, however, for fracture analyses where many consecutive load-steps must be applied this may often lead to computational problems and one has to revert to the largest available computers (see also Appendix 1).

Figure 5.2 shows some examples of material structures analyzed by Van Vliet (2000). In Case I a continuous particle distribution has been used (Fuller distribution; see Section 4.3) and the total volume of particles has been varied. Case II is based on the same particle distribution, but the particle volume has been reduced by systematically removing the particles of the smallest size fraction. Thus, in Figure 5.2d (Case II) all particles are present ($2 \leq d \leq 16$ mm); in the other structures in the same row, going from right to left the smallest particle fractions are omitted thereby reducing the total particle volume. In Case III (bottom row) a random mix of aggregate and particle elements has simply been used, and the amount of particle beams increases from left to right.

Note that particle volume in these 2D analyses must be interpreted as particle area. In all cases the thickness of the structures is constant and equal to $t = 1$ mm. Along the top boundary of the structure a uniform displacement is applied in the vertical direction; the lower row of nodes is fixed also in the vertical direction, and the middle node of the bottom and top row in the horizontal direction. The length of the lattice beams is 1 mm, which is smaller

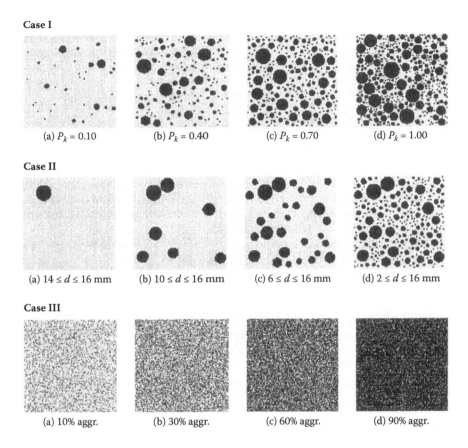

Case I

(a) $P_k = 0.10$ (b) $P_k = 0.40$ (c) $P_k = 0.70$ (d) $P_k = 1.00$

Case II

(a) $14 \leq d \leq 16$ mm (b) $10 \leq d \leq 16$ mm (c) $6 \leq d \leq 16$ mm (d) $2 \leq d \leq 16$ mm

Case III

(a) 10% aggr. (b) 30% aggr. (c) 60% aggr. (d) 90% aggr.

FIGURE 5.2
Case I (upper row): continuous (Fuller) distribution with P_k varying between 10% and 100% (for
an explanation of P_k, see main text). The effective aggregate content $P_{k.latt}$ is 0.03, 0.19, 0.34, and
0.47, respectively. Case II (middle row): Continuous Fuller distribution where the variation in
particle content has been achieved by removing the smallest particle fractions (from right to
left; middle row). The effective aggregate content $P_{k.latt}$ is 0.02, 0.11, 0.22, and 0.35, respectively.
Case III (lower row): Random mixture of aggregate and matrix beams; the fraction of aggregate
beams increases from 10% (left) to 90% (right). The effective aggregate content after lattice overlay is not affected in this case. (After Van Vliet. 2000. *Size Effect in Tensile Fracture of Concrete and
Rock*. With kind permission of Dr. Marcel van Vliet.)

than the smallest aggregate particle introduced, but still large enough to
cause distortion of the particle shape (see Figure 4.13c). The regular triangular lattice has a size of 80 × 92 cells to form a structure of approximately
80 × 80 mm² (note that in a regular triangular lattice $y = l\sqrt{3}/2$ when $x = l$,
which explains the larger number of cells in the y-direction). In Figure 5.2 P_k
denotes the aggregate fraction, as defined in Equation (4.4).

Due to the lattice overlay and the way the aggregate distribution is generated (neglecting the smallest particles) there are some losses and the effective
value of P_k is always smaller than the value indicated. Thus, for example, for

Case I with $P_k = 1$, the effective value $P_{k,\text{eff}} = 0.80$. After lattice overlay, with a 1-mm triangular lattice in the end only $P_{k,\text{latt}} = 0.48$ remains. There is in practical mixtures also a good reason why an aggregate content of 100% can never be reached. As a matter of fact, the maximum possible amount lies around 50%, but this value is to some extent dependent on the aggregate size. In experiments Wittmann, Sadouki, and Steiger (1993) showed that with increasing sand fractions ($d < 4$ mm), at a particle content of 50% or higher suddenly a higher porosity developed, which can be explained from the simple fact that not enough cement matrix is left to enclose all particles. The cement ribbons around the grains become thinner than the smallest cement grain and the composite structure is destroyed. As a result there is a clear effect on the global Young's modulus of the composite as was already shown in Figure 5.1. It is always nice to see confirmed that many aspects of material behavior simply relate directly to geometrical constraints in the material structure.

Now let us discuss some of the numerical outcomes. We limit ourselves to loading applied in the vertical (y-) direction; Van Vliet also considered loading in the horizontal (x-) direction, which makes sense because a regular triangular lattice has threefold symmetry that is broken in a Cartesian coordinate system. In Figure 5.3 results of two sets of numerical analyses are shown. In Figures 5.3a and 5.3c the Young's modulus of the aggregates is 25 GPa, of the matrix 10 GPa; in Figures 5.3b and 5.3d the situation is reversed and the Young's modulus of the aggregates is smaller than the matrix. The latter situation resembles, for example, lightweight concrete.

In all figures the open and closed symbols are the results of the numerical analyses. For the continuous particle distributions (Figure 5.3a,b) the maximum achieved P_k after lattice overlay does not exceed 48%. However, for the random mixtures of aggregate and matrix elements (Case III, Figures 5.3c,d), as expected, the whole aggregate fraction range (from 0–100%) is possible. The continuous lines in Figure 5.3 are the bounds calculated with simple analytical models such as the series and the parallel models and Hashin bounds (Equation (5.5); see Section 5.1). All numerical results are well between the upper and lower bounds predicted by these analytical models, which shows that elastic properties are correctly "predicted" from these relatively simple lattice models.

At a 0% aggregate fraction, the Young's modulus of the matrix material is found; at a 100% aggregate fraction the modulus of the aggregate grains is found. In between, the numerical results follow the trend of the analytical models. Note that it is not very important how the aggregate content is varied in particle structures (compare Cases I and II, Figures 5.3a,b). The results of random material structures of Case III are consistently higher than those of the other two cases. It should be mentioned that the results shown here are valid for the chosen lattice geometry only, that is, the threefold symmetric regular triangular lattice. For other lattice geometries, for example, a random lattice (Vervuurt et al. 1995), the same analyses have to be repeated for calibrating the model.

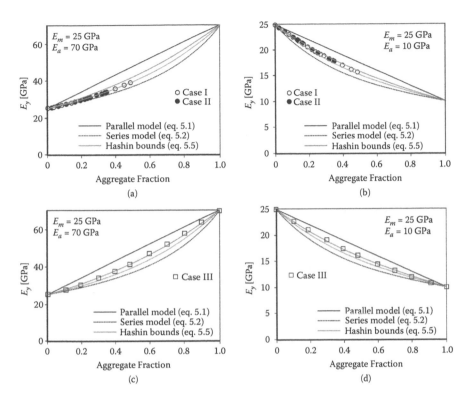

FIGURE 5.3
Effective Young's modulus for a regular triangular lattice with varying aggregate content: (a) and (c) aggregate modulus larger than the surrounding matrix; (b) and (d) low modulus aggregates in a high modulus matrix. (After Van Vliet. 2000. *Size Effect in Tensile Fracture of Concrete and Rock*. With kind permission of Dr. Marcel van Vliet.)

5.3 Effective Elastic Properties in Three Dimensions

For a 3D lattice Lilliu (2007) made similar analyses as those shown in the previous section for two dimensions. Again a complete set of new analyses is required, but now in three dimensions. Lilliu used a three-dimensional lattice, with randomness $A/s = 0.5$ (with $s = 1$ mm). The 3D random lattice is an extension of Vervuurt's methods based on the Mourkazel and Herrmann (1992) random lattice. The analyses were done for a Fuller distribution, with 2 mm $\leq d \leq 14$ mm. The maximum aggregate density that could be reached after lattice overlay ($P_{k, latt}$) is equal to 0.30. Again, as in the previous section, the effect of the ITZ was neglected, and basically a two-phase material was considered. Of course, refinements are possible, and one could decide to include the ITZ, which means that the entire series of analyses must be repeated once more. It should be mentioned again that depending on how the ITZ is handled in the lattice, an enormous refinement of the model may

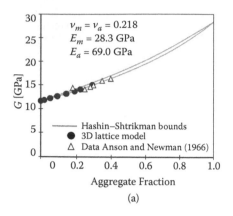
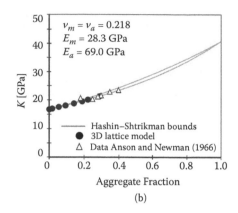

(a)   (b)

FIGURE 5.4

Effective global shear modulus G^* (a) and effective global bulk modulus K^* (b) for a three-dimensional random lattice ($A/s = 0.5$, $s = 1$ mm) with different aggregate density (spherical particles with a Fuller distribution). (After Lilliu. 2007. *3D Analysis of Fracture Processes in Concrete*. With kind permission of Dr. Giovanna Lilliu.)

be required, but for a single analysis needed to determine the global elastic properties this may not prove to be an insurmountable problem, at least, not in comparison to fracture analyses (see Chapter 6).

Figure 5.4 shows the results obtained by Lilliu. The basic properties of the matrix and aggregate particles were selected such that a comparison was possible with experimental data obtained by Anson and Newman (1966), namely $E_m = 28.3$ MPa and $E_a = 69$ GPa. The same Poisson ratio, $v = 0.218$, was assigned to both the aggregate and matrix phases. It will be obvious that such a comparison would be useless for the two-dimensional case of Section 4.2. The two diagrams in Figure 5.4 are for the effective global shear modulus G^* (Figure 5.4a) and effective global bulk modulus K^* (Figure 5.4b). As mentioned, a comparison is made with the Anson–Newman data, as well as with the Hashin–Shtrikman bounds, which are mentioned in Equation (5.6) for the bulk modulus only.

The comparison among the numerical lattice analyses, the experimental data, and the Hashin–Shtrikman bounds is quite satisfactory, which may indicate that the lattice may turn out to be a viable tool for analyzing the behavior of disordered materials such as concrete and rock. In the following chapters we return to fracture.

6

Fracture of Concrete in Tension

The tensile strength of concrete is about 8–10 times lower than its compressive strength and tensile cracks are present in almost every reinforced concrete structure. Mode I fracture of concrete is therefore considered most important. For the fictitious crack model the tensile stress–strain curve and the softening diagram are the essential input parameters. As mentioned in Section 2.4 direct tests are preferable, but due to several experimental difficulties are considered among the most difficult to perform (see also Appendix 3 on test stability). In Section 6.1 we analyze the behavior of uniaxial tension tests, and explain the fracture process from 2D and 3D analyses. Next, we elucidate the effect of a number of experimental issues including the use of notches and the effect of the rotational freedom of the loading platens. Subsequently, in Section 6.2 two types of indirect tension tests are discussed, namely the Brazilian splitting test and the 3-point bending test. Although these experiments are easier to conduct, the softening parameters can only be determined via an inverse analysis, where the uniqueness of the parameters remains an issue.

6.1 Analysis of Uniaxial Tension Experiments

In Figure 2.10 the classical results of Evans and Marathe (1968) were shown. Their results suggest that microcracking starts well before the maximum tensile stress is reached, in the prepeak regime, whereas the words by the developers of the fictitious crack model (mentioned in the beginning of Section 2.4) appear to suggest that microcracks develop in the postpeak regime. The confusion is not helped with the excellent overview paper by Mindess (1991), which shows there is no consensus regarding the size of the cohesive zone. The size of the process zone depends on the adopted measurement technique, the loading situation at the crack-tip and the specimen layout. Consensus is, however, quite essential for the FCM. Thus, it makes sense to study the fracture process in concrete specimens subjected to uniaxial tension in some detail, in an attempt to resolve the confusion. Since the introduction of the lattice model in 1990 simulations of tensile fracture have been carried out repeatedly, with or without accompanying tensile experiments. These tensile experiments were always aimed at the specific phenomenon studied. In the very beginning the experiments were carried out before the simulations were done, but occasionally the situation has been successfully reversed, which may indicate that the lattice has some predictive qualities. Of course there are weaknesses, and also in a number of cases there are obvious problems. These are always identified, and where possible, solutions for repair are suggested.

6.1.1 Fracture Process in Tension

One of the first simulations with particle overlay is shown in Figure 6.1 (Schlangen and Van Mier 1992a). Equation (4.4) was used to generate the particle distribution. The specimen geometry exactly resembles that of experiments done at the same time (see Van Mier 1991a). The specimen length is 200 mm, the width 100 mm, and the notch is 15-mm deep and 3-mm wide. Figure 6.1 contains two results, namely the σ–w diagram for normal concrete and the result for lightweight concrete (the dashed line). The stress is the (nominal) axial stress applied to the far ends of the specimen, the displacement w is taken over the same measurement length as used in the experiment, namely 35 mm. This means that w is not the pure crack opening, but some (minimal) elastic unloading is included. Only the part of the specimens where cracks are expected to develop is modeled by means of a regular triangular lattice (see inset of Figure 6.1); the length of the individual lattice elements is 1 mm.

The lightweight concrete shows a steeper postpeak stress drop immediately after peak than the normal concrete, which is in agreement with experimental observations, where in general larger brittleness for lightweight concrete

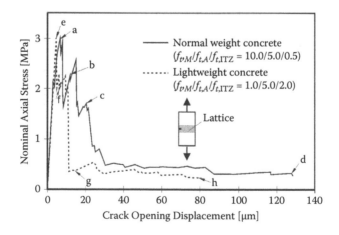

FIGURE 6.1
Stress-crack-opening diagrams for simulations of tensile fracture in normal weight and lightweight concrete. The respective crack stages are shown in Figure 6.2. The ratios of the matrix to aggregate to ITZ strength for both simulations are indicated in the inset. (From Schlangen and Van Mier. 1992a. *Cem. Conc. Comp.*, 14(2): 105–118. With permission from Elsevier.)

is found (Van Mier 1991b). The particle distribution in both specimens is identical; only the local properties have been changed. The lightweight concrete contains aggregates that are weaker compared to the grains in the normal concrete, and also the interfacial transition zone (ITZ) strength is increased. The argument for this is that the lightweight aggregates are generally quite porous, and in addition to the normal bonding mechanisms of adhesion and frictional restraint, one can rely on larger interlock between the rough aggregate surface and the cement matrix; see, for example, Zhang and Gjørv (1990). Clearly the use of identical particle distributions is an advantage in such numerical simulations in comparison to real experiments where this cannot possibly be achieved. At peak stress in both concretes some limited microcracking is observed (see Figures 6.2a and 6.2e). In normal concrete the prepeak cracking is distributed over the specimen width, whereas for the lightweight concrete just a single crack appears near the notch. The "starter crack" in the lightweight concrete appears to be within the weak aggregate particle just above the notch.

After the peak larger cracks start to grow. For normal concrete the main crack starts at the unnotched side of the specimen. By coincidence four larger aggregates were located there, close to one another, and the resulting stress concentrations must have favored macrocrack initiation at this specific location (see Figure 6.2b). Obviously the local stress concentration was larger near those four large aggregates and not at the notch. In the lightweight concrete sample the initial crack near the notch starts to grow, but clearly the crack is not continuous but contains smaller bridges (see Figure 6.2f). These may have been the result of the lattice geometry, where the crack jumps from one row of elements to the next, but may also represent the bridging observed in experiments, which takes the form of handshake cracks; see Appendix 4 (Figure A4.2) and Van Mier (1991a,b).

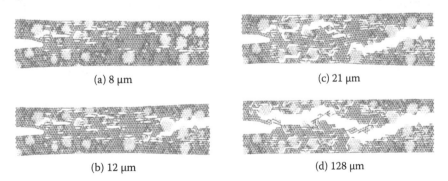

(a) 8 μm   (c) 21 μm

(b) 12 μm   (d) 128 μm

FIGURE 6.2(a)–(d)
Crack growth in single-edge notched (SEN) tensile specimen subjected to uniaxial tension; normal concrete is modeled. The tensile stress is applied in the vertical direction. Only the area where cracks were expected to grow is modeled as a regular triangular beam lattice (see inset of Figure 6.1). The total specimen width of 100 mm is shown. (From Schlangen and Van Mier. 1992a. *Cem. Conc. Comp.*, 14(2): 105–118. With permission from Elsevier.)

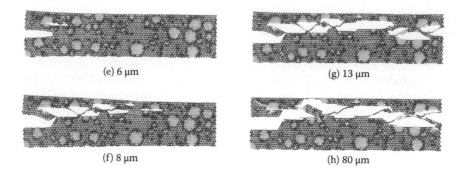

(e) 6 µm

(g) 13 µm

(f) 8 µm

(h) 80 µm

FIGURE 6.2(e)–(h)
Crack growth process in lightweight concrete. The specimen is loaded in the vertical direction, all conditions being identical to those for the normal concrete simulation in Figure 6.2a–d. (From Schlangen and Van Mier. 1992a. *Cem. Conc. Comp.*, 14(2): 105–118. With permission from Elsevier.)

Upon further loading, the main cracks extend and have more or less crossed the entire specimen width (Figure 6.2c for normal weight concrete and Figure 6.2g for lightweight concrete). Looking to the stress crack opening diagram in Figure 6.1 one can see that the major stress drop occurred in the softening diagram at this stage of the fracture process. The results suggest that a direct relationship exists between the aggregate size and the remaining load-carrying capacity (see Figure 10.5 in Chapter 10 where the four-stage fracture model is presented).

The final stage of the fracture process is then characterized by a long tail of the softening diagram; see Figure 6.1. In those final stages, the macro-crack is fully developed, but so-called crack-face bridges remain, which are capable of carrying some limited load. In experiments the same phenomena have been observed (see Van Mier 1991a,b), and the tail part of the softening diagram appeared to be quite stable. As a matter of fact the crack opening was increased manually and no instabilities occurred in the control loop of a displacement-controlled experiment. Moreover, it was possible to correlate the size of the bridges, as well as the bridging stress, to the size of the coarse aggregates used in the concrete mixture; see Van Mier (1991b). In Section A4.1 and Section 10.1.4 a more detailed account of the bridging phenomenon is given. Note that heterogeneity of the material is a prerequisite to obtain crack-face bridging. Several micromechanical models have shown the mechanism; see for instance, Vonk (1992), Wang (1994), Bolander and Kobayashi (1995), and Tijssens (2001), as well as a very early example from 1986, which was included in Van Mier and Man (2009).

As mentioned, in these analyses, which date back to 1991–1992, only the part of the specimen where cracks were expected to grow was modeled as a lattice. The remainder of the specimen and the loading platens were modeled using normal isoparametric shell elements available in the finite element package DIANA that was used. It was expected that microcracks would be concentrated near the notch area. Later analyses have shown that this assumption is not always correct, and generally it is better to model the entire structure as a lattice. Obviously this has an important consequence for the computational cost of a simulation (see Appendix 1).

6.1.2 Effect of Particle Density on Tensile Fracture

Equations (4.3) and (4.4) have frequently been used in the past to produce a computer generated particle distribution for "numerical concrete," in particular for 2D analyses. Equation (4.3) can also be the basis for a 3D particle structure where random placement of the more advanced method "space" (developed by Stroeven 1999) can be used. One of the parameters is the particle content, which is also an important starting point when real concretes are produced. With that in mind it is of interest to see how the lattice model performs when the particle content is varied. Due to the specific way the material structure is built up, an interesting transition occurs in the mechanical behavior. The ITZ plays an important role in all this. The effect of particle density on tensile fracture was investigated in the doctoral theses of Van Vliet, Prado, and Lilliu. The first two deal with 2D fracture only, whereas Lilliu developed a completely 3D model. In Figure 6.3 results are shown for several 2D analyses carried out by Prado and Van Mier (2003). Comparing the results of Figures 6.1–6.2 with those in Figure 6.3 is indicative of the progress made over the years. In 12 years the number of beam elements in a lattice analysis increased by a factor of 100–1,000, allowing for more detail in the particle structure, shorter beams, and more important, modeling the entire test specimen (in three dimensions if deemed necessary).

In the earlier simulations only the parts where cracks were expected to grow were modeled as a lattice; see, for example, Figure 6.2. The specimen analyzed is a square 60×60 mm² plate of unit thickness. The length of the beams in the regular triangular lattice is 0.25 mm; particle sizes vary between 1.0 to 11.55 mm. In Figure 6.3 for each particle density (P_k = 35, 51, and 83% before particle overlay) the load displacement diagram is plotted, with to the right, two crack patterns: stage **A** is at the peak load, stage **B** represents the crack pattern after the steep drop of load in the descending branch. The two stages are always indicated in the load displacement diagrams. The colors in the crack diagrams have meaning: they indicate the deviation from average stress at the specific loading stage. The stresses cannot be compared between the different loading stages because the average stress depends on the external load. Nevertheless the color coding (light yellow/green/blue indicates stresses lower than average, toward orange/red indicate higher than average tensile stress concentrations) is helpful for a better understanding of the fracture process. The color coding is identical to the image appearing on the book cover. In Figure 6.3 the yellow appears as light gray, red in dark gray.

The difference in behavior between the two extremes, that is, P_k = 35% and 83%, is quite telling. At 35% the crack pattern at peak, stage **A** shows

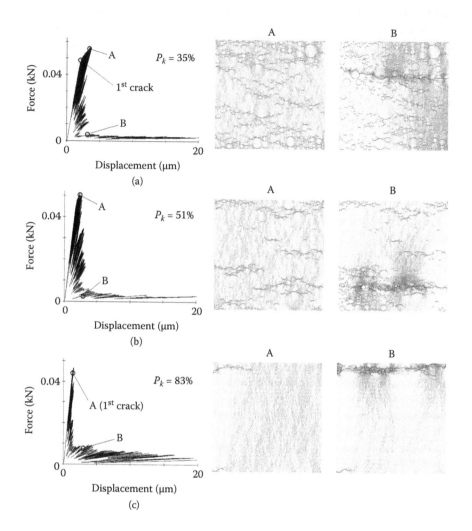

FIGURE 6.3

Effect of particle density on the fracture process in uniaxial tension. The particle density is given by the parameter P_k in each load-deformation diagram, and varies between 35, 51, and 83% (before lattice overlay). The load-deformation diagrams are shown in original zigzag format which is the result of subsequent loading cycles after each beam removal. To the right of each diagram are two stages of crack growth indicated by the letters **A** and **B** in each diagram. The colors in the crack diagrams refer to the deviation from average tensile stress at the specific loading stage **A** or **B**: yellow means average stress, toward light yellow/green/blue implies locally lower stress, and toward orange/red indicates higher tensile stress concentrations. The color codes are qualitative only, but are helpful to identify various stages in the fracture process. The color coding is visible on the cover image. Yellow translates here to light gray, red to dark gray. (From Prado and Van Mier. 2003. *Engng. Fract. Mech.*, 70(14): 1793–1807. With permission from Elsevier.)

considerable prepeak microcracking, which is absent in the 83% sample. The reason is rather straightforward: the first microcracks appear in the weakest zone, that is, the ITZ. In the sample with 35% aggregates the ITZ around the various aggregates are separated from each other by the stronger matrix phase. When the particle content increases to 83% all ITZ zones are interconnected, and percolating paths of weak elements are present in the specimen. As a result the first ITZ crack is also the critical one leading to softening and complete rupture of the specimen. The load-displacement curves show the difference in the prepeak regime. For an aggregate content of 35% there is a small, but quite recognizable "hardening regime" just before the maximum load is reached, whereas this does not appear in the 83% analysis. The intermediate case with 51% also shows some prepeak cracking, but it is significantly less compared to the 35% analysis. Note that the percolation of the ITZ can also be achieved by increasing the lattice beam length; even at sparse particle content, a thicker ITZ may cause percolation, see Lilliu and Van Mier (2007).

In the postpeak regime the behavior is quite comparable and there are hardly any differences. A single catastrophic macrocrack propagates through the specimen's cross-section and separates each specimen in two parts. In the 35% specimen the main crack clearly initiated at the left side, and propagated toward the right side. Stresses are lower than average where the main crack first developed, that is, at the left side as indicated by the light yellow/greenish color. The right side of the sample shows redder and is therefore more stressed. It is interesting to see that in all three cases the main crack is rather red along the entire length. This indicates bridging: intact pieces of material remain in the wake of the propagating macrocrack. The so-called crack-face bridges have been observed in experiments as well, especially near large stiff aggregates (see Appendix 4, Figure A4.2, and Van Mier (1991a,b)). The size of the aggregates appears to decide the size of crack-face bridges, and with that the carrying capacity of a cracked sample in the tail of the load displacement diagram. We further elaborate on these matters in Chapter 10.

The load-displacement curves are quite "spiky" and are in fact more brittle than the curves obtained from physical experiments; see, for example, Figure 6.4 where a direct comparison is made between a lattice analysis and a tensile experiment. In the computed diagram, the spikes have been smoothed by simply connecting the subsequent maximum stress levels. By doing so a higher fracture energy results; one should be cautious not to draw quick conclusions about fracture energy in this case. The initial stiffness and the maximum stress are in good agreement, but the postpeak curve is more brittle in the lattice simulation. Several reasons for the extreme brittle behavior of the lattice model have been put forward as follows:

1. The fracture law used in the lattice is an elastic, purely brittle fracture law, leading to immediate load-drop after the maximum strength of a lattice beam is exceeded. It has been argued that

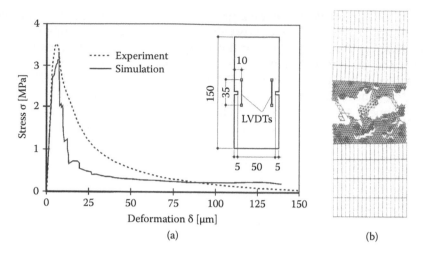

(a)   (b)

FIGURE 6.4

Comparison of experimental stress–displacement diagram from a uniaxial tension test on a double-edge notched concrete prism and from a lattice simulation. (After Schlangen and Van Mier. 1992b. In *Proceedings of the 1st International Conference on Fracture Mechanics of Concrete Structures (FraMCoS-1),*

the cement matrix between the large aggregates behaves far from brittle, and resembles the softening behavior of mortar. This has led some researchers to include a softening fracture law in lattice, for example, Ince et al. (2003). This is actually a curious decision because the lattice model was there in the first place to explain the effect of a material structure on softening, and to come to a better understanding of fracture phenomena in concrete and similar materials (Van Mier 2004b).

2. Related to the above comment is the simplification in the particle structure. Small aggregates are for computational reasons removed from the lattice structure. This allows an increase in the lattice beam length and a reduction of computational effort. This becomes less of an obstacle with advances in computer technology, both hardware and software. In the next subsection a simplified analysis is included that shows the effect of removing small particles.

3. The above analyses were all 2D simulations of fracture behavior. Laboratory-scale specimens are usually on the order of 50 × 50 ×100 mm, and are truly 3D experiments. Crack propagation is a three-dimensional phenomenon, and this is clearly not well captured in a 2D analysis. The lattice is easy to extend to full 3D, and this has been done in recent years; see Lilliu (2007). The effects of choosing a 2D simplified simulation can therefore be estimated. Some results are shown in the next section as well.

6.1.3 Small-Particle Effect

Including every small detail of the material structure into a lattice simulation is impossible with current computational facilities. If the structure of cement were included, lattice beams would have to be μm size or even smaller and only very small specimens can be analyzed, for example, the tiny cylinders used in the microtensile test of Figure 11.9. Therefore a simulation is always a trade-off between computational possibilities and resolution. If the analysis becomes 3D the problems are even larger. In order to estimate the effect of small material structural elements in lattice simulation different types of idealized simulations were carried out. One of these was the so-called small-particle effect. In Figure 6.5 the small particle effect is shown in 2D. In one example a large centrally placed aggregate is embedded in an otherwise homogeneous and brittle matrix, whereas in the second example many small particles have been included around the centrally placed large aggregates. The absolute size of the particles is not of interest; rather the relative effect of the small particles is the focus of these analyses.

The load-displacement diagram clearly shows a much increased post-peak ductility as soon as the small particles are present. The reason for the increased ductility is now quite obvious: bridging of cracks becomes more important as soon as the small particles are included, which can be seen from the respective crack patterns. Note also that due to the use of a regular triangular lattice cracks become rather straight when no small particles are included. Such straight cracks usually do not appear in experiments: materials such as cement and concrete are heterogeneous to the smallest size/scales, and the cracks show a corresponding roughness at all size/scales.

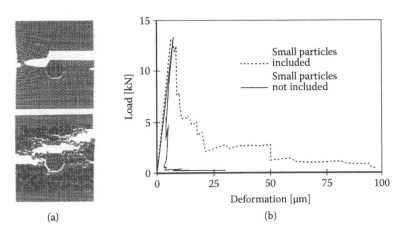

(a)   (b)

FIGURE 6.5
Effect of adding small particles to an otherwise homogeneous and brittle cement matrix surrounding a large centrally placed aggregate particle: (a) crack patterns at the end of the simulations; (b) load-displacement diagrams (After Schlangen and Van Mier. 1992c. In *Proceedings of the 1st Bolomey Workshop on Numerical Models in Fracture Mechanics of Concrete.*)

In 3D the same analysis was repeated by Van Mier and Lilliu (2001) using a random lattice in a prismatic sample of 24 × 24 × 12 mm. The average beam length in the lattice was 1.27 mm, which set a limitation on the size of the particles. The central aggregate had a diameter of 8 mm, whereas the small grains, 39 in total, each had a diameter of 4 mm. The load-displacement diagrams are shown in Figure 6.6a, and an exploded view of the crack patterns, at peak load and at 20-μm axial displacement is shown in Figure 6.6b. The conclusion is almost similar to that from the 2D simulations. The postpeak behavior is less brittle. One important difference appears though, namely,

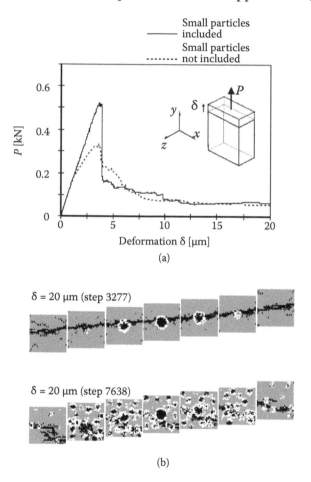

FIGURE 6.6
Load-displacement curves for two 3D simulations with different particle distribution (a). In one analysis a centrally placed particle is embedded in a homogeneous matrix, whereas in the second analysis the large particle is surrounded by many small grains. (b) Exploded view of the 3D crack patterns at peak and at 20-μm crack opening. (After Van Mier and Lilliu. 2001. In *Proceedings of the 4th International Conference on Analysis of Discontinuous Deformation, ICADD-4*.)

the analysis with small particles included shows a lower peak strength, which may have been caused by small-scale microcracking also in the ITZs surrounding the smaller particles. As a result macrocrack propagation may have started at an earlier stage compared to the one-particle analysis where microcracks would only grow along the ITZ of the single particle.

It is difficult to make a direct comparison between a 2D and 3D analysis, because the crack process is truly three-dimensional. In a 2D simulation out-of-plane crack growth is not captured, but this may have a substantial influence on postpeak behavior as well. Comparison of the relative postpeak diagrams from 2D and 3D simulations shows less brittle behavior in the 3D case; see Van Mier and Lilliu (2001). Thus it appears that not only ignoring the small-scale material structure leads to increased brittleness in lattice analyses, but also changing from 2D to 3D has some (limited) effect. In general, for multiscale heterogeneous materials such as concrete, a full 3D analysis would be required, but then of course a great deal of computational effort is required.

6.1.4 Boundary Rotation Effects and Notches

In physical experiments one always has to consider specimens of finite size. Periodic boundary conditions do not exist; the only remedy would be reverting to extremely large specimens. If the size of a specimen becomes too large, one may experience stability problems in fracture experiments; see, for example, Carpinteri and Ferro (1994) and Van Vliet and Van Mier (2000). The specimen size is therefore often chosen to be relatively small, that is, on the order of about 100 mm for concrete and mortar, which not only limits the use of maximum aggregate sizes up to 16 mm, but also results in noticeable boundary condition effects. In some of the macroscopic fracture models for concrete, tensile tests carried out between nonrotating loading platens are required (see Section 2.4, the fictitious crack model). Because some peculiar behavior is observed in tensile tests between fixed (nonrotating) loading platens, namely the development of two major cracks in the postpeak regime, a variation of the boundary conditions in a tensile test is a fine example for the lattice model. Not only is this a fine example, but the analysis may help to understand and interpret the result from displacement-controlled experiments.

An additional experimental requirement for obtaining stable softening curves is making one or more notches in a concrete specimen. The stress concentration around the notches confines the macrocrack development to a certain region of limited size in the tensile specimen, provided the stress concentration is more important compared to that around single aggregate particles, or clusters of large aggregates (see, e.g., Figure 6.2). Notches are for the same reason used in bending experiments, that is, to avoid snap-back behavior that would occur when the length of the control extensometer becomes too large and too much elastic energy is released in the postpeak regime (see Appendix 3 on test stability, and the more elaborate text in Van Mier 1997). The notches are quite often cut by means of a rotating diamond saw in a block of concrete. The notch has to be cut quite precisely, with accuracy comparable to that used for manufacturing the entire specimen. It pays to use a very accurate specimen preparation, even though this may sometimes be quite difficult to achieve for concrete specimens. The lattice may also in those cases be quite helpful because the effect of misaligned notches may be studied. These are two among many experimental issues that must be considered in doing fracture experiments on concrete, rock, and other quasi-brittle materials. In this section we discuss the effect of boundary rotations on the tensile load-deformation diagram and the ensuing fracture patterns. Next, the influence of misalignment of notches is presented.

6.1.4.1 Boundary Rotation Effect

In linear elastic fracture mechanics theory the specimen geometry and the boundary conditions both have an effect on the stress intensity factor. Corrections to the general equation for the stress intensity factor for a crack in an infinite plate subjected to farfield tension are needed when finite size samples are considered. When nonlinear fracture models were introduced for concrete, specifically the cohesive crack model, also referred to as the fictitious crack model, it was suggested that the closing pressure over the crack-tips should be derived from stable (displacement-controlled) uniaxial tension tests between fixed-end conditions (see Section 2.4). This means that it should be prohibited for rotations of the loading platens to occur during the entire experiment. Considering therefore the effect of boundary conditions on fracture response seems of eminent importance in judging whether cohesive crack models should be accepted. In Figure 2.11 we showed the effect of the rotational stiffness of the loading platen on the stress–deformation behavior of cylindrical concrete specimens. Fixing the platen has a significant effect on the shape of the softening curve and a pronounced bump is observed. In the test with freely rotating platens a very smooth softening curve is measured in contrast. The bump can be predicted using simple LEFM, as shown in Appendix 1. The flexural stiffness of the specimen itself plays a role, and enters the solution through the specimen slenderness L/W (see Figure A2.2)

Let us now focus our attention on the fracture mechanisms. In Figure 6.7 the crack patterns obtained from three analyses of the Carpinteri and Ferro (1994) dog-bone-shaped test specimen are shown. Loading is always tensile, in the vertical direction. The boundary conditions were varied in the three analyses, namely, (A) freely rotating loading platens, (B) fixed (nonrotating) loading platens, and (C) uniform deformation over a measuring length of 50 mm over the central part of the specimen. Uniform displacements should be interpreted as the same displacements at the left and right side of the specimen while cracking proceeds. In theory this is easy to achieve, but in

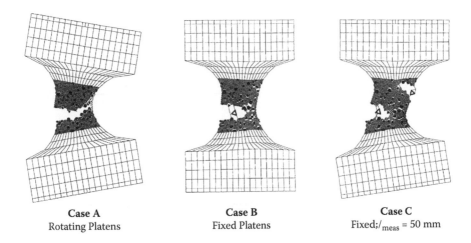

Case A	Case B	Case C
Rotating Platens	Fixed Platens	Fixed;/$_{\text{meas}}$ = 50 mm

FIGURE 6.7

Effect of boundary rotations on the fracture behavior of dog-bone-shaped tensile specimens. In Case A the far ends of the specimen can freely rotate around a center point on the respective end surfaces; in Case B the top and bottom edges are kept parallel throughout the analysis; and in Case C the deformations are kept uniform over a zone of 50-mm length in the center of the specimen. (After Van Mier, Schlangen, and Vervuurt. 1995. In *Continuum Models for Materials with Microstructure.*)

laboratory experiments the heterogeneity of the material, the specimen manufacturing, as well as alignment of a specimen in the loading frame are not always 100% controlled.

Cases B and C are quite similar in that the cracks appear more distributed over the entire specimen length, at least there where the cross-sectional area is reduced. In the case of freely rotating platens, Case A, the main crack almost immediately develops from the weakest side of the specimen (note that the material is heterogeneous and a statistical strength distribution is present as in real concrete specimens), and then propagates unhindered to the other side. As a result a relatively sharp crack zone develops, with much less side cracking as observed in Cases B and C. The analyses were done using two different grain structures, and in Table 6.1 the fracture energies for the three boundary conditions A, B, and C are listed. The fracture energy, or actually the work of the fracture, resembles the area under the load-displacement diagram up till a deformation of 100 μm. The lower fracture energy for the hinged boundary condition (Case A) is obvious in comparison to the two other boundary conditions. The result can be related to the lesser amount of cracks in Case A, that is, when fewer cracks are formed likewise a decrease of fracture energy is observed.

In Figure 6.8 the fracture mechanism is explained schematically for all three boundary conditions. For an understanding of the observed phenomena it is absolutely essential to recognize that the heterogeneous material structure plays an important role. The specimens are certainly not uniformly

TABLE 6.1

Fracture Energies for the Three Analyses of Figure 6.7,
Carried Out for Two Different Grain Distributions

	Grain Structure [–]	Fracture Energy [N/m]
Case A: Rotating platens	1	29.8
	2	11.6
Case B: Fixed platens	1	35.3
	2	32.5
Case C: l_{meas} = 50 mm	1	40.6
	2	43.8

Source: After Van Mier, Schlangen, and Vervuurt. 1995. Lattice
type fracture models for concrete. In H.-B. Mühlhaus
(Ed.), *Continuum Models for Materials with Microstructure*,
John Wiley & Sons, Chichester, UK, 341–377.

built up in reality, and this is reflected in the heterogeneous particle structure that was adopted in the lattice analyses. In Figure 6.8 the heterogeneous particle structures and the related statistical strength distributions in the samples have not been drawn. It is easy to see, however, that crack nucleation will start at a location where the weakest material strength is found; see, for example, also the tensile simulation of Figure 6.2a–d. In that specific case it was not where the notch was made, but at the other side where a concentration of large aggregate particles existed.

If the loading platens can rotate freely (Figure 6.8a), the main crack can simply propagate immediately after it nucleates. There is no restraint, no mechanism that can stop the crack from propagating. Consequently the load-displacement curve shows a very smooth and gradual load-drop in the softening regime. At the side where the main crack opens large positive deformations are measured, whereas at the other side of the specimen compressive displacements are found to occur. Clearly the hinges operate as they should and the crack is not restrained to open from the side where it started. This situation changes considerably when the end platens cannot rotate; see Figure 6.8b. The nucleation of an initial crack at the left side of the specimen, for example, triggers the development of a bending moment as sketched in Figure 6.8b. The moment develops because the end-platens are forced to remain parallel. As the axial deformation keeps increasing, a second crack may nucleate from the right side of the specimen. The first crack is thus temporarily arrested by the bending moments, which results in a plateau (often referred to as a "bump") in the softening curve. The individual displacements at the left and right sides of the sample have been included in the stress–displacement diagram of Figure 6.8b, and clearly indicate that first the left crack opens, showing large positive deformations, and later, after the plateau the right-side crack opens thereby partly closing the left crack.

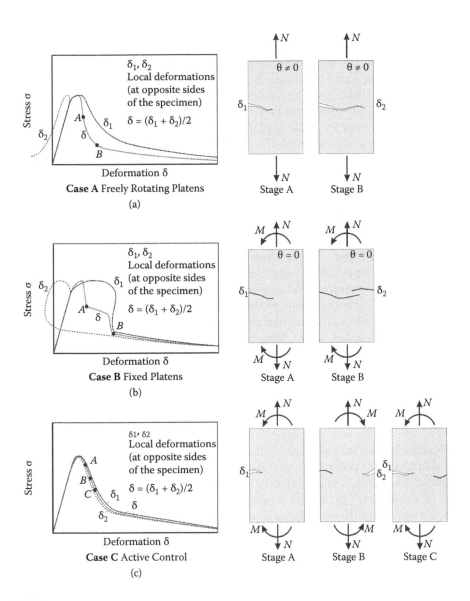

FIGURE 6.8

Effect of boundary rotations on the stress–displacement behavior and fracture mechanism under uniaxial tension: (a) the situation sketched for freely rotating end platens; (b) crack development for fixed end platens; (c) for active control. Crack stages A, B, and C in the diagrams to the right are at the locations indicated along the curves. (After Van Mier. 2004a. In *Proceedings of the 5th International Conference on Fracture of Concrete and Concrete Structures (FraMCoS-V)*.)

The third case is of interest inasmuch as the debate about boundary rotations triggered some researchers to develop a tensile experiment with uniform deformations around a crack zone, as was required in the fictitious crack model. For experiments with active control three axial actuators are needed: one for the axial load, and the two others for controlling the in-plane and out-of-plane bending when nonuniform crack growth develops in either of these directions. Tensile tests with active control were, for example, carried out by Carpinteri and Ferro (1994), and more recently with a manually operated device by Akita et al. (2003). In active control it is tried to keep uniform deformations around the specimens' circumference by constantly adjusting the control actuators for the bending moments. The axial deformations along the two specimen sides in the 2D example shown in Figure 6.8c are almost identical in such an experiment. The corresponding numerical analysis (Table 6.1) shows that more microcracks develop under these boundary conditions and that the fracture energy is correspondingly higher (in comparison to Case A).

It is quite clear that when the fracture process can be influenced so easily by changes in the boundary rotations, one can hardly hope to interpret the outcome of tensile fracture tests as real material properties. Yet, this is done by those favoring applications of the fictitious crack model. On the basis of the above it is rather obvious that predicting fracture behavior of structures by means of the fictitious crack model must be cumbersome. Indeed, a good example that showed the problems was the international round-robin analysis of anchor pull-out; see Elfgren (1992). In Section 7.1 we show how well the lattice model performs for the anchor pull-out problem.

6.1.4.2 Notches

Notches are a not very pleasant consequence of the demand to perform a stable displacement-controlled fracture experiment (see Appendix 3). By using notches, the location of the fracture zone is more-or-less fixed. The terminology "more-or-less" is used because the material structure may be reason for crack growth in different places in a specimen; see, for example, the extreme case of the large grain concentration in the analysis of Figures 6.2a–d. It would be more realistic to perform fracture experiments using unnotched samples, and in that case dog-bone-shaped specimens would be the obvious choice in uniaxial tension tests. In Figure 6.7 dog-bone-shaped specimens used by Carpinteri and Ferro (1994) were shown. In their experiments it proved to be difficult to achieve stable results for 400-mm large specimens. The test method was substantially improved by Van Vliet (2000), who developed a method where the electronic system, that is, the regulation amplifier in the closed-loop test control system, would automatically switch to another LVDT as soon as it recorded larger deformations (see Appendix 3 on the stability of fracture experiments). This so-called max-control proved to be very effective, and large dog-bone-shaped specimens with a total length of

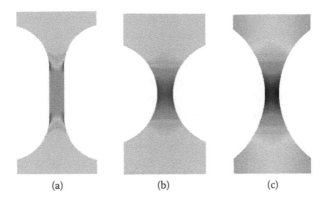

(a)   (b)   (c)

FIGURE 6.9
Stress distributions in dog-bone-shaped specimens subjected to vertical tension from linear-elastic FEM. The geometry changes between straight bays (left, (a)) to curved bays with varying radius (b) and (c). The highest tensile stress concentrations appear in the neck region and are black. Note that the specimen with straight bays at the left (a) shows stress concentrations where the straight part connects to the curved ends. If tests are conducted using such specimens, cracks mostly appear at this transition. (From Rieger. 2010. *Micro-Fiber Cement: Pullout Tests, Uniaxial Tensile Tests and Material Scaling.* With kind permission of the author.)

2,400 mm could be controlled in this way. This would then require a total of 16 LVDTs, so the experiments were not really cheap.

The shape of the dog-bone is very important, and over the years it has been observed that the smallest tensile stress concentrations may lead to quite unsolicited responses. In Figure 6.9 the stress distributions from simple linear elastic finite element analyses are shown for three different dog-bone-shaped specimens. The obvious advantage of the geometry to the left is that the central section has straight sides and thus a constant cross-sectional area. This is not the case in the other two geometries, both with curved bays, which of course result in a varying cross-sectional area over the length of the specimen. One of the reasons for favoring dog-bone-shaped specimens is that the area for gluing the specimens to the (steel or aluminum) loading platens of the test machine is increased. With that stresses are lower in the contact zone and chances for glue failure decrease substantially. In particular for very large test specimens ([m]-size) this is quite desirable.

The specimen geometries with curved bays are most suited, because of one small, but quite important disadvantage of the dog-bone geometry with straight sides. As can be seen from the results in Figure 6.9a, small tensile stress concentrations appear where the curved neck changes to the straight sides of the specimen. This would suggest that cracking may occur there first, and when experiments with this geometry are performed, indeed, over 90% of the samples will fracture at the location of these small stress concentrations. The other dog-bone geometry, with curved bays (Figures 6.9b-c), does not show this disadvantage: the specimen will fracture there where the material strength is most critical (lowest). When the curvature of the bays is increased the variation in cross-sectional area in the central part of the specimen becomes less. The geometry at the right side has proven to be quite useful for testing fiber-reinforced concrete and phenomena such as multiple cracking can be observed. It is obvious that with a variation of the curvature of the bays, the part of the specimen where cracks may develop can be regulated. Thus, the influence of the specimen shape cannot be ignored, and the outcome of a fracture experiment necessarily must be judged against the experimental boundary conditions.

In the limit case a notch is sawn (or cast) into a concrete sample, and the location of a crack is with that more or less fixed, unless the material shows large-scale heterogeneity. Most common is to produce prismatic specimens with two notches at the same height. Basically, one notch would suffice, because the heterogeneity of cement or concrete tested usually leads to crack propagation from one of the notches anyway. The idea that a symmetric loading situation is created by sawing two identical notches at the same height is an illusion because of the materials' heterogeneity. The lattice model has been used in the past to judge the error that is allowed in the alignment of the notches (Shi et al. 2000).

In Figure 6.10 results of four different analyses with the same grain structure, but with varying notch offsets are shown. The notch offset is defined as the vertical distance between the center lines of the left and right notches. The loading is tension in the vertical direction. The specimen width is 60 mm; with two notches each 10-mm deep a central area of 40 mm remains. The results can be summarized as follows:

1. With 0-mm offset the crack forms in the plane between the left and right notch.

2. At 5-mm and 10-mm offset, the crack initiates from the left notch and does not connect to the right notch.

3. At 15-mm offset many microcracks develop between the two notches, as can be seen from the three load-steps in Figure 6.11, but eventually two larger cracks are found. The particle structure in the concrete has some influence; see Figure 6.12.

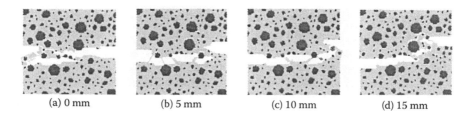

    (a) 0 mm   (b) 5 mm   (c) 10 mm   (d) 15 mm

FIGURE 6.10
Effect of notch offset on the fracture patterns in uniaxial tension simulations with the lattice model. Note that in these analyses the bond strength was set to 1.7 MPa.

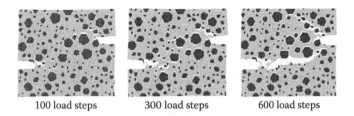

100 load steps   300 load steps   600 load steps

FIGURE 6.11
Crack growth in a specimen with 15-mm notch offset; particle structure #2 (Figure 6.12). At stage 300 numerous microcracks have developed in a wide zone between the notches. At 600 steps the main crack has propagated from the left notch and a secondary crack appears to grow diagonally upwards from the right notch. The same behavior was found in tests by Shi and Van Mier (2000).

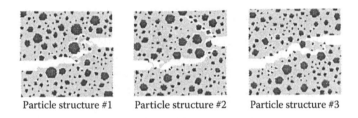

Particle structure #1   Particle structure #2   Particle structure #3

FIGURE 6.12
Three analyses with different particle structure; notch offset is 15 mm. In these analyses the bond strength was set at 1.25 MPa. (After Shi and Van Mier. 2000. In *Proceedings of Meso-Mechanics 2000.*)

Experiments by Shi and Van Mier (2000) show similar behavior. There is a tendency to develop two cracks when the notch offset is large, whereas for a notch offset smaller than 10 mm a single crack zone is always observed. The experiments were carried out on a mortar with 2-mm max aggregate size, and a concrete with 8-mm max aggregate. The experimental results were quite comparable. The conclusion is that a relatively large error can be made in sawing the notches in concrete specimens. For creating an unwanted/uncontrolled offset of 15 mm one has to work rather sloppily. From these results it could be concluded that the greater heterogeneity would allow for less accurate specimen manufacturing. However, experience shows that in general this is not quite true and it will pay to work with the highest possible accuracy.

6.2 Indirect Tensile Tests

For practical purposes one often reverts to indirect determination of tensile properties of concrete. The main reasons cited are that gluing a specimen in a tensile loading frame is tedious; aligning the specimen is difficult, and frequently the test will fail in the glue connection and much time and money are lost without obtaining useful results. In some cases a proper tensile loading frame is also lacking and one is forced to determine the tensile properties in another way. Two tests spring to mind, namely the Brazilian splitting test and the simple (3-point or 4-point) bending test. In all cases the loading on the specimen is far from uniform, but through careful analysis it is possible to derive values close to the tensile strength obtained in uniaxial tension tests. What the "true" value is will likely always be unknown, but for structural engineers it is normally sufficient to rely on approximate values. In this section a more detailed analysis is made of a splitting experiment on cylindrical discs and a 3-point bending beam.

6.2.1 Brazilian Splitting Test

When only a simple compression machine is available the tensile strength of (quasi-) brittle materials can be estimated using the so-called Brazilian splitting test. Nilsson (1961) mentions that the test was originally proposed in 1943 by the Japanese Akazawa, and later by the Brazilian Carneiro in 1949. A cylindrical specimen is loaded to failure via the sides between load-distributing strips. Model codes exactly prescribe what the size of the load-distributing strips should be, as well as the material (often triplex). The state of stress in a cylinder loaded between two opposite line loads can be calculated from the theory of elasticity; see, for example, Timoshenko and Goodier (1970). Figure 6.13 shows the stress distribution between the line loads from the theory of elasticity for three different widths of the load-distributing strips. The plotted stresses act perpendicular to the line connecting the two line loads P. In the center part of the cylinder horizontal tension (σ_h) prevails; just below the loading strips a confined compressive state of stress develops, quite similar to the stress-distribution under the loading platen below rigid steel loading platens in a cube compression test; see Chapter 8. The width of the loading strips has some limited effect on the stress distribution. Note that when cubes are used instead of cylinders and the loading strip width increases, the failure mode gradually moves toward the compressive failure modes discussed in Chapter 8. Here we concentrate on the loading on cylindrical specimens over a narrow loading strip. The horizontal tensile stresses between the loading strips are almost evenly distributed and can be approximated through

$$\sigma_h = \frac{2P}{\pi Dd} \approx 0.64 \frac{P}{Dd} \tag{6.1}$$

where D is the diameter of the cylinder, d is the depth, and P is the line load. In Figure 6.13 the vertical stress σ_v is also plotted. This is a compressive stress,

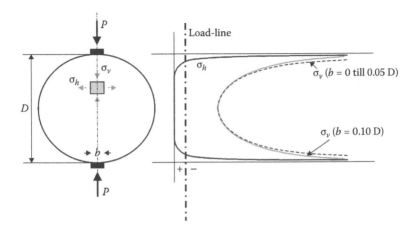

FIGURE 6.13
Stresses between the two line loads in a Brazilian splitting test for two different widths *b* of the load-bearing strips. (From Nillson. 1961. *RILEM Bull.*, 2(11): 63–67. With permission from RILEM.)

and the resulting state of stress in the crack plane is biaxial tension compression, which becomes important in later stages of the fracture process. With the development of the fictitious crack model (see Section 2.4) some researchers were tempted to use the Brazilian splitting test for determining the fracture energy of concrete and rock, as well as for concrete–rock or old-concrete/new-concrete interfaces (Hassanzadeh 1995). The latter applications may likely lead to useful approximations of the interfacial fracture energy, yet, the application on solid concrete or rock may be debatable when the experiment is run in displacement-control because more than one crack surface may develop, which must be accounted for somehow.

Note that displacement control is essential because the complete stress-crack-opening diagram is required for the determination of the fracture energy as defined by Hillerborg et al. (1976) (see also Section 2.4). Detailed analyses of Rocco et al. (1999a,b,c) and Olesen, Ostergaard, and Stang (2006) indicate that when the diameter of the cylinder increases, or the width of the loading strips decreases, a better estimate of the tensile strength is obtained (see Figure 6.14). Also it was concluded that cylinders perform better than cubes for estimating tensile strength; compare Figures 6.14a and 6.14b. These conclusions are true for intermediate concrete qualities. If the material has a deviating ratio between tensile and compressive strength, that is, deviating from the common value of roughly $f_t \approx 0.10 |f_c|$, different behavior may emerge. When tensile and compressive strength of the material are more or less equal, crushing of the material below the loading strips may prevail. Moreover, for plastic materials, for example, when fibers are added to the cement-based matrix, the split-cylinder test is not suited for determining the tensile strength, not even in an approximate manner (see, for example, Olesen et al. 2006).

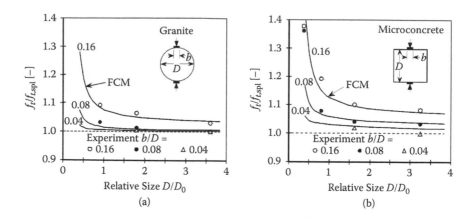

FIGURE 6.14

Ratio of tensile strength f_t and splitting tensile strength $f_{t,spl}$ for cylinder tests on cylinders (a) and cubes (b). The size of the loading strip is indicated with the measure b/D, whereas D/D_0 denotes the relative specimen size. The solid lines indicate the results from analyses with the fictitious crack model. (From Rocco et al. 1999b. *Mater. Struct. (RILEM)*, 32: 210–217 and 1999c. *Mater. Struct. (RILEM)*, 32: 437–444. With permission from Springer.)

Several years ago we attempted to get some idea of the influence of the stiffness of the loading strips (restraint) on the failure of a concrete disk subjected to vertical splitting loads (Lilliu and Van Mier 1999). Another reason to run displacement-controlled experiments and to perform lattice analyses was to see whether some way could be found to adapt the fracture law in the lattice model in such a way that compressive failure could be handled in just the same simple and straightforward manner used in tension. Changing the restraint between the loading strips and the concrete disks leads to an important change of the failure of the disks. Some results are shown in Figure 6.15. These experiments were conducted on concrete disks ($d_{max} = 8$ mm) of two different sizes ($D = 75$ and 150 mm) loaded between plywood strips ($b = D/6$) or simply attached to steel platens of the same width by means of a two-component epoxy adhesive. The thickness of the disks was small: $d = 10$ mm for both specimen sizes. The tests were performed in displacement-control using the average horizontal displacement (perpendicular to the loading direction) measured at the front and back sides of the disk as a feedback signal. Figure 6.15 clearly shows that stable load-displacement diagrams are obtained. The crack opening refers to the crack width of the central section of the splitting crack.

The failure process was in all cases more or less similar with a small but significant difference between plywood platens and glued steel platens. In all experiments a central splitting crack developed, which gradually opened beyond peak. During opening of the main crack, the regions of the specimen close to the crack were loaded in vertical compression as mentioned before. The lattice analysis of Figure 6.16a clearly shows that the magnitude of these compressive stresses is about eight times larger than the tensile stresses that

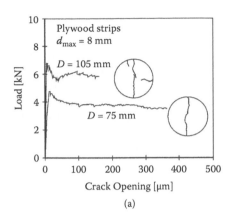
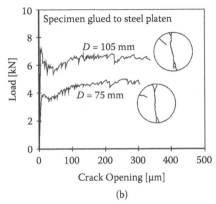

FIGURE 6.15
Load-crack opening diagrams for Brazilian splitting tests on concrete disks of two different sizes (a) loaded between plywood strips or (b) between glued steel loading platens having the same width as the plywood insert. (Results after Lilliu and Van Mier. 1999. In *Construction Materials: Theory and Application*.)

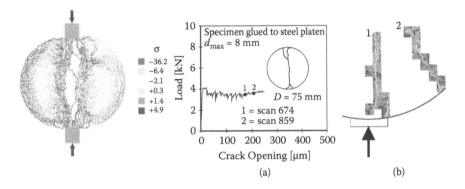

FIGURE 6.16
(a) Stresses in a Brazilian splitting test after the formation of the main splitting crack and one of the radial (secondary) cracks, which is the result from the tensile stresses along the outside of the specimen. Note that the stresses in the vertical direction along the main crack are compressive and along the disk's circumference tensile; and (b) crack development near the lower loading platen in Brazilian splitting tests on a specimen with $D = 75$ mm. (After Lilliu and Van Mier. 1999. In *Construction Materials: Theory and Application*.)

develop along the outside of the specimen. The tensile stresses along the outside lead to the growth of radial cracks, which will go undetected when the test is done in load control because of the sudden collapse of the specimen in that case. This underlines the importance of performing stable tests when dealing with fracture. Dynamic failure, as would occur under load control, may result in several secondary cracks that might be missed in the analysis of the results.

The differences between the failures of the two different specimen sizes are small, if nonexistent. The difference between plywood and glued steel platens is that in the latter case small wedge-shaped elements develop near the loading platen, whereas when plywood inserts are used the splitting crack extends toward the center of the loading strips. Obviously, the restraint friction leads to the wedges, quite similar to the mechanisms observed under uniaxial compression (see Chapter 8). In Figure 6.16b some results from microscopic observations of the specimen surface during loading are shown. Using a long-distance microscope part of the specimen surface near the lower loading platen was scanned, and the images where cracks appeared are gathered in the diagrams. In the example shown in Figure 6.16b the loading platens were glued to the specimen. Clearly visible is that the inclined cracks (numbered with "2") develop in the second rise in the softening regime. In the larger samples two more-or-less parallel cracks developed just at the edge of the loading platen, but ultimately a wedge-shaped part developed there as well. In the larger specimens the second rise in the load-crack opening diagrams was more pronounced (see also Figure 6.15).

Finally, in Figure 6.17 the failure sequence from a simple lattice analysis is shown. The failure mechanism as observed in the experiments with glued platens is simulated, thus no relative movements between the loading platens and the concrete disks are allowed. At 450 steps the main splitting crack is more or less complete, and gradually widens until the bottom of the post-peak valley is reached (point (c)). The splitting crack stops a short distance away from the loading platen. Next a widening of the fracture zone near the loading platen occurs, slightly resembling the wedges observed in the experiments with glued platen, but not good enough as was also observed in simulations of compressive fracture (Chapter 8). The failure criterion used is Equation (3.36), with $\alpha = 0.005$. Thus, the flexural component in the fracture law has been switched off almost completely. When another failure criterion is selected, for example, one based on the Mohr–Coulomb criterion, the wedge can develop, but as soon as it is fully detached from the remainder of the structure, the computation will stop for obvious reasons. A friction criterion in the crack is what seems to be lacking, and one of the main questions is whether a simple mechanism, as proposed here for tensile failure of the lattice beams (simple removal of a beam when a stress criterion is exceeded), can be found to model frictional sliding, or whether perhaps a completely different type of model would be needed.

It is interesting to note that the splitting mechanism evolves quite naturally from the lattice analysis, in particular when plywood inserts are used. This gives further confidence that, at least for tensile fracture, the lattice model is capable of capturing true failure mechanisms, and as such the model can be used to "predict" certain situations, such as those where no restraint applies.

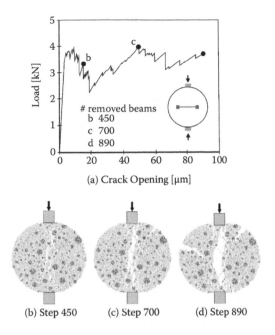

(a) Crack Opening [μm]

(b) Step 450   (c) Step 700   (d) Step 890

FIGURE 6.17

Failure process in a Brazilian splitting test from a lattice simulation: after the peak in the load-crack opening diagram (a) a vertical splitting crack develops (stage b, 450 elements removed). Subsequently a secondary mechanism develops comprising a number of radial splitting cracks (stage c, 700 elements removed and stage d, 890 elements removed). The crack patterns shown in Figures (b)–(d) correspond to these three stages. (From Lilliu. 2007. *3D Analysis of Fracture Processes in Concrete*. With permission.)

6.2.2 Bending

Flexural tests are also performed to avoid (presumed) difficulties in a uniaxial tensile test. Sometimes a flexural test is chosen because only a compressive machine is at hand, and an estimate for the tensile strength must be available. The splitting tensile test on relatively large specimens with narrow loading strips is then probably the best choice, but somehow people keep reverting to the so-called bending tensile strength of concrete. The bending strength is simply determined from the maximum bending moment divided by the section modulus. For a simple 3-point bending test this results in

$$\sigma_{fl} = \frac{6Pl^2}{4bh^2} \tag{6.2}$$

where P is the maximum load the specimen can sustain, l is the span between the supports, and $A = b \times h$ is the sectional area. It is quite obvious that this can only be an indication of the tensile strength. The same problems with boundaries occur as in the four-point shear beam discussed in detail in Section 7.1.2. If frictional restraint becomes too large, this may probably still be all right at peak load, but when it progresses one step further to determine the fracture energy following the fictitious crack model (Hillerborg 1985; RILEM TC 50-FMC 1985) the supports and load-point need to be free of restraint. The pendulum bar system will work (Figure 7.7), although this will require the construction of a special loading frame. Specially designed frictionless roller-systems might also be used. In a recommendation for the determination of the fracture energy from 3-point bend tests Planas et al. (2007) proposed the use of normal steel roller supports, but the slightest plastic deformation at the point contacts of such roller bearings may significantly affect the frictional restraint and with that the postpeak regime of the load-deformation diagram.

When the 3-point bending test is used for determination of the flexural tensile strength only, the support conditions are less critical. Because the test is only indicative (notably the stress and strain gradients during the entire loading history are not quite convenient), deviations on the order of 5–10% can be considered negligible. If the idea is to extract more advanced information, such as the fracture energy, the test is quite unreliable. The prisms used in standard 3-point bending tests are quite stubby: dimensions varying from $100 \times 100 \times 600$ mm^3 to $150 \times 150 \times 700$ mm^3 ($b \times h \times l$) are found in model codes. Obviously, such beams do not meet requirements from Bernoulli beam theory. In spite of all these drawbacks the test made it to become a standard, but it is clear that, as mentioned before, the results are indicative only and can only be used in comparison to results obtained with the same test method. The main reason for using a 3-point beam test is for sheer convenience; the test can be classified as quick and dirty, but as a comparative means can suit some goals.

It is interesting to use the same tools as before to examine in a bit more detail what would happen in a 3-point bending test if the aggregates were modeled realistically. Realistic aggregate shapes and aggregate size distributions from CT scans at University Hospital in Zurich were included in the lattice model (see Figure 4.7, where two different aggregate structures from CT scans are shown). For the exact procedure the reader is referred to Man (2009) and Man and Van Mier (2008b). Figure 6.18 shows the particle structure in a two-dimensional section of a scanned prism and the structure after lattice overlay, where a distinction is made between three phases: matrix, aggregate, and interfacial transition zone as explained in Section 4.5.

The lattice is completely three-dimensional; the lattice beam length is 0.25 mm. The resulting material structure is quite realistic, but obviously the price to pay is the extended computer time (see also Appendix 1). From a scanned prism, specimens of different size were cut and numerical simulations were carried out. In Figure 6.19 an example is shown of a simulation of a so-called C-type specimen (containing 1,155,549 elements; prism dimensions $9.3 \times 9.5 \times 24.88$ mm^3 and crushed aggregate size between 1.5 and 3

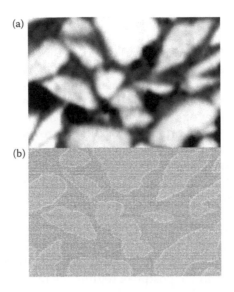

FIGURE 6.18
(a) 2D section from a three-dimensional CT scan of a concrete prism containing basalt aggregates (light gray) in a porous cement matrix (dark gray), and (b) the same structure after lattice overlay. (From Man and Van Mier. 2008b. *Int. J. Fract.*, 154(1–2): 61–72. With permission from Springer.)

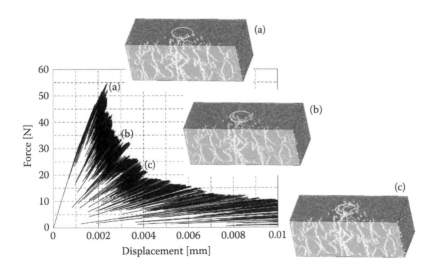

FIGURE 6.19
Example of fracture process in a numerical concrete prism containing crushed aggregates (1.5–3-mm aggregate size). The three fracture stages are shown at the indicated locations along the load-displacement curves. (From Man and Van Mier. 2008b. *Int. J. Fract.*, 154(1–2): 61–72. With permission from Springer.)

mm). The analysis shows that (micro-) cracking is not limited to the final fracture plane, but appears distributed along the lower side of the beam, in particular in the ITZ as this was defined as the weakest part of the concrete. At peak (stage (a) in Figure 6.19) a localized (macro-) crack starts to grow in the central part of the specimen, and thereafter gradually grows toward the central load point (stages (b) and (c)). In the circled area the main crack hits upon a larger aggregate, but because of the irregular shape (and thus the high stress concentrations) the crack grows straight through the aggregate particle, in spite of its much higher strength. When rounded or oval-shaped aggregates are modeled, aggregate fracture seldom occurs. In concrete containing crushed aggregates with quite irregular forms aggregate fracture is more common, also in real physical experiments. The main issue here is that the amount of cracks, that is, the total crack area, is much larger than the area of the main localized crack. Thus, if it is assumed that the fracture energy of the main crack can be calculated from the area under the load-displacement diagram (RILEM TC 50-FMC 1985) it is quite important to check whether in the postpeak regime the main (macro-) crack is the only propagating crack. Using a notch may have a positive effect, and this is actually proposed in many of the suggested standard tests for measuring the fracture energy of concrete. Yet, real structures have no notch, and will always show multiple cracking.

One final remark should be made at this point. In the scanning and lattice overlay technique sometimes it is found that aggregates are so close together that real separation is difficult to detect; compare, for instance, Figures 6.19a and 6.19b. Consequently some of the aggregates appear as rather irregular lumps which may then actually represent a cluster of two or more separate particles. If the resolution of the scanners improves it becomes easier to distinguish between separate aggregates.

7

Combined Tensile and Shear Fracture of Concrete

So far attention has been given to mode I, or tensile, fracture of concrete. This is undoubtedly the most important fracture mode, but shear can have influence as well. Consider, for example, the case of bending, where along the tensile part of the beam shear forces must be active too. For a material such as concrete, and many similar materials such as rock, one could easily argue that tensile fracture will always prevail. The tensile strength of normal strength concrete is no more than 10% of its compressive strength. Because shear can be interpreted as equal-biaxial tension-compression (i.e., $\sigma_1 = -\sigma_2$), the enormous imbalance between tensile and compressive fracture strength will inevitably lead to premature tensile fracture. Shear, or mode II or III fracture, would only be possible if through some means the imbalance between the tensile and compressive strength is restored, as argued in Van Mier (1997, 2004c). In this chapter we explore fracture of concrete subjected to combined tension and shear. We distinguish two shear modes (see Figure 2.3), namely mode II or in-plane shear and mode III or out-of-plane shear, therefore we discuss both modes. In both sections, Section 7.1 devoted to in-plane shear and tension, which, in fracture mechanics terminology would be referred to as mixed mode I and II fracture, and in Section 7.2 debating torsion, or mode III fracture, the loading combination refers to the externally applied stress. The pure fracture mechanics modes, as well as the mixed-modes refer to the crack-tip loading, which may differ substantially from the external loading on the specimen or structure. This normally may lead to confusion, so, once more: the loading cases mentioned always refer to the external loading on the considered structure.

7.1 Tension and In-Plane Shear

In this section we present three different experiments: one for measuring the mixed-mode I and II fracture properties of concrete, one for elucidating whether mode II fracture is possible, and a third more practical example where combined tension and shear fracture might prevail.

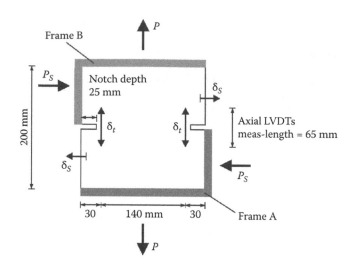

FIGURE 7.1

Double-edge notched plate (200 × 200 × 50 mm) subjected to axial tension (*P*) and lateral shear (*P_s*). The loading platens for applying axial tension can hardly rotate, which leads to the development of bending moments as soon as asymmetric crack propagation occurs, similar to the phenomenon discussed in Section 6.2.1. The shear displacement δ_s is the relative displacement between the two separate loading frames A and B. (After Van Mier and Nooru-Mohamed. 1989. In *Fracture Toughness and Fracture Energy: Test Methods for Concrete and Rock.*)

The first experiment concerns plates subjected to biaxial tension and shear. An interesting device to study fracture of concrete under these circumstances was developed by Reinhardt, Cornelissen, and Hordijk (1989) at Delft University of Technology. In the developed apparatus it was possible to subject platelike specimens to any possible combination of biaxial tension/shear. Using this device various experiments were carried out by my doctoral student Nooru-Mohamed (1992). It was decided to adopt a slightly modified specimen shape as was originally foreseen by Reinhardt and coworkers. The specimen was simplified to a double-edge notched plate as shown in Figure 7.1.

The second example is the well-known 4-point shear beam, which leads to more or less the same loading situation as shown in Figure 7.1. This example is of importance because several researchers have claimed that by using a double-edge notched version of the 4-point shear beam pure mode II fracture can be achieved. Others have argued against that, and we show here that it is easy to misinterpret the results. Basically the 4-point shear beam fails in tension, and pure mode II fracture can only be achieved under rather extreme circumstances that are normally not encountered in reinforced concrete structures.

The third example is the pull-out of a steel anchor from a concrete substrate. In this case one might suspect that a combination of tension and shear may lead to failure. Using both experiments and numerical simulations the case is explained. The pull-out problem formed the basis of an extensive round-robin analysis to show the validity of numerical simulation tools for concrete fracture, which makes this example of more general interest. So, let us first explore the structure of Figure 7.1 and see what the effect can be of changing the external loading on the plate.

7.1.1 Biaxial Tension Shear Experiments

Two examples from the various experimental series are included here: (1) shear after tensile loading up till a prescribed crack width, and (2) proportional tension/shear displacement path (viz. constant δ_t/δ_s). Figure 7.2 contains results from the first series that shows the development of secondary tensile cracks inclined to the shear direction (i.e., the horizontal direction in Figure 7.1) when the crack width after tensile load is small enough (viz. smaller than 225–250 μm). The results indicate that sliding occurs at an approximately constant shear-load of 2–2.5 kN when $\delta_t > 225$–250 μm, and

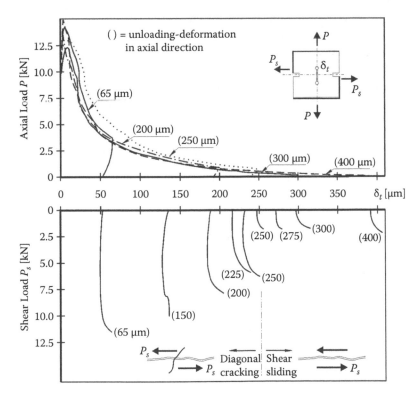

FIGURE 7.2
Shear resistance of precracked mortar (d_{max} = 2mm). The failure mode changes from secondary inclined cracking to shear sliding at a normal crack width of approximately 225–250 μm. (Result after Van Mier and Nooru-Mohamed. 1989. In *Fracture Toughness and Fracture Energy: Test Methods for Concrete and Rock.*)

that a higher fracture load $P_s > 5$ kN is required to create the diagonal cracks for $\delta_t < 225$–250 µm. The material tested here was a 2-mm cement mortar. It is very likely that for coarser grained materials the transition crack width and shear-load will have different values. The roughness of the crack will certainly affect the crack width at which substantial shear transfer is still possible. The second example concerns constant displacement-ratio paths δ_t/δ_s. In addition to the tests in Delft, similar experiments were performed by Hassanzadeh (1992) at Lund University in Sweden, employing a device with manual gears for applying shear and using circumferentially notched specimens of rather small size (Hassanzadeh used 70-mm cubes with a circumferential 15-mm deep notch resulting in a (too) small effective specimen cross-section of 40×40 mm; the effective cross-section in the Delft experiments was 50×200 mm as shown in Figure 7.1). Direct comparison with Hassanzadeh's results is not without problems because linear and parabolic displacement paths were investigated with slightly different displacement ratios from those tested by Nooru-Mohamed. Some results from displacement paths $\delta_t = \tan \alpha \cdot \delta_s$ with $\alpha = 45$ degrees are shown in Figure 7.3.

For this δ_t/δ_s-ratio the axial stress changes from tensile to compressive as can be clearly seen in Figure 7.3a. This means that initial cracking develops due to the relatively large tensile component, which has to change to compressive stress at relatively small crack openings ($\delta_t < 50$ µm) in order to keep crack propagation during the remainder of the experiment stable. Note that at higher ratios of δ_t/δ_s tensile failure seems to prevail and the effect of shear is minimal; see Nooru-Mohamed (1992). The behavioral trends of the tests by Hassanzadeh were the same as shown in Figure 7.3. It is interesting to look at the failure patterns. This was quite difficult in the tests from Lund because the circumferential notch did not allow for an unrestricted

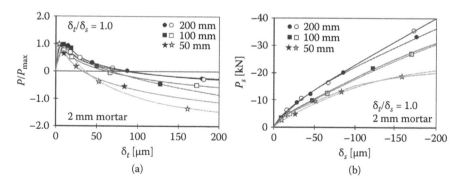

(a)   (b)

FIGURE 7.3

Behavior of concrete under biaxial tension-shear for specimens of three different sizes: constant displacement ratio tests ($\delta_t/\delta_s = 1.0$) on 2-mm mortar. (a) Axial load P/P_{max} versus axial crack opening δ_t; and (b) shear-load P_s versus crack shear displacement δ_s. (From Nooru-Mohamed, Schlangen, and Van Mier. 1993. *Adv. Cem. Based Mater.*, 1(1): 22–37. With permission from Elsevier.)

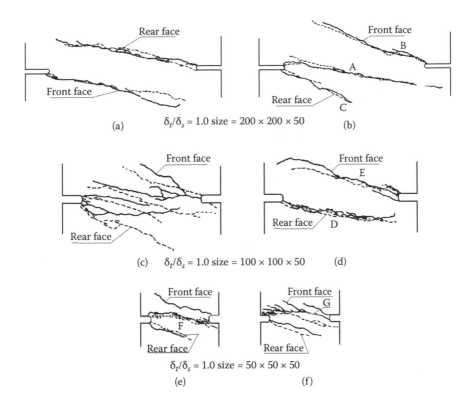

FIGURE 7.4
Crack patterns for the constant displacement-ratio tests ($\delta_t/\delta_s = 1.0$) of Figure 7.3. (From Nooru-Mohamed, Schlangen, and Van Mier. 1993. *Adv. Cem. Based Mater.*, 1(1): 22–37. With permission from Elsevier.)

view of the fracture zone. In this respect the geometry chosen in the Delft experiments appears to be more useful. Results of constant displacement-ratio paths of specimens of three different sizes are included in Figure 7.4. The smallest size of 50 × 50 mm is quite close to the specimen size used by Hassanzadeh; the two others (100 × 100 and 200 × 200 mm) are large in comparison. It is interesting to note that a transition from diffuse cracking to a few localized cracks occurs when the specimen size increases. The consequence of this transition in failure modes is that the size effect model developed by Bažant (1984; see Chapter 9) cannot be unrestrictedly applied under biaxial tension-shear loading. Namely, one of the basic assumptions in the size effect model is that the failure mode is the same for all considered sizes. Another interesting observation is that for one of the two largest specimens a diagonal splitting crack developed between the two inclined cracks that nucleated from the left and right notch. Such behavior is reminiscent of the fracture observed in 4-point shear beams when fixed boundary supports are used; see Section 7.1.2.

The biaxial tension-shear experiments can be simulated by means of a lattice and a simple tensile strength criterion. The role of friction in all the experiments appears to be rather minimal, and the main role of the (horizontal) shear load is to drive the cracks away from the plane between the notches; hence the inclinations of the cracks as shown, for example, in Figure 7.4. For the smallest specimen sizes shown in Figure 7.4 the role of the material structure appears to be more significant than in the larger sizes. Various results of simulations have been published in the past, for example, in Nooru-Mohamed, Schlangen, and Van Mier (1993).

In Figure 7.5 two different results are shown, namely the shape of the cracks that develop when the shear-load P_s = constant = –10 kN throughout the test while the axial deformation increases, and when $P_s = P_{s,max}$ = constant (i.e., the maximum shear-load the specimen could sustain). Preliminary experiments showed that a small crack would sometimes develop close to the vertical loading platens that was interpreted as "glue-failure," as can be seen in Figure 7.5b (top-right corner). This crack would disappear when at locations A and B additional steel platens of sufficient thickness were attached

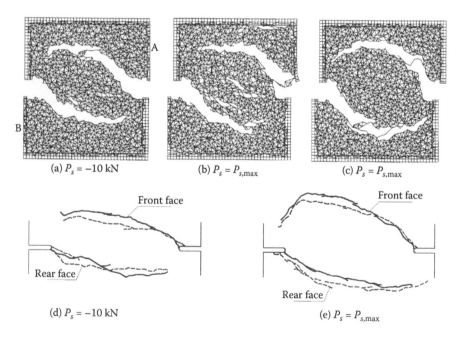

$\qquad$ (a) P_s = –10 kN $\qquad\qquad$ (b) $P_s = P_{s,max}$ $\qquad\qquad$ (c) $P_s = P_{s,max}$

Front face

Rear face

(d) P_s = –10 kN

Front face

Rear face

(e) $P_s = P_{s,max}$

FIGURE 7.5
Simulated crack patterns in biaxial tension-shear with P_s = constant = –10 kN (a,d) and P_s = $P_{s,max}$ = constant (b,c,e). The axial deformation is increased up till complete failure after the shear load has reached the level of –10 kN (a,d) or its maximum value (b,c,e). (a,b) A single row of steel elements was modeled at locations A and B as indicated; (c) a double row of steel elements was modeled; (d,e) the crack pattern observed in experiments is shown and compares well to the simulated patterns. (From Nooru-Mohamed, Schlangen, and Van Mier. 1993. *Adv. Cem. Based Mater.*, 1(1): 22–37. With permission from Elsevier.)

to the specimens' sides (compare the top-right corner of the simulations in Figures 7.5b and 7.5c, which are both at maximum shear-load. The difference is in the additional row of steel elements attached at locations A and B in Figure 7.5c). Two large curved cracks develop in the specimens, one from each notch, which is similar to the experimentally observed crack patterns. The curvatures become larger when the shear-load increases from −10 kN to its maximum value (Figures 7.5d and 7.5e). The simulations also show an increase of curvature with increasing shear; compare Figures 7.5a and 7.5c. An important point is actually the curvature of the two main cracks, which may help to identify values of some of the parameters in the lattice simulation. As can be seen, a random lattice was used in the simulations, which is quite essential to avoid alignment of the cracks along the mesh lines. This would occur when a lattice with regular geometry (triangular or square, for instance) was used; see Schlangen and Garboczi (1996). As an alternative a regular lattice with material structure overlay could be used. However, in that case the problem of mesh alignment would be overcome only partially.

The simulations of Figure 7.5 are carried out with a simple tensile strength criterion, and the commonly adopted complicated mixed-mode fracture laws are quite unnecessary. The same is true for the 4-point-shear beam test discussed in the next section. That loading situation was claimed to generate pure mode II cracks by Bažant and Pfeiffer (1986), but as shown, reality is somewhat different.

7.1.2 4-Point-Shear Beam Test

The loading situation in a 4-point-shear beam resembles the DEN biaxial tension/shear plates discussed in the previous section. Figure 7.6a shows the loading scheme in the biaxial tension/shear tests, and in Figure 7.6b the loading situation in a double-edge notched 4-point-shear beam is indicated. The biaxial loading rig used by Nooru-Mohamed was quite stiff in order to suppress possible rotations of the loading platens. Although infinite rotation stiffness is hard to achieve, the tests showed quite satisfactory stiffness (see Nooru-Mohamed 1992). As soon as asymmetric crack growth occurred (which is the rule rather than exception in coarse-grained heterogeneous materials like concrete) a bending moment developed as sketched in Figure 7.6a. This bending moment restrains the crack from growing; as we have seen in uniaxial tension this leads to the aforementioned bump in the softening diagram (see Figure 6.8, Case B). The moment distribution and shear-load distribution in the 4-point-shear beam of Figure 7.6b lead to similar loading of the midsection of the beam. Note that the situation is rotated 90 degrees in comparison to Figure 7.6a.

Bažant and Pfeiffer (1986) claimed that between the two notches of the 4-point shear beam a vertical "shear" crack develops, which is built up from small inclined tensile cracks. Although their test setup was quite questionable, and no real proof was provided that stable crack propagation was

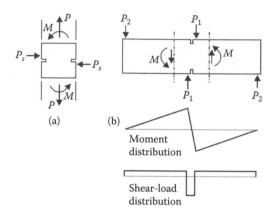

FIGURE 7.6
Similarity in loading between (a) biaxial tension/shear tests and (b) 4-point-shear beams. (After Van Mier. 1997. *Fracture Processes of Concrete: Assessment of Material Parameters for Fracture Models.*)

assured, they strongly defended their conclusion. In fracture experiments it is quite essential that stability of crack propagation is assured; if crack propagation is not stable dynamic effects may become important and disturb the crack path that would otherwise be found under quasi-static conditions. Thus, quite inevitably Bažant and Pfeiffer's claim was questioned by many. The first indication that there might be something wrong came from numerical simulations carried out by Ingraffea and Panthaki (1986). They showed that after the formation of two short curved cracks from the two notches, a diagonal splitting crack would develop in between these two curved cracks. Therefore, the conclusion should be that the behavior observed by Bažant and Pfeiffer had nothing to do with shear, but again, was merely governed by local tensile crack growth. Because the outcome of simulations can never be conclusive, new experiments were carried out by Schlangen and Van Mier (1992b, 1995). The identical specimen geometry was used as proposed by Bažant and Pfeiffer.

One improvement was that the test was now controlled over the average value of crack mouth opening displacement (CMOD) and the crack mouth sliding displacement (CMSD). The CMOD and CMSD were measured at the tip of the lower and top notch, both at the front and back sides of the specimen. The average value of the CMOD and CMSD resulted in a continuously increasing displacement throughout an experiment, and thus stability was guaranteed when the test was done in closed-loop displacement control. A second improvement was that in the test setup used by Van Mier, Schlangen, and Nooru-Mohamed (1992) the supports could either rotate freely, or could be fixed. A sketch of the test setup is shown in Figure 7.7. Through the addition of diagonal bars (shown as dashed lines in Figure 7.7a) the free rotation of the pendulum bars is prevented and the supports act as fixed supports,

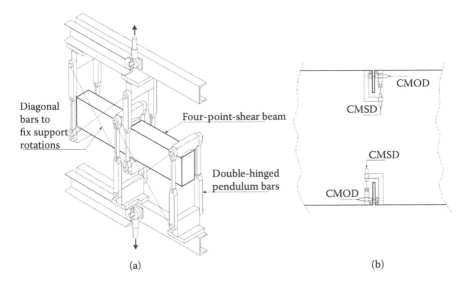

(a)   (b)

FIGURE 7.7
(a) Loading frame for 4-point shear tests on SEN of DEN beams. The beams are loaded verti-
cally, in tension. Depending on the specimen dimensions a certain load distribution develops.
For the DEN beam geometry used by Bažant and Pfeiffer (1986) the load at the outermost supports is $P/15$ when a load P is applied at the supports that are close to the notch. In the test
set-up of (a) the pendulum bars that allow for rotation of the supports can be fixed by adding
the diagonal bars; (b) measurement of CMOD and CMSD. (From Van Mier et al. 1992.)

which resembles the experimental boundary condition used by Bažant and
Pfeiffer in their tests. The difference between loading a DEN beam between
freely rotating supports and between fixed supports can be seen in Figure 7.8.
Two sizes of beams were tested; the larger ones had twice the dimensions of
the smaller ones, except for the thickness t. For the large beams the difference in behavior is not very large when either of the support conditions is
used. The small beams show a longer plastic plateau for the fixed support
condition, whereas the peak load is not affected. Also included in Figure 7.8
are results from 4-point shear tests on Felser sandstone (Schlangen and Van
Mier 1995), but now only for $d = 150$ mm. The initial stiffness of the sandstone is a factor $1/3$ smaller than the concrete, as can be clearly seen from
the load-deformation diagrams. The sandstone diagrams follow the same
trend as observed in the concrete tests: approximately the same peak-load is
measured, but postpeak ductility increases when rotations at the supports
are fixed. Note that the diagrams show a continuously increasing CMOD +
CMSD, which is a prerequisite for stable crack growth; that is, there is absolutely no doubt about the stability of these experiments.

The crack patterns initially show the development of two curved cracks, as
predicted by Ingraffea and Panthaki (1986). When fixed supports are used an
inclined splitting crack develops between the two curved cracks in the case
of 8-mm concrete; when freely rotating supports are used one of the curved

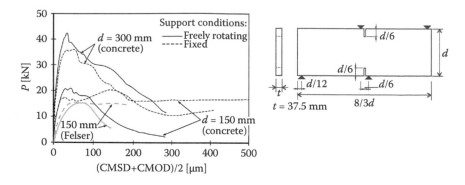

FIGURE 7.8

Effect of boundary conditions on the load-deformation behavior of DEN concrete beams loaded in 4-point shear: results are included for 150-mm and 300-mm concrete beams and 150-mm beams of Felser sandstone. (Adapted from Schlangen and Van Mier. 1995. *Rock Mech. Rock Engng.*, 28(2): 93–110.)

cracks keeps growing until it reaches the opposite side of the beam (see Figure 7.9a). The large beams did not show the growth of the inclined splitting crack as clearly as the 150-mm beams. Also the sandstone behaved a bit differently: the inclined splitting crack did not develop when fixed supports were used. Instead the sandstone seemed to fail under the support platens, which may be related to the lower compressive strength of the sandstone in comparison to the concrete (33.4 MPa and 46.6 MPa for Felser sandstone and 8-mm concrete, respectively). The lattice model is capable of simulating the two crack patterns using a simple tensile fracture criterion, Equation (3.36). In Figure 7.10 the computed failure modes are shown, which compare very favorably to the experimental results (at least for concrete). It should be mentioned that the load-deformation response calculated using the lattice model is too brittle, as observed before in the tensile simulations (compare to

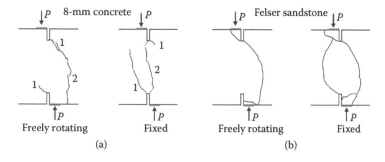

FIGURE 7.9

Crack patterns observed in DEN 4-point beam tests between freely rotating and fixed support conditions for (a) 8-mm concrete and (b) Felser sandstone. The numbers along the cracks indicate crack nucleation (#1) and crack leading to complete rupture (#2). (From Schlangen and Van Mier. 1995. *Rock Mech. Rock Engng.*, 28(2): 93–110. With permission from Springer.)

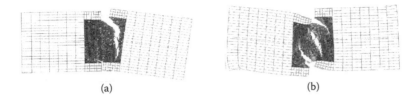

(a)   (b)

FIGURE 7.10
Crack patterns from lattice model simulations: (a) freely rotating supports, and (b) fixed supports. (After Schlangen. 1993. *Experimental and Numerical Analysis of Fracture Processes in Concrete*. With permission of Dr. Erik Schlangen.)

Figure 6.4). The general impression is that this is caused by a lack of detail in the simulations; clearly more research is needed at this point.

One final remark should be made. Swartz and Taha (1990) also performed quite a number of 4-point-shear tests. In some experiments axial compression was applied in the direction of the beam axis, which led to a closer spacing between the two curved cracks. This observation is in agreement with the biaxial tension/shear experiments mentioned before (see Section 7.1.1), where in some cases during a constant displacement-ratio path compressive constraint would also develop perpendicular to the crack plane; see Nooru-Mohamed (1992).

From the results shown to this point, shear failure seems to be nothing more than lower-scale tensile failure, which does not make the situation easier. Scale dependency may be quite important as, for example, is visible in the failure modes observed in the constant displacement-ratio experiments of Figure 7.4. Shear failure may develop under confined compression (shear bands; see Van Mier (1984), Van Geel (1998), and others), in fiber-reinforced concrete (Van Mier 2004c), and perhaps under dynamic loading (high strain-rate regime). The uncontrolled 4-point-shear beam tests of Bažant and Pfeiffer (1986) might actually be seen as an indication that under dynamic loading shear failure may occur. Contrary to the claim of the authors they are certainly not proof for the existence of mode-II failure under quasi-static conditions.

7.1.3 Anchor Pull-Out

The pull-out of a steel anchor from a concrete substrate is a rather practice-oriented example. One would suspect that shear plays an important role under these conditions. To demonstrate the effectiveness of the Fictitious Crack Model by Hillerborg and co-workers in 1976 it was decided to organize a round-robin analysis of a two-dimensional pullout problem. A 2D problem was chosen because in those days not everyone had a fully operative 3-dimensional nonlinear finite element code available; a 2D analysis would likely attract more contenders. We speak of the years around 1990. For those willing to perform a 3D analysis an axial-symmetric version of the same pull-out problem was suggested as well.

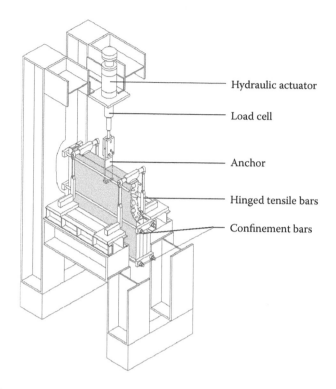

FIGURE 7.11
Experimental setup for the two-dimensional anchor pull-out problem. A T-shaped steel anchor is pulled-out from a concrete slab with the same thickness (100 mm). The support-loads are carried via pendulum bars that can either rotate or be fixed against rotations. In addition horizontal confinement can be applied via three horizontal confinement bars attached to steel beams that are glued to the sides of the concrete plate as sketched. (After Vervuurt et al. 1993a. In *Fracture and Damage of Concrete and Rock (FDCR-2)*.)

About 20 groups submitted a solution to the 2D pull-out problem; the variation of predicted failure loads varied by a factor of 200. No experiments were available at the time of most analyses, but were later conducted at ETH Zurich (Helbing, Alvaredo, and Wittmann 1991) and at TU Delft (Vervuurt, Schlangen, and Van Mier 1993a; Vervuurt, Van Mier, and Schlangen 1994). The test setup used in Delft is shown in Figure 7.11. The loading device is built in the same frame where the DEN 4-point-shear beams were tested. The support conditions in the anchor pull-out tests could also be varied by adjusting diagonal bars between the pendulum bar supports. In Figure 7.12 the result from a number of lattice analyses is compared to experimental results. In the experiments the stiffness of the horizontal confinement was either $k = 0$ or $k = 500$ MPa. The support span was $a = 2d = 300$ mm, where $d = 150$ mm is the embedded depth of the anchor. When confinement is applied the softening part of the load-displacement curves rises in comparison to the unconfined situation. Moreover, the crack pattern changes: when

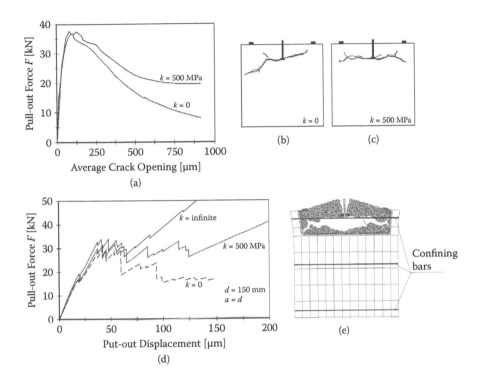

FIGURE 7.12
Experimental results from pull-out experiments of a T-shaped steel anchor from a concrete slab. (a) Load-crack opening diagrams for $a = 2d$ with and without horizontal confinement. (b,c) Two experimental crack patterns ; (d) results from numerical lattice simulations with a random lattice for support span $a = 2d$; (e) simulated fracture behavior of confined pull-out ($k =$ 500 MPa) for the case $a = 2d$. (Results adapted from Vervuurt et al. 1993a. In *Fracture and Damage of Concrete and Rock (FDCR-2)*, and Vervuurt et al. 1994. *Mater. Struct. (RILEM)*, 27(169): 251–259.)

confinement is applied the crack is forced to grow in a horizontal plane, whereas without confinement the crack can "escape" in different directions (see Figure 7.12b,c). The influence of horizontal confinement is reflected in the same way in the outcome of simulations, as can be seen in Figures 7.12d. In this figure the load crack-opening diagram is plotted for support span $a = 2d$. Although the confinement effect is the same in the simulations, the comparison with the experimental curves (Figure 7.12a) shows that the lattice analyses are too brittle. Some reasons for the brittleness of the lattice analyses were mentioned before in Sections 6.1.2 and 6.1.3, and relate to too-limited detail in the material structure and the 2D representation of (basically) a 3D fracture problem (among others, in 2D aggregates are simulated as cylinders instead of spheres in full 3D). Note that the analyses were done using a random lattice, but here the same objection of a 2D analysis of a 3D situation can be made because under the removal of a lattice beam the structure actually fractures over the entire concrete slab thickness, which is not realistic. A simulated crack patterns for $a = 2d$ is finally shown in Figure 7.12e. The comparison with the experimental pattern in Figure 7.12c is quite favorable.

The round-robin analysis (Elfgren 1992) showed that the application of advanced fracture models in combination with any type of numerical program is not quite straightforward. A thorough understanding of fracture processes in concrete is essential, and estimating the various parameters in the models in combination with good engineering judgment is very important in order to make useful predictions. Often the best check is still in performing the physical experiment, even for the seemingly simple example of the two-dimensional anchor pull-out problem.

7.2 Torsion (Mode III Fracture)

Out-of-plane shear can best be studied in torsion experiments. The third fracture mode can also be important for concrete, especially when a premature tensile fracture is suppressed through the application of confinement. This was mentioned in the introduction to this chapter, and the matter is further debated in Chapter 8 (compressive fracture). In the case of torsion, confinement can easily be applied by simply preventing elongation along the longitudinal axis of the specimen subjected to torque. The axial constraint can be (a) free axial deformations, (b) active loading in the axial direction, or (c) suppressed axial deformation of the cylinder. The failure mode changes from a clearly mode-I-controlled process (tensile fracture) for free axial deformations (Case (a)) to a confined shear mode (Cases (b) and (c)). The latter may be helped by machining a circumferential notch in the test cylinder at the place where the shear fracture should develop (see Figure 7.13b). Mostly, however, the failure mode takes all kinds of twists and tilts, which actually refer to mixed-mode situations (Hull 1993). Torsion tests on plain concrete were carried out, among others, by Xu and Reinhardt (1989), Yacoub-Tokatly, Barr, and Norris (1989), Bažant, Prat, and Tabbara (1990), and Van Mier and Lilliu (2002).

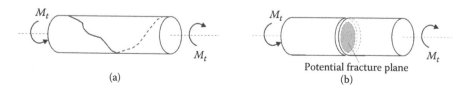

M_t M_t M_t M_t

Potential fracture plane

(a)   (b)

FIGURE 7.13

Fracture in concrete cylinders subjected to torsion: (a) spiral crack in a solid cylinder and (b) predefined crack plane in a notched cylinder. (After Van Mier. 1997. *Fracture Processes of Concrete: Assessment of Material Parameters for Fracture Models.*)

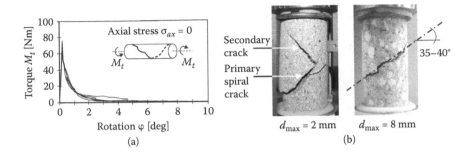

(a)   (b)

FIGURE 7.14
(a) Torque-rotation diagrams for mortar cylinders (34-mm diameter, 68-mm length, d_{max} = 2 mm) without axial confinement (σ_{ax} = 0), and (b) formation of spiral cracks in a 2-mm mortar specimen and an 8-mm concrete cylinder. Note that in the mortar specimen a secondary crack developed perpendicular to the spiral. The spiral crack growth is completely around the cylinders; here only a front view is provided. (After Van Mier and Lilliu. 2002. In *Structural Integrity and Fracture*.)

Various examples of the three mentioned loading cases (a) through (c) are shown in Figures 7.14 and 7.15 for 2-mm cement mortar (results are almost identical for 8-mm concrete). When axial deformations are free (note: the end surfaces of the cylinder are always glued to a stiff steel platen and are forced to remain planar) and the axial force is constant zero, a spiral crack develops as shown in Figure 7.13a. The torque rotation diagram shows a smooth softening branch when this crack develops (Figure 7.14a). Eventually a secondary crack may grow, perpendicular to the main spiral as shown in Figure 7.14b for the mortar specimen (d_{max} = 2 mm). The secondary crack is caused by bending when the two parts of the specimen separated through the spiral crack touch again in an advanced stage of the fracture process; see Van Mier and Lilliu (2002).

When a small axial load is applied basically a similar failure mode is observed (Figure 7.15). The difference with zero axial loading is mainly in the tail of the torque-rotation diagram: the load remains constant, yet small, and does not reduce to zero (compare Figures 7.14a and 7.15a). The reason is that the axial load causes frictional restraint in the spiral crack and thus leads to a higher residual stress in the softening branch. In the third case, when the axial deformation is kept constant and equal to zero, an extended hardening behavior is observed (Figure 7.15b). The axial load (upper diagram of Figure 7.15b) starts from zero at the beginning of the experiment but as restraint builds up a maximum axial stress of approximately –4 MPa is reached. This is almost the same as in the second loading case, although here because of the different boundary condition it is not constant. In the last case, zero axial displacement (Figure 7.15b), the failure mode is identical to failure of a specimen loaded under constant axial load (Figure 7.15a). There are two important differences between unconfined and confined tests, namely, first, the spiral crack is fully completed and makes a complete path around the

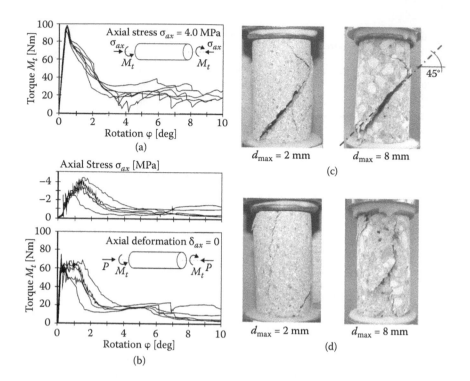

FIGURE 7.15

Torque-rotation diagrams for confined torsion tests: (a) with constant axial stress $\sigma_{ax} \approx 4.0$ MPa, and (b) with zero axial deformation ($\delta_{ax} = 0$); (c) and (d) the front and back sides of failed mortar and concrete cylinders for the tests of (a) with constant axial stress. Note that under confined torque the spiral does not loop around the complete perimeter; final failure is now through the vertical splitting crack at the back side of the cylinder, see (d). (After Van Mier and Lilliu. 2002. In *Structural Integrity and Fracture*.)

specimens' perimeter when no confinement is applied. Second, the inclination of the spiral crack changes from 35–40 degrees for no confinement to a steeper >45 degrees for confined torsion. Under axial confinement, the spiral crack is not complete and a vertical splitting crack appears where the spiral is not complete. Basically this would hint at compressive failure of the last intact part of the cylinder after partial development of a spiral crack. Perhaps this should be expected to occur: after the cross-sectional area of the specimen has been reduced by the developing spiral crack, the last remaining intact part may turn out to be too small to carry the confining axial stress. In Figure 7.15c,d the various images for confined tests are gathered: both the front and back sides of 2-mm mortar and 8-mm concrete specimens are shown. They clearly show the above-mentioned mechanisms.

All results seem to indicate that local mode I fracture is mostly responsible for the observed failure modes. When the axial confinement increases the residual stress level in the softening regime increases, but in order to

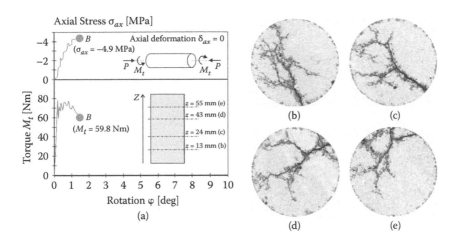

FIGURE 7.16
Crack pattern in a mortar cylinder (d_{max} = 2 mm) under zero axial displacement (δ_{ax} = 0). (a) axial load variation while applying torque to the specimen following the shown torque-rotation curve. (b)–(e) sections of the cracked cylinder at z = 13, 24, 43, and 55 mm, respectively, where z is measured from the bottom of the cylinder as indicated in the inset of (a). (After Van Mier and Lilliu. 2002. In *Structural Integrity and Fracture*.)

avoid softening altogether and have a fully plastic response the confinement would need to increase by a factor of 6–7 (and likely even more). According to the criterion mentioned in the first paragraph of this chapter, the expected behavior for antiplane (mode III) fracture would basically be the same as the conditions for pure mode II fracture to occur.

In Figure 7.16 the results are shown in a slightly different way. Now the specimen has been impregnated after loading, and carefully sawn in thin slices perpendicular to the cylinder axis to reveal the interior cracks. For a confined test on 2-mm mortar cracking is visualized just after the maximum stress level in the first part of the softening curve. The fracture pattern is not yet fully developed, but clearly shows how the main crack(s) take a different orientation in the various sections, and thus how the spiral crack (visible along the specimens' perimeter from the outside) is built up. When images of all sections are placed sequentially in a simple movie the spiral shape is clearly visible.

As mentioned, without axial constraint (σ_{ax} = 0), or with just minor axial stress, the fracture process seems to be dominated by local mode I cracking. This is confirmed in analyses by Lilliu (2007) who performed a lattice analysis of the unconfined situation, which actually is a nice example of an analysis with the three-dimensional version of the lattice model. In Figure 7.17 some results are shown. The parameters in the 3D random lattice were: l_{avg} = 1.24 mm, randomness A/s = 0.001, cell size s = 1 mm (see Figure 4.11b), $E_A/E_M/E_{ITZ}$ = 70/25/25 and $f_{t,A}/f_{t,M}$ = 2 and $f_{t,ITZ}/f_{t,M}$ = 0.25 (i.e., the suggested parameters from Table 4.1). The 3D lattice contained a total of 449,179 elements. A particle distribution following Equation (4.4) was used

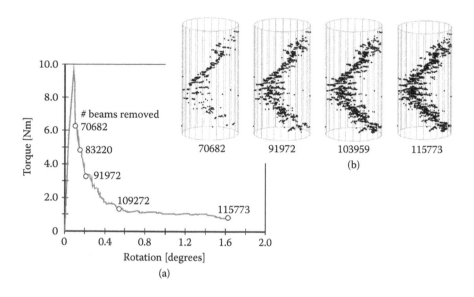

FIGURE 7.17
Lattice analysis of the unconfined torsion experiments of Figure 7.14; (a) torque-rotation diagram from the analysis; (b) fracture propagation in four postpeak stages. (From Lilliu. 2007. *3D Analysis of Fracture Processes in Concrete.* With kind permission from Dr. Giovanna Lilliu.)

with particle diameters between 2 and 8 mm. The effective particle density, after lattice overlay, was equal to $P_{k,latt} = 0.22$ (thus, a very sparse particle content was used). In Figure 7.17a the torque-rotation diagram is depicted; in Figure 7.17b are four stages of postpeak cracking. These stages are identified in the torque-rotation diagram. The formation of the spiral crack is very clear from this analysis. The crack starts from the perimeter of the specimen, where stresses are highest, and gradually grows toward the center of the cylinder. Lilliu (2007) also presented the results as planar cuts, like those shown from the experiment in Figure 7.16. The (qualitative) resemblance is extremely good. The shape of the torque-rotation diagram in the simulation resembles those from experiments with unconfined torque (Figure 7.14a), but the values of the torsion moment and the rotation are much smaller. The setting of the parameters in the 3D lattice, and likely also the chosen fracture law are the cause for these deviations. Nevertheless, the lattice again is capable of capturing qualitatively the correct failure mode, which is actually by far the most important aspect here, namely understanding fracture under a variety of loading conditions.

All examples of lattice analyses shown to this point are qualitatively correct (tension, bending, splitting, biaxial tension/shear, anchor pullout, and torsion) and in good agreement with experimentally observed failure modes. Moreover there is a size effect, and there appears no need for a separate size-effect law. The main difficulty now arises, namely modeling of fracture involving frictional restraint in cracks. For that purpose, before turning to fracture in compression, a last, but quite interesting result from the torsion experiments is discussed. Failure of a notched specimen subjected to confined torsion showed differences compared to the unnotched specimens. This case was explored earlier by Bažant et al. (1990). In the present experiments the focus was on understanding the failure mode, and not trying to "prove" that mode III fracture of concrete exists (which seems a waste of time anyway in view of the simple theoretical consideration presented in the introduction to this chapter).

In Figure 7.18a the torque-rotation diagrams are shown for confined torsion ($\delta_{ax} = 0$ = constant) on notched specimens (L = 68 mm, D = 34 mm, 8-mm deep notch at half height) made of 2-mm mortar. The global shapes of the diagrams are identical to those shown in Figure 7.15b, except that maximum torque is smaller, caused by the reduced cross-section at the notch. The torque-rotation diagram shows hardening behavior and the confining axial stress increases. All specimens failed in an asymmetric mode; that is, mostly only one half of the cylinder, for example, above the notch, showed clear signs of cracking, whereas the bottom part remained seemingly intact. The asymmetric failure mode must be related to the heterogeneity of the specimen, and perhaps to some asymmetry in the loading device, for example, caused by stiffness distribution in the loading frame. The part of the cylinder that failed gave the impression that compressive failure occurred. Two conical parts, which remained more or less intact, were found in the cylinder as

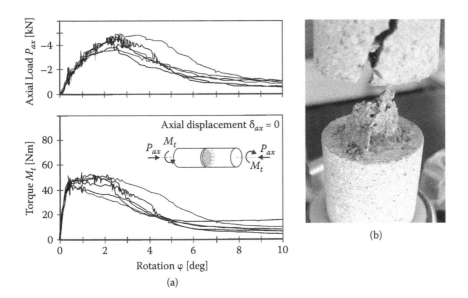

(a)

(b)

FIGURE 7.18
(a) Torque-rotation diagrams for 2-mm mortar notched cylinders loaded under constant zero axial displacement (δ_{ax} = 0); (b) example of a failed cylinder. (After Van Mier and Lilliu. 2002. In *Structural Integrity and Fracture*.)

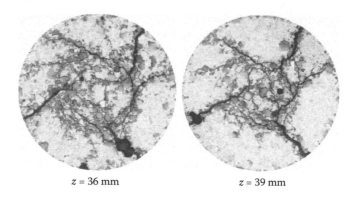

$z = 36$ mm　　　　　　　　　　$z = 39$ mm

FIGURE 7.19

Internal cracking in two slices from an impregnated notched cylinder after loading into the softening regime. The slices are close to the plane of the notch, which is at $z = 34$ mm. The coordinate z is measured from the bottom of the cylinder as indicated in Figure 7.16a. (After Van Mier and Lilliu. 2002. In *Structural Integrity and Fracture*.)

can be seen from Figure 7.18b. There were no differences when the maximum grain in the mixture was increased to 8-mm; see Van Mier and Lilliu (2002).

Using again vacuum impregnation of a fluorescent epoxy dye after the specimen was cracked, more detail about the failure mode was found. Figure 7.19 shows two sections from a cracked specimen, that is, loaded into the softening regime. The slices are located close to the notch (the notch appears at half height, viz. at $z = 34$ mm). The conical central part can clearly be identified; in addition a number of radial cracks can be recognized. These radial cracks are arranged in a spiral pattern. It seems that after the weakening of one part of the specimen conditions are favorable for compressive fracture to occur. The compressive fracture mode resembles that when high-friction loading platens are used, such as dry steel platens (see Figures 8.5a and 8.8 in Chapter 8). The observed "conical" failure mode in these notched specimens clearly involved quite some mode I cracking, but friction very likely plays an important role as well. In compressive fracture the role of friction can be systematically studied by simply changing the loading conditions of the specimen, that is, by changing the frictional restraint caused by the loading platens. We turn our attention now to compressive fracture.

8

Compressive Fracture

In comparison to tensile failure, compressive fracture is at least one step up in the degree of complexity. Shear, and probably also buckling instabilities close to the surface of specimens/structures subjected to compression are two additional mechanisms that cannot be neglected. The lattice model used thus far has shown excellent capability of simulating the fracture process in concrete subjected to external tension or combined tension/shear, as shown in Chapters 6 and 7. Always, it seems the cracks tend to follow the direction of the major principal stresses. When cracks are confined to grow in a certain direction they will escape in another direction where the specimen/ structure allows for that. Shear fracture (pure mode II or mode II according to the classical fracture mechanics definition) may occur under some peculiar conditions that are not regularly met in ordinary (reinforced) concrete structures. In Section 7.1.2 we alluded to possible circumstances under which shear failure might occur.

When shear fracture occurs (modes II and III) and confinement is large enough to prevent local tensile fracture, frictional restraint in cracks starts to become more important and may eventually have an effect on the fracture process. One might hope that by modeling the mechanical behavior at the meso-level the local mechanism can again be reduced to mode I fracture, and friction can be neglected. Unfortunately this appears not to be the case. In lattice type models it is difficult—if not impossible—to incorporate friction; particle models such as those developed by Cundall and Strack (1979), Iwashita and Oda (2000), Thornton and Antony (2000), and Luding (2004) might be a more useful alternative concept to include friction. This chapter has been subdivided into five parts. First a variety of possible mesomechanisms underlying fracture in compression are discussed. These mechanisms have been proposed mainly on the basis of experimental observations, and to some extent from theoretical considerations. Next, the focus is on macroscopic fracture, in particular the observed influences from boundary conditions (frictional restraint) and specimen geometry (slenderness, size), with a limitation on uniaxial compression. In the third section attention turns to softening as a crack propagation phenomenon and the effect of (external or active) confinement. Next, it is shown what can be achieved with a simple lattice approach. Some limitations have been observed, and extending the model by including some additional local failure mechanisms appears to be unavoidable. Inasmuch as for engineering applications these meso-level models are hardly applicable, there is a need for macroscopic approaches for dealing with compressive fracture. In Section 8.5 we review some of the proposed models. Unfortunately there remain some constraints in these macromodels that might be resolved by returning to classical fracture mechanics, which are, however, delayed till Chapter 10. It will be important to consider a compression test as a small-scale structural test, which, as a matter of fact indicates that for fracture we cannot design an experiment free of boundary effects and geometry (size and shape) influences. So, let us now first consider fracture of concrete at the scale of the aggregate particles, the mesoscale.

8.1 Mesomechanisms in Compressive Fracture

In Figure 8.1 various mesomechanisms of fracture in concrete under compression are summarized. Most simple is perhaps to represent concrete as a stack of equal-sized balls that are in contact, as shown in Figure 8.1a. Under compression the force lines are not vertical any more but take a certain inclination following the contact points between the balls. As a consequence horizontal splitting forces must develop to ensure equilibrium. When external confinement is applied (biaxial and triaxial compression), the internal splitting forces may be partly counteracted and microcracking along the weak ITZ may be delayed as depicted in Figure 8.1b. As a result the material can undergo larger deformations and sustain higher stresses (as shown in Section 8.3). The particles in real concrete do not have equal size, and although the mechanism of Figure 8.1a also will develop in a stack of particles of different size, alternatively concrete can also be envisioned as being built up from large aggregate particles embedded in a more or less homogeneous cement matrix, as discussed in Section 4.1, more specifically Figure 4.1b (meso-level).

The cement matrix, which is assumed to contain small sand particles, has a lower stiffness than the large solid aggregates ($E_M/E_A \approx 20/75$ GPa). Because the lateral expansion of the cement matrix will eventually exceed that of the aggregate particle, restraint develops at the top and bottom of the particle. As a consequence two triaxially compressed matrix cones develop above and below the aggregate particle. After the formation of tensile interface cracks along the side of the particle, shearlike cracks, in the form of "en-echelon" tensile cracks, will develop along the uncracked matrix cones as indicated in Figure 8.1c. This mechanism was first put forward by Vile (1968), and later supported through observations by Stroeven (1973). The typical mortar cones were actually found among the rubble left behind after a compression test on a concrete specimen was carried out. Once the idea has taken root, one becomes suddenly aware of the shape of the rest of the pieces. The mechanism proposed by Vile can only develop when the cones have a chance to develop, which implies that the stiffness of the grains should be larger than that of the surrounding cement matrix. If the aggregate stiffness

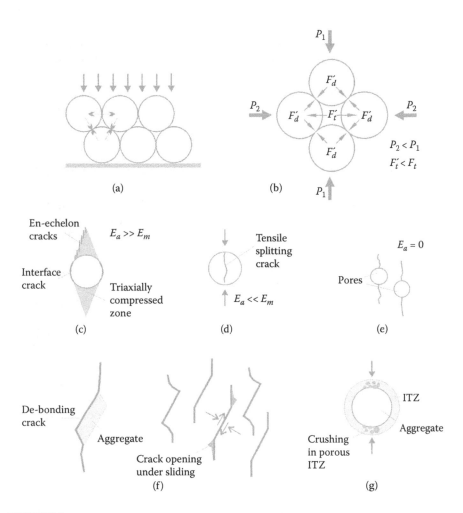

FIGURE 8.1

Mesomechanisms of fracture in concrete: (a) concrete represented as a stack of equal sized balls; horizontal tensile forces develop under external compression; (b) the internal splitting forces may be balanced by external confinement to delay crack growth; (c) triaxially confined areas develop in the "soft" cement matrix when the concrete contains stiff aggregate particles; (d) in "soft" aggregates splitting cracks may be enforced by the surrounding matrix; (e) large pores and air entrapments may attract splitting cracks; (f) wing cracks may develop when the material in which the cracks grow is more or less homogeneous; and (g) crushing of highly porous interfacial transition zone may occur under compression. (After Van Mier. 1998. *Fracture Mechanics of Concrete Structures: Proceedings FraMCoS-3.*)

is substantially smaller than the cement matrix (Figure 8.1d), or if the aggregate is replaced by a large void (Figure 8.1e), splitting cracks will develop. The shear mechanism (i.e., similar to Figure 8.1c), at the level of the aggregate particles, will not occur.

The first case (Figure 8.1d) leads to intraparticle fracture as, for example, observed in lightweight concrete (see Figure A4.2b in Appendix 4 where an example is shown of particle fracture in lytag concrete; although this example relates to tensile fracture, particle fracture will also occur in compression). The mechanism is quite similar to the Brazilian splitting tests discussed in Section 6.2.1. The second case (Figure 8.1e) leads to splitting cracks in the matrix above and below the void. The latter can easily be seen from a simple elastic analysis of the stresses around a circular pore in a plate of unit thickness, as can be found in Timoshenko and Goodier (1970); see also Van Mier (1997, pp. 272–274).

In a compressive stress field, a crack oriented parallel to the compressive loading direction will only grow when the external load is increased (contrary to cracks in a tensile stress field that tend to be unstable). This is the basis of many models that try to capture compressive fracture, for example, the model developed by Sammis and Ashby (1986). Note that as early as 1924 Griffith pointed out that tensile stress concentrations around cracks may cause premature failure of solids. Using a simple fracture-mechanics-based analysis he argued that in tension and torsion the failure strength would be approximately the same, whereas for compression the strength would be eight times higher than the tensile fracture strength (Griffith 1924). It is interesting to compare this to the fracture laws for the lattice model presented in Section 3.5. But let us not digress too much and return to the Sammis and Ashby (1986) analysis. The starting point for the analysis is the observation that tensile stress concentrations occur around the perimeter of holes (pores). Cracks develop in a vertical direction, which is, as mentioned, the outcome of the Timoshenko and Goodier (1970) analysis assuming perfect linear-elastic behavior. After an initial pop-in these vertical cracks are stable and can only grow when the external load increases.

Using the model of Figure 8.2 Sammis and Ashby continue and include interactions between neighboring pores and cracks. The deformation of the ligaments (one has been shaded in Figure 8.2) between neighboring cracks, and summing up various contributions to the stress-intensity factor they manage to calculate the shape of the compressive stress–strain curve of porous solids. The mesoscale tensile cracks and the presence of the pores appear to be a sufficient mechanism for energy dissipation before the maximum compressive load is reached, and thus lead to a curved relationship between external stress and global strain just before peak.

When the initial cracks are inclined to the compressive stress wing cracks develop from the crack-tips as sketched in Figure 8.1f (see, for example, Horii and Nemat Nasser (1986), Kemeny and Cook (1991), Ashby and Hallam (1986), and Bobet and Einstein (1998) in two dimensions and Dyskin, Germanovich,

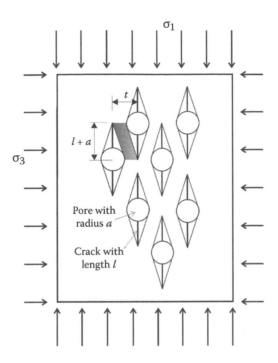

FIGURE 8.2
Model based on the growth and interaction of vertical splitting cracks. (From Sammis and Ashby. 1986. *Acta Metall.*, 34(3): 511–526. With permission from Elsevier.)

and Ustinov (1999) for analysis of wing crack growth from elliptical cracks in three dimensions). Sliding along the inclined initial cracks leads to the growth of curved wing cracks from the tips. This mechanism is observed for cracks growing in a homogeneous material; most of these experiments were carried out using PMMA, which is brittle at low temperatures. Machining the initial cracks is most difficult, in particular in 3D. Ashby and Hallam (1986) show in a series of photographs the progression of wing cracks. The external load must be increased continuously to keep the cracks growing, which makes it a very stable mechanism. Sometimes secondary cracking is reported, as, for example, the "comb-cracks" reported to occur in the fracture of ice (like PMMA also translucent, which allows us to visualize internal cracking more easily than in concrete) by Schulson and Gratz (1999) and in rock-type materials by Bobet and Einstein (1998). One may wonder whether the wing-crack model makes much sense in materials with a coarse micro-structure. In Figure 8.3 the wing-crack model is shown again, but now add-ing the possible effects of a regular or more random material structure. In a homogeneous solid, such as the PMMA so often used to demonstrate the mechanism, wing cracks developed as shown in Figure 8.3a. When the mate-rial consists of a regular packing of hexagons, and the interfaces are weaker than the elements, the same fracture shapes may appear. However, when the

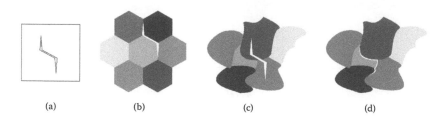

FIGURE 8.3
Growth of wing cracks in (a) a homogeneous isotropic continuum and in (b)–(d) granular materials with increasing randomness of the materials' particle structure and local material properties. (From Van Mier. 2008. *Engng. Fract. Mech.*, 75: 5072–5086. With permission from Elsevier.)

grains become distorted, that is, have a random geometry, and on top of that random properties, the wing crack mechanisms may be seriously hindered and not appear at all (Figure 8.3b–d).

It all seems related to the scale of the material microstructure in comparison to the scale of the cracks. Thus, grain boundaries may affect crack growth, and they are not always neatly arranged to allow for an undisturbed growth of the wing cracks, which usually have a slightly curved shape (see Figure 8.1f, right sketch). The growth of wing cracks is delayed even more when lateral confinement is applied, as was clearly shown in the analysis of Horii and Nemat-Nasser (1986). In all, it is quite questionable whether the wing-crack model applies to concrete, at least at the scale of the specimens discussed here, and the mechanisms depicted in Figures 8.1a,c–e and 8.1g seem to be more appropriate, although the result of crack arrest in the compressive stress field certainly applies to concrete as well. The last possible mechanism depicted in Figure 8.1g concerns the compaction of the porous interfacial transition zone above and below aggregates in the major compressive direction. This mechanism was proposed by Van Geel (1998).

What emerges is a rather complex crack mechanism, dictated very likely by the complex microstructure of concrete. Both the geometry (size and shape) of the microstructural elements and the distribution of strength and stiffness will affect the overall result. If the lattice model were successful in coping with all the above mechanisms, it seems obvious that friction should play some role, but let us first take a closer look at the final stage of fracture during softening in (uniaxial and triaxial) compression in the next section.

8.2 Softening in Compression

After the peak stress has been reached, a softening regime is found in compression tests quite similar to tensile softening (see Figure 2.10), providing, of course, the test is run in displacement-control (see Appendix 3 "Stability of

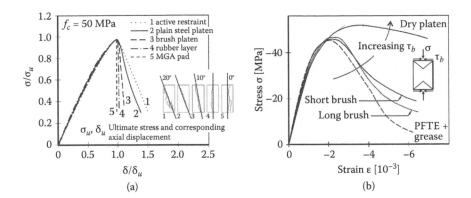

FIGURE 8.4

Effect of boundary restraint on the uniaxial compressive stress–strain curve. (a) Results on cylinders with h/d = 2.5 by Kotsovos (1983, *Mater. Struct. (RILEM)*, 16(1): 3–12) and (b) on 100 mm cubes by Van Mier and Vonk. 1991. With permission from Springer.)

Fracture Experiments"). In Figure 8.4 results from uniaxial compression tests between different types of loading platens are shown. In Figure 8.4a experimental results of Kotsovos (1983) are presented, who performed experiments on cylinders with an aspect ratio h/d = 2.5. Van Mier and Vonk (1991) repeated the experiments of Kotsovos, using 100-mm cubes (h/d = 1.0) instead, which confirmed the earlier findings. The cube results are included in Figure 8.4b. The results of Kotsovos are shown in dimensionless form; those by Van Mier and Vonk in absolute values of stress and strain. In both representations it is clear that the frictional restraint of the loading system has a significant influence on the softening behavior. The softening curve tends to become steeper as frictional restraint decreases.

For uniaxial compression tests the restraint is usually not considered as an important factor because by simply using more slender specimens the end-zone effects are reduced and it is generally assumed that the middle part of the specimen is under a true uniaxial compressive state of stress. Reducing the friction, for example, by using lubricants or friction-reducing pads (such as aluminum or PFTE-foil (Teflon)), the frictional restraint can be reduced from an inaccurately defined 15–49% (see discussion in Ottosen 1984) when solid steel platens are used to a mere 1–2%. The development of triaxially compressed end-zones in the specimen parts in contact with the loading platens is prevented, and the failure mode of the specimen changes. This is clarified in Figure 8.5, where two impregnated specimens after loading into the softening branch are shown. Figures 8.5a and b represent two vertical cuts of a specimen loaded between rigid steel platens and between steel platens with a Teflon (PFTE) -grease insert between the platen and the specimen, respectively. Quite clear is the development of the restrained triaxially compressed end-zones in Figure 8.5a. When friction-reducing pads are used, the cracks are more vertical as depicted in Figure 8.5b, although at some

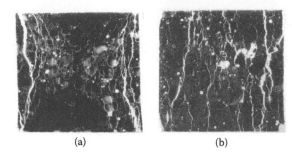

(a)   (b)

FIGURE 8.5
Cracking in two specimens after loading in uniaxial compression (loading direction is vertical) between (a) rigid steel platens and (b) rigid steel platens with friction-reducing Teflon grease pads. Cracking is visualized by means of fluorescent epoxy impregnation. (After Vonk et al. 1989. In *Fracture of Concrete and Rock: Recent Developments*.)

locations cracks show some inclination, which may have been caused by the internal stress distribution from the heterogeneous material structure. Note that in the latter case, with Teflon, cracks are distributed throughout the specimen's volume, whereas when dry steel platens are used, the conical end-parts are virtually uncracked.

Using a quite different crack detection technique, namely, digital image correlation, similar results are obtained; see Figure A4.8 in Appendix 4. The difference is that the crack process can be monitored continuously during loading with the latter technique, whereas with impregnation we can only show cracking at a single loading stage. On the other hand, Figure 8.5 shows the internal crack growth; the images from digital image correlation are limited to showing surface cracks only. In Figure 8.4a the change of the final crack plane observed in Kotsovos's experiments is depicted. The slope of the final "shearlike" crack changes from approximately 20 degrees to approximately zero degrees (vertical splitting crack) with decreasing boundary restraint (the crack angles were measured from photographs of fracture specimens that appeared in the original paper by Kotsovos). The loading systems used by Kotsovos are indicated in the inset of Figure 8.4a, and range from a rather large restraint (referred to as "active restraint" when a steel ring is enclosing the outer parts of a specimen,) via plain steel platens, brushes, to rubber and MGA-pads.

When different deformation measurements are taken on a concrete cube under uniaxial compression it appears that failure occurs from outside to inside (very much like peeling an onion). This means that the first outer surface layers are split off from the core of the specimen, and later the core will fail; see Van Mier (1984). In Figure 8.6a the stress–strain diagrams are shown from strain gauges glued to the surface of the specimen and an overall measurement with LVDTs between the loading platens (in this case brushes, similar to those originally developed by Hilsdorf (1965); brushes are an effective medium to reduce frictional restraint). The difference between overall

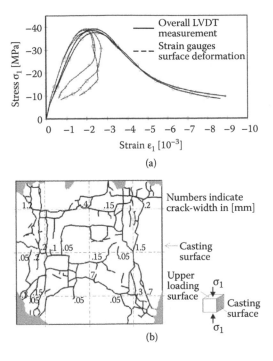

FIGURE 8.6
Difference between surface strains and overall strains in a concrete cube loaded between brushes (a), and surface crack pattern of one of the specimen's top-surface that was in contact with the brush platen (b). In (b) the numbers refer to the crack width in [mm]: wider cracks appear consistently closer to the outer edges. (Results after Van Mier. 1984. *Strain-Softening of Concrete under Multiaxial Loading Conditions.*)

strains and surface strains is quite clear; beyond peak the surface strains stay significantly behind the overall strains. The experiments were controlled using the overall strain as control parameter, and would have failed when the surface strains were chosen instead. What happens becomes clear when the fracture pattern is considered.

Figure 8.6b shows the top surface of one of the cubes after the test was terminated. Cracks are visible on the surface, also in the interior parts, indicating that the brushes indeed allowed a failure mode similar to the experiment with Teflon-grease pads shown in Figure 8.5b. The width of the cracks is largest toward the edges of the specimen, indicating that first the surface layers are split off and subsequently the core material will fail. More recently, Meyer (2009) showed by means of x-ray tomography of foamed cement specimens loaded in uniaxial compression that surface cracking already starts far before peak stress is reached. Figure 8.6 shows that the process has become quite dominant in the postpeak regime. Figure 8.5 also shows that the crack widths are larger toward the sides of the specimen. Additional results by Vonk (1992) clearly confirm this trend. In a way this mode of failure is no surprise. The surfaces are the easiest to move sideways; there is no lateral support. One might even expect that after the vertical cracks developed buckling instabilities cause the outer layers to fail completely. For the core material the situation is quite different and in order to have cracks in the core sufficient lateral deformations must have occurred, which implies surface cracks to appear first.

The straightforward conclusion might be that the loading on the specimen is not as uniform as one would prefer. The question is whether uniformity can be achieved at all. When steel platens are used, one may apply uniform boundary displacements, but the stress distribution is, due to the frictional restraint between platen and specimen, not quite uniform, as is clearly observed from finite element analyses by Ottosen (1984). He showed the effect of boundary restraint on the normal and shear stress distribution along the surface of cylinders ($h/d = 2$) in contact with the (steel) loading platens. The analyses predicted a rather nonuniform normal stress distribution (σ_z/σ_{nom} in Figure 8.7) and shear stress distribution (τ_{rz}/σ_{nom}, where σ_{nom} is the nominal axial stress). A linear analysis shows the largest stress concentrations in the corner, which diminishes somewhat when nonlinear material behavior is assumed. The analyses also indicated, similar to the

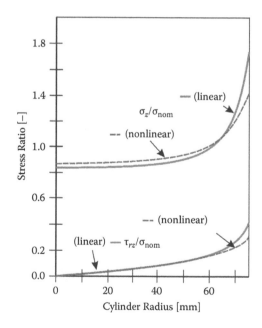

FIGURE 8.7

Normal (σ_z/σ_{nom}) and shear stress (τ_z/σ_{nom}) distribution along the end surface of a cylinder ($h/d = 2$, $d = 150$ mm) just before failure. One set of results shows the stress distribution assuming linear material behavior, and a second set the result from using a nonlinear constitutive law. σ_{nom} is the nominal axial stress. (From Ottosen. 1984. *J. Engng. Mech.*, 110(3): 465–481. With permission from ASCE.)

above-mentioned experimental results, that failure started from the outside. The heterogeneous structure of the concrete has some additional effect on the stress distribution. It appears therefore that the specimen/material behavior cannot be completely separated from the used boundary conditions, and that the stress–strain diagram could perhaps better be formulated in terms of load and displacement: the P–δ curve shows the failure of a small-scale structure (at lab-scale) rather than a material property (as was concluded in Van Mier 1984 and 1986a). Figure 8.4 shows how the P–δ (or σ–ε) diagram is affected by variations in boundary restraint; Figure 8.6 shows clearly how the core fails later than the surface layers. These two observations are sufficient to dismiss the result of such (routinely performed) compression tests as a "true property" of the material. One solution might be to use prismatic specimens with a slenderness $h/d > 2$. In higher specimens one might expect that the effect of the triaxially constrained parts of the specimens becomes increasingly less important, as sketched in Figure 8.8. However, the situation is actually even more complicated as the following results underscore.

When experiments are carried out on specimens of varying slenderness it appears that large differences in postpeak behavior occur. The prepeak curves are quite similar, but the postpeak behavior becomes gradually more brittle when the slenderness increases (Van Mier 1984), and gradually even may develop into a snap-back situation, especially for high-strength concrete at very high slenderness ($h/d > 5.0$; see Jansen and Shah 1997). In Figure 8.9 the results by Van Mier are shown, who was the first to show that deformations localize under uniaxial compression, indicating that a fracture mechanics approach might be needed to explain the softening behavior in compression. As a matter of fact, such a solution may be quite similar to the model developed by Hillerborg, Modéer, and Petersson (1976) for tensile fracture,

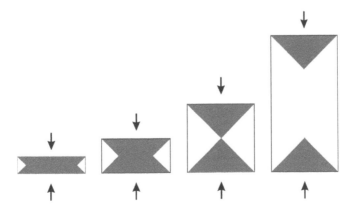

FIGURE 8.8
Triaxially constrained end-zones (gray) in specimens of different slenderness loaded between rigid steel platens. Especially for the shorter specimens the triaxial constraint may eventually extend over the largest part of a specimen's volume. (Reprinted from Van Vliet and Van Mier. 1996. *Mech. Coh.-Frict. Mater.*, 1(1): 115–127. With permission from Wiley.)

although some changes are needed; see Section 8.5. The compressive variant of FCM still assumes continuumlike behavior in the softening regime (stress is used as a state variable), which is in contradiction to many observations, and an alternative approach has been proposed in Van Mier (2009). These matters are discussed in detail in Chapter 10.

Figure 8.9 shows stress–strain curves for compression tests between brushes for specimens with three different slendernesses, h/d = 0.5, 1.0, and 2.0. The stress-axis has been normalized to the maximum stress in each experiment. This eliminates minor differences between the tests (note that the tests gave about the same maximum strength because brushes were used; see Van Mier 1984). Presenting results in terms of strain shows an almost identical prepeak behavior but in the postpeak regime the softening curve becomes gradually steeper when slenderness increases.

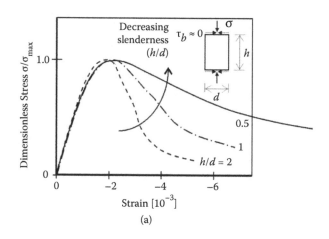

(a)

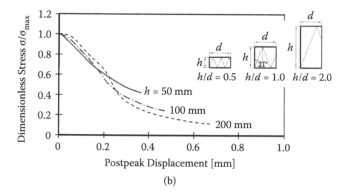

(b)

FIGURE 8.9

(a) Stress–strain curves for concrete prisms with varying slenderness (h/d = 0.5, 1.0, and 2.0), and (b) stress postpeak displacement curves of the same tests. (After Van Mier. 1984. *Strain-Softening of Concrete under Multiaxial Loading Conditions.*)

This result has been confirmed by many others, in particular by the work of the RILEM Technical Committee 148-SSC; see Van Mier, Shah et al. (1997). Tests at both ends of the slenderness spectrum were carried out by Van Vliet and Van Mier (1996) for $h/d = 0.25$ to 2.0, and by Jansen and Shah (1997) who considered $h/d = 2.0$ to 5.5. The results basically all give the same result, that is, similar to the results shown in Figure 8.9b. Here the postpeak stress–displacement curves are shown of the same three experiments of Figure 8.9a. The postpeak displacement is calculated following:

$$\delta_{soft} = \delta - \delta_{pp} = \delta - \varepsilon_{pp}l \qquad (8.1)$$

where δ_{soft}, δ, and δ_{pp} are the softening displacement, the total displacement, and the prepeak displacement, respectively, and l is the measurement length which is in these tests is equal to the specimen length. Quite clearly the curves now fall in a narrow band, and a unique response is observed. Basically the same displacement prevails, which points toward a localized fracture mode.v

The conclusion is clear: a fracture mechanics solution must be applied in order to describe softening in compression realistically. This finding was a drawback for continuum-based models in the 1980s. Actually it should not have been a surprise but the desire to model everything in continuum theory appeared to be stronger than simply accepting the observation that after peak the specimen fractures and separates into several parts. In Chapter 10 we return to these matters; here we continue and show some further experimental results.

Equation (8.1) is a simple first-order approximation and some further improvements can be made. For example, deducting the elastic deformation following

$$\delta_{soft} = \delta_{tot} - \frac{\sigma}{E}l \qquad (8.2)$$

leads to a small clockwise rotation of the three curves and the similarity is slightly improved. We return to these matters in Section 8.5.

Figure 8.9b also indicates the failure modes: an inclined shear crack (shear band) appears to develop. In specimens of large slenderness this localized mode is clearly visible (see, e.g., Schickert 1980 and Jansen and Shah 1997); in specimens with small slenderness the localized crack seems to "fold up" as sketched in Figure 8.9b. Van Mier (1984) shows a clear photograph of a fold-up (zigzag) crack in a low prism. Similar observations were reported for soils (see, e.g., Desrues 1998, viz. Figure 10.4 in his paper shows results obtained by Mokni 1992).

The question is now whether this localized failure mode would appear only for situations/tests where low-friction loading platens are used. The answer to this question appears to be negative as indicated with the results

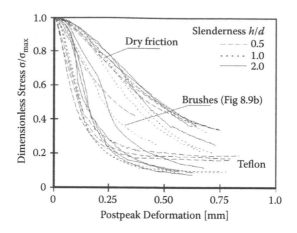

FIGURE 8.10
Post-peak stress–displacement curves for normal strength concrete specimens with varying slenderness. Specimens were loaded between Teflon (PFTE)-grease sandwiches, brushes, and rigid steel platens. (From Van Vliet and Van Mier. 1996. *Mech. Coh.-Frict. Mater.*, 1(1): 115–127. With permission from Wiley.)

shown in Figure 8.10. The localized failure mode appears irrespective of the frictional restraint of the loading system. This means that there is an effect of the loading system, in particular when relatively small specimens are tested, but the material characteristic, that is, the localized failure mode, is preserved. The tests carried out by Kotsovos also were reported to give localized cracks as indicated in Figure 8.4, and many examples can be found in Schickert (1980) as well; just the inclination of the localized crack seems to vary with frictional restraint

8.3 Softening as Mode II Crack-Growth Phenomenon

The end-restraint due to friction at the interface between specimen and loading platen is a type of confinement, albeit of the outer parts of a specimen only. When confinement is applied over the full length of a specimen the stress–strain diagrams change dramatically. How much they change depends on the actual level of the lateral stress applied, whether the lateral stress is applied symmetrically or asymmetrically in the second and third principal loading direction. By adding confinement, we enter the realm of multiaxial loading, which is very important, especially when considering fracture. Under sufficiently high confinement the quasi-brittle failure mode that we have seen thus far for concrete subjected to uniaxial compression may change to show much more ductility. Under hydrostatic loading it may turn out that a material specimen cannot be failed; pressure may increase so much that phase transformation of the material will become possible (see also Section 3.5 in Van Mier 1997). We do not enter into that aspect of multi-axial compressive behavior, and restrict our attention to the low-confinement regime, up to the brittle-to-ductile transition.

So, let us first start with symmetric confinement. We apply a lateral stress of –1 MPa to the side surfaces of a prism or cube subjected to uniaxial compression. This simple measure allows an increase of the axial compressive failure load with approx. 4–5 MPa; see Van Mier (1984). As early as 1929 Richardt et al. concluded that the strength of (symmetrically confined) concrete is equal to the uniaxial compressive strength plus four times the confining stress. This is valid for tests with constant confinement, but more recently it has been shown that the same is true for proportional loading paths, that is, where the ratio of axial and lateral stress increase is kept constant throughout the test. More elaborate equations have been developed, for example, by Newman (1979), but the result is about the same as the original result by Richardt, Brandtzaeg, and Brown (1929). In Figure 8.11 the nominal stress ($\sigma_1 - \sigma_3$) versus axial strain (ε_1) curves from triaxial compression tests on cylinders ($\sigma_2 = \sigma_3$) by Jamet, Millard, and Nahas (1984) are shown.

When the confinement increases we clearly see an increase of the nominal axial failure stress. The unconfined concrete, which is actually a coarse mortar with 5-mm maximum aggregate size, has strength of approx 35 MPa. Under uniaxial compression the behavior is quasi-brittle, comparable to the results shown in the previous section. With increasing confinement

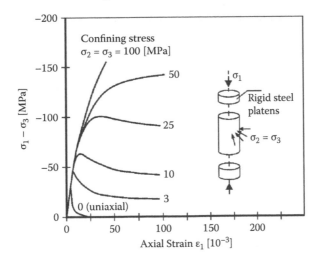

FIGURE 8.11
Nominal stress $\sigma_1–\sigma_3$ versus axial strain ε_1 curves from triaxially confined cylinder tests with constant confining stress $\sigma_2 = \sigma_3$. Note that first hydrostatic pressure was applied, and after reaching the required confinement only the axial load was increased further. (Adapted from Jamet, Millard, and Nahas. 1984. In *Proceedings of the International Conference on Concrete under Multiaxial Conditions.*)

we observe a gradually increasing residual stress level, and the quasi-brittle behavior appears to change into ductile behavior at 50 MPa confining stress. This so-called brittle-to-ductile transition takes place probably between −25 and −50 MPa confinement, that is, at relative stress-levels of $\sigma_3/\sigma_1 \approx 0.2$–$0.25$. So, what mechanisms are at play? In Figure 8.1b it was suggested that the wedging stresses due to particle interactions in the concrete are relieved by the confinement; consequently the axial stress must increase to achieve failure. Another mechanism is the delay of surface instabilities. In the tests by Van Mier (1984), loading was applied in a true multiaxial machine, that is, with three actuators any stress combination in stress space could be handled. The load was applied through steel brushes, which forces the surfaces to remain (more or less) flat, and the large surface cracks shown in Figure 8.6 can simply not develop.

In the triaxial tests of Jamet et al. (1984) cylinders were tested using oil pressure to achieve the lateral pressure (thus $\sigma_2 = \sigma_3$); the same effect must occur, even though the flexible membrane and the oil fluid allow differential deformations along the length of the cylinder. These large splitting cracks will likely develop at and beyond peak, and simply delaying the reduction of the cross-sectional area through the splitting off of surface layers (like an onion) causes the material specimen to act in a more ductile manner. In addition, because crack widths are smaller, friction between the crack faces may become increasingly more important, also enhancing the postpeak carrying capacity and improving the postpeak ductility. The various mechanisms and their probable interactions can perhaps best be studied using mesomechanics models such as the lattice model used for analyzing tensile fractures in Chapter 6. We return to these matters in Section 8.4. One last remark should be made. Figure 8.1b suggests that the splitting cracks derive from the interactions between large aggregates. In the tests by Jamet et al. no real large aggregate particles were present, therefore the results suggest that other sources of heterogeneity are the cause of the microcracks, such as pores and pre-existing flaws, which may have developed during casting and curing the specimens. As mentioned before, any flaw inclined to the major loading direction will grow under increasing external compression, and the tips will rotate in the direction of the principal compressive stress; see also Horii and Nemat-Nasser (1986).

If the confinement is no longer symmetric, the majority of cracks will develop in planes perpendicular to the minor principal stress. Actually these are the majority of stress situations; what can be achieved in a classical triaxial cell is rather limited (mostly tests such as those shown in Figure 8.11 are done). In Figures 8.12 and 8.13 two examples of asymmetrically confined tests are shown. Both sets of tests are performed on cubical specimens in a true triaxial machine fitted with servohydraulic actuators to allow for full measurement of the softening behavior. Contrary to the tests shown in Figure 8.11 the load-paths are proportional; that is, $\sigma_2 = \alpha \cdot \sigma_1$ and $\sigma_3 = \beta \cdot \sigma_1$, from the very beginning of an experiment. In Figure 8.12 we see the effect of varying the magnitude of

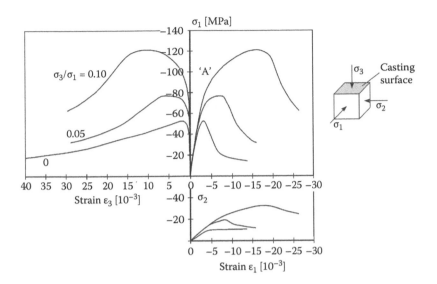

FIGURE 8.12
Effect of minor principal stress $\sigma_3 = \beta \cdot \sigma_1$ ($\beta = 0$, 0.05, and 0.10) on the stress–strain curves of plane–strain tests. Note that because the deformation in the intermediate stressed direction is kept constant and equal to zero, we have a case of passive confinement (see Appendix 5). Consequently σ_2 fluctuates with the variation of σ_1. (After Van Mier. 1984. *Strain-Softening of Concrete under Multiaxial Loading Conditions.*)

the minor principle stress while $\varepsilon_2 = 0$ (plane strain). In Figure 8.13 the effect of varying σ_2 is shown keeping in all experiments $\sigma_3 = 0.05 \cdot \sigma_1$. The maximum strength points all seem to relate to the same failure envelope; see Van Mier (1984), thus with these moderate changes in load-paths, in particular the plane–strain tests deviate from the linear stress-ratio tests in all other cases, no stress-path dependency seems to occur. In order to achieve that some quite bold variations are needed (see Chapter 7 in Van Mier 1984).

But, let us not deviate too much from the main issue at hand. The results in Figure 8.12 show that the σ_1–ε_1 curves increase in size: the first bend-over points (marked "A" for the highest curve only) occur at a higher level of σ_1 when the confinement in the minor direction increases. Again, because cracks will open in the direction of the smallest compressive stress, this indicates that the level of confinement regulates when these cracks can nucleate and widen. The shape of the curves is approximately identical; the slope of the softening curves just after peak stress is approximately the same in all three tests. The main difference lies in the level of the residual stress, which is substantially higher for the highest confinement level $\sigma_3 = 0.10 \cdot \sigma_1$. The cubical specimens will fail as shown in Figure 8.14a for a comparable experiment, namely a linear stress-ratio test with $\sigma_2 = 0.33 \cdot \sigma_1$ and $\sigma_3 = 0.05 \cdot \sigma_1$. This proportional stress-path leads to approximately the same failure stress as a plane–strain test with the same confinement σ_3 because ε_2 fluctuates just a

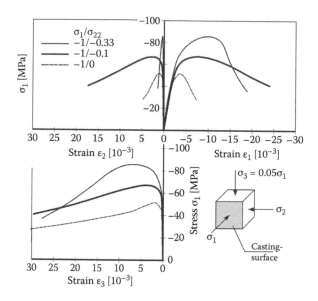

FIGURE 8.13
Effect of varying the $\sigma_2 = \alpha \cdot \sigma_1$ ($\alpha = 0$, 0.05, and 0.33) for true triaxial tests with $\sigma_3 = 0.05 \cdot \sigma_1$. Note that for the tests with $\sigma_1/\sigma_2/\sigma_3 = -1/-0.10/-0.05$ the intermediate strain ε_2 fluctuates around zero, which makes the experiment almost the same as a plane–strain test. (After Van Mier. 1984. *Strain-Softening of Concrete under Multiaxial Loading Conditions.*)

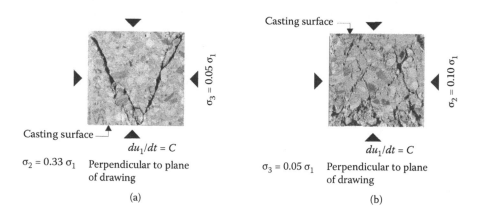

FIGURE 8.14
Examples of planar (a) and cylindrical (b) failure modes in cubes subjected to multiaxial compression. Stress–strain curves of the two tests were shown in Figure 8.13. These images were obtained after the test was completed, and after carefully grinding down the capping that was applied to the surface of the cubes. (After Van Mier. 1984. *Strain-Softening of Concrete under Multiaxial Loading Conditions.*)

little around zero; see also Figure 8.13. The different cases are shown here on purpose in this way to provide the widest possible view of failure of concrete under multiaxial compression. Thus, it appears that all these specimens fail through the development of cracks that open in the direction of the minor principal stress. As long as the intermediate principal stress is larger, fewer or even no cracks will develop in that direction. Thus, in spite of the fact that we are dealing with a complex three-dimensional state of stress, the failure mode is relatively simple. So-called shear bands form in the plane formed by the major and minor principal stress as indicated in Figure 8.14a.

Before drawing some conclusions and discussing the failure mechanisms, let us have a brief look at Figure 8.13. In this case the minor confining stress is the same: $\sigma_3 = 0.05 \cdot \sigma_1$, whereas the intermediate stress $\sigma_2 = \alpha \cdot \sigma_1$ varies, with $\alpha = 0$, 0.10 and 0.33, respectively. We now immediately notice the difference between the softening curves (σ_1–ε_1 curves, right top diagram). When the intermediate and minor principal stresses are almost the same (test with $\sigma_2 = 0.10 \cdot \sigma_1$ and $\sigma_3 = 0.05 \cdot \sigma_1$, solid line in Figure 8.13) we see that the softening curve indicates more ductile behavior compared to the two other, more asymmetric cases. The largest deformations occur in the direction of the minor principal stress (σ_3) and the intermediate principal stress (σ_2), with the strains in the intermediate direction just a bit smaller than those in the smallest stressed direction. Thus, the failure mode is almost symmetric, as can also be seen in the photo of the fractured sample in Figure 8.14b. Short inclined cracks are found in all directions, and, in comparison in the failure mode of the tests in Figure 8.12 a fully three-dimensional crack pattern now emerges. The terms "cylindrical" and "planar" mode have been coined for these two crack modes, respectively; see Van Mier (1984). The cylindrical mode appears to be a form of distributed cracking, yet as we show, localized cracks are clearly visible when longer prisms are used instead of cubes.

The experiment with the distributed crack pattern of Figure 8.14b has approximately the same (positive) deformations in the intermediate and minor loading directions. This then makes the test comparable to classical triaxial tests in a Hoek cell. The tests by Jamet et al. (1984) shown in Figure 8.11 indicate a brittle-to-ductile transition at confinement levels between 20 and 25% of the major principal stress. In the test of Figure 8.13 and 8.14b we are at 5–10% confinement level. The shapes of the curves are similar to those obtained by Jamet et al. (1984) using cylinders. So, what could explain the clear shear band observed in the cylinder tests, also at the confinement levels applied in Figure 8.14b? It might be speculated that the specimen length has some influence too, as a matter of fact quite similar to what was shown in Figure 8.9 for uniaxial compression.

The localized failure modes are most clearly visible when a classical triaxial test is done using cylinders with a slenderness ratio of 2, or when asymmetric stress combinations in a true multiaxial machine are tested. Under such loading regimes the development of shear bands and compaction bands has been extensively documented, for many different materials. For example, Nádai (1924) reports on the development of shear bands in iron. Also in this classic paper there appear examples on shear bands in wax and marble. For the group of materials usually referred to as "geomaterials," for which marble is just one example, it is now generally accepted that under low confining stress clear localized failure modes appear (Hallbauer, Wagner, and Cook 1973; Lockner and Byerlee 1992; Moore and Lockner 1995, for rock; Desrues and Viggiani 2004, for soils; Van Mier 1984 and Van Geel 1998, for concrete; and Fortt and Schulson 2007, for ice). The localization zone can be visualized by means of different techniques, for example, stereophotogrammetry, impregnation, acoustic emission monitoring, and so on.

In Figure 8.15 some further results are shown, obtained by Van Geel (1998) on the same multiaxial machine used by Van Mier (1984). The tests are plane–strain tests, comparable to those shown in Figure 8.12, except that prismatic specimens with a slenderness ratio $h/d = 2$ ($d = 100$ mm) were used. Using the fluorescent-epoxy impregnation technique explained in Appendix 4, six specimens loaded to different levels of axial deformation were treated after unloading and cut open. The crack patterns are shown in Figure 8.15a–f; the different stress–deformation curves can be identified in the diagram next to the crack patterns. With the applied crack detection technique, in stage

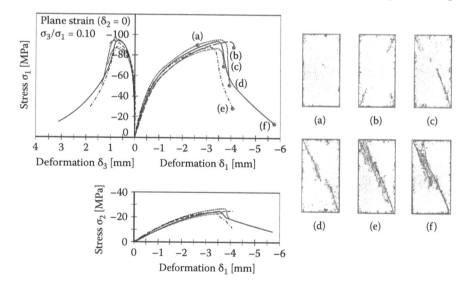

FIGURE 8.15
Shear-band propagation in concrete prisms ($100 \times 100 \times 200$ mm³) loaded under plane–strain ($\varepsilon_2 = 0$), with $\sigma_3/\sigma_1 = 0.10$. The diagram to the left shows the σ_1–δ_1 and σ_1–δ_3 curves, as well as the σ_2–δ_1 curves for six specimens loaded to a prescribed loading level in the prepeak and postpeak regimes. (a)–(f) The corresponding crack patterns are shown in the prisms that were cut after vacuum impregnation with an epoxy resin containing fluorescing particles. (After Van Geel. 1998. *Concrete Behaviour in Multiaxial Compression*. With permission from TU Eindhoven, Faculty of Architecture, Building and Planning, Unit Structural Design and Construction Technology.)

(a), in the prepeak regime, hardly any cracks are found. In the next stage (b) three small cracks emanate from the corners of the specimen. This is in agreement with the numerical analyses shown in Figure 8.7 that predict a stress concentration in the corners of a specimen. In stage (c) it appears there is a preferential direction, and a single larger inclined crack develops from the lower-right corner. In stage (d) the main inclined cracks extend from the lower-right corner and the upper-left corner, almost completely traversing the specimen. Stage (d) is almost at the end of the steep part of the softening curve as can be seen from the σ_1–δ_1 diagram. Stages (e) and (f) then show a "widening" of the shear zone, which may occur due to frictional restraint in the main cracks as well as aggregate interlock. Large aggregates often obstruct the path of the inclined cracks, which somehow will have to go around such stiff obstacles (see also Figure 8.14a: two large aggregates disturb the left lower portion of the V-shaped crack). In stage (f) it appears that two inclined cracks overlap in the center of the specimen. Similar observations were reported by Desrues (1998), who observed the growth of localization zone(s) including overlapping of multiple zones in sand samples. The results would actually point toward initiation of the inclined localization zone (or shear band) at a location where the highest stress concentrations are found (in these specimens obviously the corners), and a subsequent propagation process. The similarity to observations during tensile fracturing is quite striking; see, for example, Figure A2.1 in the Appendix 2. In that case also crack growth seems the dominating mechanism during the steep part of the softening curve. In the tail regime different mechanisms then take over, like the crack-face bridging in tension (see Section 6.1, and Appendix 4, Figure A4.2), and most likely Coulomb friction in compression.

So, what can be concluded from the results presented in this section? First of all, fracture of concrete under triaxial compression is a localized phenomenon: shear bands grow through a specimen's cross-section during the steep part of the softening curve. The shear band appears to remain straight while it is propagating; locally widening of the band may occur due to the presence of obstacles such as large aggregate particles (see, for instance, Figure 8.14a), but the main direction seems unaffected. Thus, the shear band growth can likely be interpreted as a mode II crack growth phenomenon, which would hint toward a classical fracture mechanics approach as the best theoretical solution. We return to this result in Chapter 10 where the 4-stage fracture model is presented. Toward the brittle-to-ductile transition the clear shear band disappears and instead a more distributed crack pattern is observed; see, for example, Newman (1979). So-called barreling of the specimens occurs: the middle part of the cylinder expands more than the end parts that are in contact with the stiff steel loading platens. Deformations in experiments over the brittle-to-ductile transition are so large, up to 40–60% that the nonuniformity of the loading situation becomes clearly visible to the naked eye. We do not further discuss stress states above the brittle-to-ductile transition and limit the discussion to stress states within the quasi-brittle field.

Just one last remark, however. For extensile loading ($\sigma_1 < \sigma_2 = \sigma_3$) the brittle-to-ductile transition occurs at higher confinement levels; see Paterson (1978) and Van Mier (1997). The higher transition is the result of the fact that crack growth occurs in the plane perpendicular to the lowest compressive stress (now σ_1), and the driving force is substantially larger (σ_2 and σ_3, against σ_1 only for triaxial compression).

Plane–strain tests, such as those shown in Figures 8.12 and 8.15 are examples of passive confinement tests. Keeping the deformation in one direction equal to zero and applying compression in the other two loading directions leads to an increase of stress also in the direction with zero-deformation. Restraint appears under many different situations, for example, the aggregate restraint during drying shrinkage cracking; see Section A4.4. The passive confinement mechanism is frequently used in practice, albeit not always consciously; see Appendix 5 for some further background information.

8.4 Lattice Approximations

It is interesting to see whether a lattice simulation, similar to the one used for tensile fracture in Chapters 6 and 9, is capable of generating realistic behavior in compression, that is, in agreement with the experimental results summarized in Sections 8.2 and 8.3. It is obvious that several of the aforementioned mechanisms are not included in the simple lattice model, but looking to the model by Sammis and Ashby (1986) in Figure 8.2, we might come quite close. This model is characterized by modeling crack growth from pores in a compressive stress field and interactions between neighboring cracks as shown in Figure 8.2. These elements also are included in a lattice simulation. When modeling foamed cement, as we discuss below, lattice simulations indicate crack growth from the top and bottom of the pores in the direction of the compressive load.

In the past we have made several attempts to model compressive fractures in the lattice model; see, for example, Schlangen and Van Mier (1994) and Margoldová and Van Mier (1994). The fracture law used is the normal force/bending rule, Equation (3.36), with α varying between 0 and 1.5. We can summarize the results from these early (1994) simulations as follows. Using a random beam-length lattice (3D, $5 \times 5 \times 5$ cells), in which all beams have the same tensile threshold strength of 1 MPa, an increase of α in Equation (3.36) from 0.1 to 1.5 leads to an improvement of the softening behavior in compression. Yet, the ratio between overall uniaxial compressive to uniaxial tensile strength decreases rapidly. At $\alpha = 0.1$ the ratio is about 20, for $\alpha = 1.5$, symmetry in tensile and compressive response is found, that is, almost the same global failure strength is found, except for the sign, of course. In an attempt to model the material more realistically 2D analyses were carried out, also with a beam lattice model, using the particle overlay method of Figure 4.13. When uniaxial compression was applied, the failure mode changed from vertical splitting when $\alpha = 0.5$ to inclined cracking for $\alpha = 1.0$. In a similar approach, but trying to model various phases and interfaces in cemented particle composites in a slightly different manner, Topin et al. (2007) ran into the same difficulties.

A 2D lattice is perhaps not the best choice for modeling compressive fracture, which is a truly three-dimensional phenomenon. Modeling the material structure realistically in three dimensions has recently become possible with the doctoral work of Man (2009). Using data from CT scans, the real particle or pore structure of concrete can be included in a lattice; see Figures 4.6 and 4.7. A step back from the complicated structure of normal aggregate concrete is to leave out all the large and small aggregate particles, use cement only, and provide heterogeneity through including a large amount of foam. The foam will dissolve in time, after the cement has hardened, leaving highly porous cement stone. The material is simpler than normal concrete: the interfacial transition zone has been excluded by removing the aggregates, and the behavior now is completely dependent on the stress concentrations from the large amount of air bubbles and the strength of the cement matrix. The results were disappointing in the sense that no real improvement over the 1994 simulations was achieved. Although there is now much more detail in material structure, as well as in the calculated crack patterns, the maximum loads are still highly dependent of α: For $\alpha = 0.0$, the largest maximum load was calculated as 38.6 N for the foamed cement containing 70% porosity. When α increased to 1.0, the maximum load was just 3.9 N. Note that from the experiments of Meyer (2009) an average maximum compressive load of 3.46 N was achieved for foamed cement with 70% porosity, using exactly the same specimen geometry and boundary conditions as in the simulations. Thus, for $\alpha = 1.0$ the computed peak load comes close to the average maximum load measured in the physical experiments; yet in this case the crack pattern is not correct. Smaller α leads to vertical splitting cracks, as in the experiments (see Meyer 2009); larger values of α lead to inclined cracks, as shown in Figure 8.16 for $\alpha = 1.0$. The presented results are representative for different initial porosities of the foamed cement, ranging from 30–70%; see Man (2009).

The conclusion drawn at this moment is that the fracture law in the lattice model must be wrong, or at best incomplete. The criterion used for simulating tensile fracture doesn't seem to matter much and excellent results have been obtained (see Chapter 6, 7, and 9), but for compression it seems either necessary to include friction, for example, through the use of Equation (3.37) as an alternative fracture law, or to move to a completely different type of model, for example, the particle models alluded to in the introduction to this chapter. At this moment such analyses are not considered very useful for further understanding, therefore we close this chapter by summarizing a macroscopic model. The model may be used for analyzing the rotational

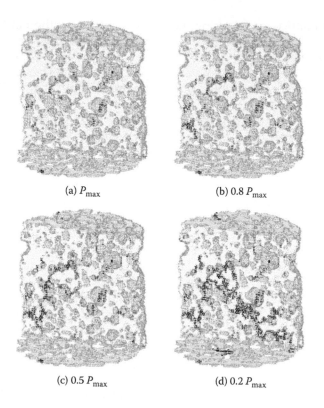

(a) P_{max} (b) 0.8 P_{max}

(c) 0.5 P_{max} (d) 0.2 P_{max}

FIGURE 8.16
Crack patterns at peak load and at three different stages in the softening regime of foamed cement with foam content of 70%. Stages (b)–(d) are postpeak at 80%, 50%, and 20% of peak load P_{max}. The model has 0.1-mm long beams; see also Figure 4.6b. The Young's modulus and tensile strength of the cement matrix were equal to 25 GPa and 5 MPa, respectively. The removed elements are colored black. Only a few elements are removed at peak load; at 20% of P_{max} in the softening regime a large inclined crack developed. (After Meyer et al. 2009.)

capacity of reinforced concrete structures failing in compression (viz. over-reinforced beams subjected to bending).

8.5 Macroscopic Models

Model codes are tools for practice engineers. Most of them, such as the CEB-FIP 1990 model code give formulas for the tensile and compressive stress–strain behavior of concrete. In the CEB-FIP model code the following compressive stress–strain relation is given, valid up to a certain maximum strain in the softening diagram. In the comment in the draft model code reference was made to Van Mier (1986a) (see Figure 8.9a,b), who showed the specimen-length dependency of the softening slope, which, similar to the Fictitious Crack Model, indicates that continuum solutions are no longer valid in the postpeak regime.

The CEB-FIP equation is purely phenomenological:

$$\sigma_c = -\frac{\dfrac{E_{ci}}{E_{c1}}\dfrac{\varepsilon_c}{\varepsilon_{c1}} - \left(\dfrac{\varepsilon_c}{\varepsilon_{c1}}\right)^2}{1 + \left(\dfrac{E_{ci}}{E_{c1}} - 2\right)\dfrac{\varepsilon_c}{\varepsilon_{c1}}} f_{cm} \qquad \text{for } |\varepsilon_c| < |\varepsilon_{c,\text{lim}}| \qquad (8.3)$$

where E_{ci} is the tangent modulus and $E_{c1} = f_{cm}/0.0022$ is the secant modulus from the origin to the peak compressive stress f_{cm} as shown in Figure 8.17. Furthermore, σ_c and ε_c are the compressive stress and strain, $\varepsilon_{c1} = 0.0022$. Such quadratic equations are the most used forms and fit well to experimental data, for example, those by Wischers (1978). If confinement is added the equation changes; see, for example, Attard and Setunge (1996) and Samani and Attard (2010). The only possible solution is to revert to a localized model for the postpeak regime, as suggested in Equations (8.1) and (8.2). These equations are the compressive equivalent to the fictitious crack model presented in Section 2.4 for tensile fracture. There is one important difference, however. The prepeak stress–strain behavior is much more pronounced in compression, and the simple approximation of a linear prepeak stress–strain diagram does not apply to compression. When confinement is applied even larger prepeak deformations are found, and deviations from linear behavior increase. Therefore an issue is whether the simple unloading behavior assumed in Equations (8.1)–(8.2) is correct. In that approach zero recovery is assumed (see Figure 8.18), whereas it seems more appropriate to expect

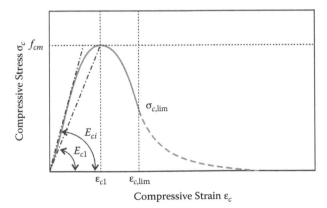

FIGURE 8.17
Stress–strain diagram for concrete subjected to uniaxial compression following the CEB-FIP Equation (8.3).

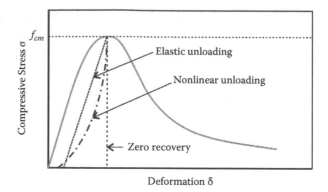

FIGURE 8.18
Possible recovery corrections for localization models in compression.

elastic unloading or even nonlinear unloading. Figure 8.9 certainly gives the impression that at least a linear unloading must be assumed.

The localization phenomenon in compression has been addressed by several researchers, including Bažant (1989), Hillerborg (1990), and Markeset (1993). The new problem, after the various equations have been worked out, is the definition of a localization length, which forms an integral part of the solution. In analyzing rotational capacity of beams one might assume that simply the height of the compression zone in the considered beam might be taken, but this length scale continuously changes as loading progresses along the softening branch. The most complete model to date is without doubt the model proposed by Markeset (1993).

Central to Markeset's model is the notion of a damage zone with length L_d. In this damage zone axial splitting cracks develop distributed over the volume (allowing the use of strains), and later on in the softening branch a localized shear crack appears (demanding the use of displacements). Outside the damage zone the material behavior is assumed to behave (nonlinear) elastic. The model is depicted in Figure 8.19. The three contributions to the overall specimen behavior are clearly distinguished and lead to the following expression for the average compressive strain ε_m:

$$\varepsilon_m = \varepsilon + \varepsilon_d \frac{L_d}{L} + \frac{w}{L} \tag{8.4}$$

that is, the elastic, damage, and localization deformations, respectively. The composite stress–strain curve, distinguishing the three contributions to the axial strain is shown in Figure 8.19b. The various cracks contribute to the compressive fracture energy. The two main contributions are the energy consumed in the damage zone and in the inclined shear band. The main problem with the model is the estimate for the damage zone length L_d.

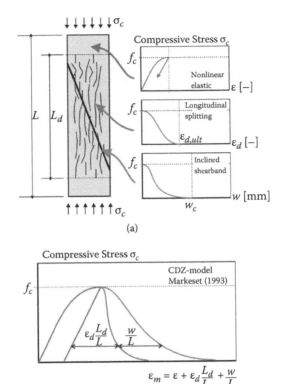

FIGURE 8.19
Compressive damage zone (CDZ) model: (a) the three contributions to the average axial compressive strain ε_m, and (b) composite diagram. (From Markeset. 1993. *Failure of Concrete under Compressive Strain Gradients*. With kind permission from Dr. Gro Markeset.)

Furthermore, there are several ongoing discussions as to whether the compressive fracture energy is a material property, and whether localization is as perfect as assumed in this chapter. For example, experiments by Vonk (1992) and Hurlbut (1985) show less perfect localization than those by Van Mier (1984) and Jansen and Shah (1997). The issue is not easy to resolve because boundary restraint changes from experiment to experiment. Boundary restraint has a significant influence on the compressive stress-deformation curve, and, as shown in Figure 8.4, affects the slope of the shear band as well. To my opinion the solution should not be sought in a model equivalent to the Fictitious Crack Model, perhaps augmented with a volume energy dissipating mechanisms such as the axial splitting cracking which is part of Markeset's CDZ-model. Rather a model based on classical linear elastic fracture mechanics seems close to reality. Such an approach would support the view that softening is a structural property, and in addition includes the notion of shear-band propagation in the postpeak regime as discussed in Section 8.3. Also one can refrain from adding further length parameters, such as L_d, which are hard to determine. We return to the resulting 4-stage fracture model in Chapter 10.

For the time being a macroscopic and phenomenological approach seems the best solution for dealing with compressive fracture. Research is under way to unravel the complex fracture mechanism of concrete under (confined) compression. Improved mesomodels are needed, capturing the essential mechanisms of mode I and mode II crack (growth), frictional restraint in cracks, and likely also buckling instabilities of (laterally) unsupported surface layers of a concrete structure/specimen.

9

Size Effects

Size effect is one of the salient characteristics of fracture mechanics. Classical linear elastic fracture mechanics exhibit a size effect on strength, which was mentioned in Chapter 2 and worked out in Appendix 1. For a long time there has been interest in size effects. In their days Leonardo da Vinci and Galileo Galilei studied the effect of structural size on strength. The issue is of considerable importance, namely, can the strength of real-size structures be predicted from small-scale laboratory experiments? It turns out that larger structures are generally weaker and their behavior is more brittle than structures that are geometrically scaled down. Testing full-scale structures in the laboratory is, however, rather impracticable and expensive. Often sufficiently large loading equipment is simply not available. Only a few exceptional labs are fitted to do really large-scale work, for example, the large structural labs at the University of San Diego. The rule is, however, that most laboratories are limited to test structures up to the 10-[m] scale (this means, for instance, beams with a depth up to 1 [m]). Figure 9.1, from Bažant and Yavari (2005), quite clearly illustrates the lack of large-scale data. The largest existing tests measuring the shear capacity of reinforced concrete beams are likely those carried out by Shioya et al. (1989) with a depth of 3,000 mm for the largest beams. Once in a while it is possible to test a full-scale structure, prior to demolishing, but in those cases one obviously has no influence on the composition of the materials used, or on the structural design and all the relevant details.

Structural disasters, such as the collapse of the Koror Babelthaup Bridge in Palau (see Burgoyne and Scantlebury 2006) are, at least partially, ascribed to misunderstanding of the size effect. The best way to cope with size effects appears to be to develop a theory, based on correct physical mechanisms of quasi-brittle fracture. In this book we have focused on full understanding of quasi-brittle fracture, and so far this has led to very good insight in tensile failure, whereas for (confined) compression further research appears to be essential to come to the required in-depth understanding of those complex failure modes. In this chapter on size effects the limitation is again on tensile fracture. We first discuss classical models for size effect on strength. Because the tensile strength is the weakest property of concrete (as well as for several other materials), it seems more than appropriate to focus on tensile fracture first. Classical approaches include Weibull's weakest link theory, and the more recent models developed for concrete fracture in particular: the "size effect law" (SEL) by Bažant (1984) and the "multifractal scaling law" (MFSL) developed by Carpinteri (1994), Carpinteri, Chiaia, and Ferro (1995), and

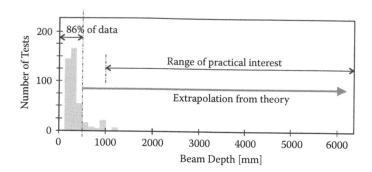

FIGURE 9.1
Range of available experimental data of beams failing in shear in comparison to the actual range of interest by practical engineers. Extrapolation from theory is needed. (From Bažant and Yavari. 2005. *Eng. Fract. Mech.*, 72: 1–31. With permission from Elsevier.)

Carpinteri, Chiaia, and Cornetti (2003). Although both models are labeled "laws," they hardly deserve that credential because in both cases curve-fitting forms an essential part of the exercise. The implication is that, in spite of the claims made, the range of applicability does not exceed the range of test data used to make the "fit." As mentioned, in Section 9.1 we review these models. In Section 9.2 recent experiments on the size effect for tensile fracture of plain concrete are analyzed. These tests are among the largest tensile tests ever conducted on plain concrete, which is partly explained by the fact that test control is quite difficult for large specimens, which, as mentioned, tend to have more brittle behavior than the usual laboratory-size specimens (with a size range up to 100 mm). Lattice approximations of size effect in tension and bending are presented next in Sections 9.3 and 9.4. Two-dimensional analyses are of course easier to perform than fully three-dimensional simulations; yet, it turns out that in the case of 2D scaling the transition from plane–strain to plane–stress is missed in most models. The lattice analyses allow the calculation of the distribution of crack sizes, which is essential input for theories with a much improved physical background such as the Weibull weakest link theory and the 4-stage fracture model outlined in Chapter 10.

9.1 Classical Models Describing Size Effect on Strength

Let us start this section by noting that a viable fracture theory should correctly predict the size effect on strength and no separate rule should be devised for just that. The Fictitious Crack Model presented in Chapter 2 is a general fracture theory that predicts size effects; see Hillerborg, Modeér, and Peterson (1976) and Hillerborg (1985). The size effect was calculated for notched and unnotched beams, and depends on the choice of the softening diagram (linear, bilinear, or any other shape). There is uncertainty about the assumption that softening is a material property. Because a valid general theory for the fracture of concrete is still lacking we have been using separate rules, such as, for example, the one for the size effect on (tensile) strength. Almost every size-effect model (theory) developed to date is phenomenological in nature and a clear physical argument is lacking. One important exception is the Weibull weakest link theory, which considers that largest volumes of the same material have a larger probability of containing a weak element and would thus have lower failure strength.

The strength of a structure of given size depends on the probability that the structure contains a weak spot. The easiest way to visualize this is a structure subdivided into small unit cells (e.g., cubes). The larger the number of unit cells contained in a structure, the larger the probability that one of the unit cells has a low strength. The weakest link concept developed by Weibull (1939, 1951) predicts the probability of failure $P_f(\sigma, V)$ as

$$P_f(\sigma, V) = 1 - \exp\left\{ -\frac{V}{V_0}\left(\frac{\sigma}{\sigma_0}\right)^m \right\} \tag{9.1}$$

where σ is the applied external stress, V is the structure's volume, V_0 is the normalizing volume (unit cell), and σ_0 is a characteristic (or normalizing) strength. The exponent m is the so-called Weibull modulus, which is considered a characteristic property of the material under consideration. Sometimes a lower strength limit σ_u is included for which the probability of failure is equal to zero. Equation (9.1) changes into:

$$P_f = 1 - \exp\left\{ -\frac{V}{V_0}\left(\frac{\sigma - \sigma_u}{\sigma_0}\right)^m \right\} \tag{9.2}.$$

The classical approach assumes that as soon as the first element fails the entire structure will fail (i.e., fracture is purely brittle), which is not the case for concrete, but, as we show, the size/scale of the structure has an effect on the apparent brittleness. As a matter of fact there are just a few materials that at a given size/scale will behave in a purely brittle manner. Glass is often mentioned as a material exhibiting purely brittle behavior, but recently some evidence at the nanoscale for precritical crack growth was found (Célarie et al. 2003), implying that at very small scales glass may behave as quasi-brittle or even plastic. The reason is likely that the amorphous structure of glass has an effect on the fracture process, actually quite like what we have seen up till now for concrete at the scale of the aggregate particles (mm scale). We will return to these matters in Chapter 11. Thus, one might conclude that the Weibull theory cannot be applied to concrete and several related materials.

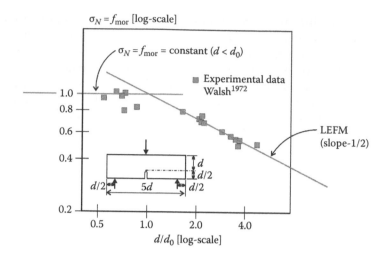

FIGURE 9.2

Normalized nominal strength σ_N/f_{mor} for beams of varying size d/d_0 (log-log diagram). The beam geometry is shown in the inset. (From Walsh. 1972. *Indian Conc. J.*, 46(11): 469-470. With permission.)

Yet, a way around may be to consider the microcrack distribution just before reaching peak load as, for example, done by Jayatilaka and Trustum (1977) and Danzer et al. (2007). We do not debate these matters here but delay that to the last section of this chapter after presenting some experimental results and insights derived from numerical (lattice) simulations.

Two phenomenological size-effect models have been extensively debated over the past decade, namely the "size–effect law" and the "multifractal scaling law." SEL is partly based on cohesive softening models and contains elements derived from it such as the length of the fracture process zone (see Chapter 2); yet it is also directly related to the early work by Walsh (1972) and Leicester (1973). MFSL is based on considerations about the fractal structure of concrete and its effect on mechanical behavior. MFSL predicts complete stress–deformation behavior and is in that respect more complete than SEL. Let us first, however, summarize some of the findings by Walsh and Leicester. In Figure 9.2 Walsh's results are shown. Notched beams of varying size were loaded in 3-point bending. The beam length was $5d$, the height $3d/2$, and the notch depth was $d/2$, with d the size of the ligament above the notch as indicated. Walsh observed that, as expected, the nominal strength of the beams (normalized to the modulus of rupture) decreased with increasing beam size. Surprisingly this decrease was found only for beams larger than a certain size threshold d_0. For beams smaller than d_0 a more or less constant strength was found. Walsh argued that the decrease of nominal strength along the inclined branch followed linear elastic fracture mechanics. In a paper published around the same time Leicester (1973) proposed a simple expression for the decreasing nominal strength with increasing size:

$$\sigma_N = a_1 D^{-s}, \quad s \geq 0 \tag{9.3}$$

Here a_1 is a constant and D is the characteristic size of the structure. It is safe to assume that the characteristic size is the smallest dimension of the cross section where a crack will appear, in most cases in the direction of crack propagation. In Equation (9.2) one can substitute the volume V by D^n, where n denotes the number of dimensions, and the relation between nominal strength and structure size can be written as (see Bažant, Xi, and Reid 1991)

$$\sigma_N \propto D^{-n/m} \tag{9.4}$$

Thus, for a material with a Weibull modulus $m = 12$, and considering three-dimensional scaling, the negative exponent in Equation (9.4) is equal to $-n/m$ $= -3/12 = -1/4$. The decrease of nominal strength with increasing characteristic size takes a certain slope, which depends both on the material and on the dimensionality of the problem. LEFM predicts a negative slope of $-1/2$, irrespective of the material, as shown in Section A2.1. Some researchers consider the LEFM limit for very large sizes is the largest possible size effect. This would imply that for very large sizes the material structure has no effect on the strength of a structure; all is dictated by the (relative) size of the notch and the geometry of the structure itself. This could be debated. There is no experimental evidence available for extremely large structures. Moreover, the LEFM analysis is highly idealized: the material is homogeneous and isotropic, and the stress-state around the notch-tip is assumed to be identical to the stresses at the tip of an actual crack. For the moment we assume that the large-scale asymptote is set by LEFM.

Let us now return to the lower size asymptote from Walsh's experiments. It appears that below a certain threshold d_0 no size effect is observed and the strength of the (notched) beams is constant, contrary to sizes beyond the threshold d_0 for which a strength decrease is found. For very small structures we have to consider the effects from the material structure, for example, the particle structure of concrete. In Chapter 6 an example was shown of a tensile test on a single-edge notched plate (see Figures 6.2a–d). Although the notch was relatively deep, larger than the maximum aggregate size, the main crack started where a cluster of large aggregates was present, by chance at the opposite side from the notch. The stress concentration at the large cluster of aggregates was obviously more important than the stress concentration from the notch. If the size of the structure becomes smaller than what is considered as the representative volume element (RVE) (the smallest volume of a material that can be considered as a continuum; usually it is assumed that the RVE for concrete is three to five times the maximum aggregate size), effects from the local material structure will become increasingly more important. The notch will be of less importance, until the point is reached where the structure has become so small that it consists of just cement matrix or aggregate material. For very small structure one should therefore expect a larger scatter in strength results. The horizontal branch in Walsh's experiments for small structures may therefore be quite indeterminate when scatter increases.

In the "size–effect law," which is based on the model by Walsh, a number of important assumptions are made. The size-effect law can be seen as a clever curve-fitting method that tries to connect the horizontal asymptote for small sizes and the LEFM-based asymptote for very large sized structures by means of a continuous function. "Asymptotic matching," known from the field of fluid mechanics is the curve-fitting technique used. Two conditions are that the failure mechanism in structures of varying size remains unchanged and that two-dimensional scaling is applied; that is, the thickness of structures of varying size is assumed to be constant. The structure always contains a notch (also geometrically scaled) from which the main crack initiates. In front of the tip of the notch a fracture process zone exists with half-length c_f. The nominal strength of a structure can then be expressed as a simple function of its size D, following

$$\sigma_N(D) = \frac{Bf_t'}{\sqrt{1 + D/D_0}} \tag{9.5}$$

The size-effect relation can be described also by considering the energy release rate, which leads to

$$\sigma_N(D) = \sqrt{\frac{EG_f}{D\hat{g}(\alpha_0, \theta)}} \tag{9.6}$$

where $\hat{g}(\alpha_0, \theta)$ is the effective dimensionless energy release rate, $\theta = c_f/D$ is a dimensionless parameter, E is the material's Young's modulus, and G_f is the fracture energy. These latter two parameters should be determined on "infinitely" large specimens; that is, they are the limit value. As we show in Section 9.2 the fracture energy of concrete from tensile tests is not constant, but increases with structure size (see Figure 9.8). D_0 is the transitional size, which was already proposed by Walsh (Figure 9.2). When $D \ll D_0$, the energy release from the structure is negligible, whereas for $D \gg D_0$, the energy release from crack growth is dominant. In the latter case the classical fracture mechanics solution is retrieved that leads to the –1/2 asymptote for very large sized structures, again, as was proposed by Walsh. The energy release caused by the growing crack is simply calculated by considering a triangular area above the total crack length, that is, the stress-free crack plus the fracture process zones with length $c_f/2$.

As we have seen in Chapter 6 the definition of a fracture process zone for concrete is by no means clear. In a panel subjected to tensile loading the fracture process develops from distributed microcracking through the growth of a single large macrocrack with substantial bridging. This whole microfracture and bridging zone could then be interpreted as the fracture process zone, but measuring the extent of it is certainly not straightforward. Moreover, the fracture process zone doesn't seem to be constant throughout the fracture process. Depending on the resolution of the measuring technique different fracture process zone lengths were reported ranging from a few µm to almost a meter; see, for example, the overview article by Mindess (1991). The extent of the fracture process zone is thus the fundamentally unknown parameter, which then leads to fixing the problem by means of curve-fitting. Doing just that, it can be shown that on a double-logarithmic scale Equation (9.5) fits the behavior of many different structures made of a variety of different materials; see, for example, Bažant (2004). It is obvious that applying results from a curve fit outside the range of observations is a haphazard endeavor, and one should refrain from trying. SEL is just that, a curve-fitting exercise, in spite of all the theoretical considerations made in dozens of papers published on the topic by Bažant. The matter becomes quite clear when one studies the values of the parameters in Equation (9.5), when it is fitted to various data sets. An overview was made by my doctoral student Nooru-Mohamed (1992), who concluded that as long as there is no physical basis for the parameters, SEL has a very limited range of applicability. Indeed, as was concluded by Mindess (1991), c_f, the fracture process zone length, does not appear to be a fundamental property of the material, but depends on specimen geometry and loading conditions. The various analyses presented in Chapter 6 confirm this view. The fracture process zone length is an essential parameter in Equation (9.6), and given its elusive character, any theory based on it is to be used with caution. There are additional problems, however. SEL applies to notched structures only. The enormous amount of papers on SEL is confusing people and has even led physicists to believe that there is interest in "notched structures" in the field of civil engineering (see Alava, Nukala, and Zapperi 2006). Luckily we all know better.

The second, more recent theory developed to describe size/scale effects in the fracture of concrete is the multifractal scaling law. This approach considers the fractal nature of the material structure, in particular the aggregate structure, and leads to the following expression.

$$\sigma_N(D) = f_t\sqrt{1 + \frac{l_c}{D}} \tag{9.7}$$

where f_t is the tensile strength of the material (for infinitely large specimen sizes) and $l_c = \lambda \cdot d_{max}$ is a characteristic length representing the influence of disorder on the mechanical behavior. In the small-scale (fractal) regime the size-effect law has a slope of $-1/2$ on a log σ_N-log D diagram, whereas in the homogeneous regime MFSL grows toward a horizontal asymptote.

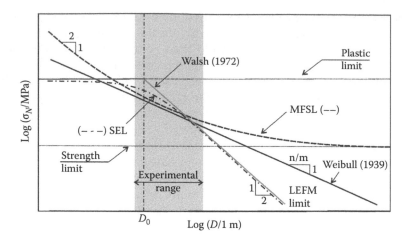

FIGURE 9.3
Comparison of the various size-effect models: SEL and MFSL. Walsh's model has been included, as well as Weibull's (a straight line with slope –n/m), and the experimental range is shown as a gray-shaded area.

The characteristic length l_c marks the transition between the fractal and the homogeneous regimes. The fracture energy behaves in the opposite way, namely in the small-scale (fractal) regime the slope is + 1/2 in a log-log representation, and again, grows toward a horizontal asymptote. This is, as a matter of fact, also the assumption in SEL, and is confirmed by experiments: the fracture energy increases with structural size, and seems to grow to a horizontal asymptote; see Figure 9.8. It is interesting to note the differences in asymptotic behavior between SEL and MFSL. In Figure 9.3 the various models are shown in a single log σ_N – log D diagram. Walsh's model has been included, as well as the Weibull approach, which reduces to a straight line with negative slope in a log–log representation. The range of available experimental data is shown as a gray-shaded area. SEL forms a continuous variant of the Walsh proposal. A horizontal asymptote for the small-scale regime and a –1/2 slope from LEFM are connected by means of asymptotic matching. MFSL goes in the opposite direction: a –1/2 slope in the small-scale regime, changing to a horizontal asymptote for large-scale structures. The two models, SEL and MFSL, overlap in the area where most test data are available, and the two models fit these data equally well. This is no surprise because in both cases a fit is always made directly to the experimental data: in the case of SEL there is uncertainty about the length of the fracture process zone, whereas in MFSL the same is true for the characteristic length and the limit value of the tensile strength for infinite size. So, which model should be used? This question is very difficult to answer. The small-scale and large-scale asymptotes are difficult, if not impossible, to measure. In the case of SEL, Bažant maintains that the asymptotes are only used to construct the model, and it should not actually be attempted to measure the behavior in the two asymptotic regimes, in particular the one at the small-size/scale end of the spectrum. Bažant and Yavari (2005) have critically assessed the MFSL, and make various comments to show that erroneous predictions are made, especially concerning the small-scale and large-scale asymptotes. The horizontal asymptote for large sizes should have a slope, at least similar to Weibull. Moreover, the –1/2 slope for the small-scale asymptote in MFSL does not follow from the hypotheses made in MFSL (personal communication with Bažant).

We do not attempt to solve the argument, or rather controversy, in this book. The discussion about the asymptotic behavior seems rather futile and not to the point; it will never be possible to validate the small-size nor the large-size regime through experiments. It should be noted that both models, in the end, need to be fitted to experimental data. In both cases it is not clear how the model parameters relate to the common parameters used in concrete structural engineering, such as the compressive and tensile strength. Rather what we intend to do in the remainder of this chapter is to come to grips with the actual fracture process in concrete structures of varying size. Experiments have been conducted, in a size range of 1:32, revealing interesting details of the behavior of tensile specimens of different size. Next to that, lattice simulations have been performed that give further insight to the fracture process. The numerical simulations relate to two loading cases: uniaxial tension and 3-point bending. In the latter case detailed analyses of the crack size distributions have been made, which forms essential input for the new model presented in Chapter 10.

9.2 Size Effect on Strength and Deformation: Experiments

Experiments will always form the basis for new models and theories for fracture of concrete. Considering this it is amazing that so few are really engaged in developing new critical experiments to test existing theories such as those presented in the last section. Truly open experimentation is not very often done, which probably is caused by the fact that in order to get some research funding one often has to "predict" what is to be found or observed. Performing size–effect experiments on concrete is a costly affair. Constructing the necessary loading equipment to handle specimens over a sufficiently large size range is time consuming; one needs to ensure that all measurements, on small and large specimens, can be done with the same accuracy. If different loading rigs are used to tests specimens of different size one needs the assurance there is no influence from the test method. In Chapters 2, 6, and 8 we have seen that the softening regime, both in tension and compression, is notably affected by the boundary conditions adopted in the test. Therefore, it is believed that softening is, at least for a large part, a structural property rather than a material property. The lowest fracture energy measured directly from a uniaxial tensile test is obtained when freely rotating supports are used: crack overlaps from bending moments in the supports cannot develop. The main crack simply grows from one side of the specimen to the other side.

In 1994 Carpinteri and Ferro published results from size effects on concrete subjected to uniaxial tension. Active control was used, meaning that in a center part of the dog-bone-shaped specimens deformations measured along the circumference of a specimen were kept uniformly distributed. This was achieved by using three actuators, one centrically placed, the two others with a certain eccentricity to counteract possible bending moments when the main crack developed asymmetrically. Active control was mentioned in Chapter 6; see Figure 6.8c. According to computations with the lattice model higher fracture energy would be found for specimens loaded using active control, whereas the lowest possible fracture energy would be found when freely supporting boundaries are used. The size range of the experiments by Carpinteri and Ferro (1994) was 1:16. This was the largest available range in 1994, which relates directly to difficulties in measuring stable softening behavior for the largest tensile specimens. Following these experiments my student Van Vliet and I embarked on an ambitious endeavor trying to extend the range of size–effect tests for uniaxial tension. The goal was developing a test technique that would allow measuring the complete stress–deformation behavior, including the softening regime. Adopting and slightly modifying a technique also used by Li, Kulkarni, and Shah (1993), it was possible to determine the softening response of 2.4-m long dog-bone-shaped panels with a thickness of 100 mm. The loading was applied with a small eccentricity, which was scaled along with the specimen dimensions. Two-dimensional scaling was applied: the thickness of all specimens was 100 mm. Special hinges were developed, based on a pendulum bar system, which allowed testing the dog-bones between freely rotating loading platens.

A sketch of the specimens is shown in the inset of Figure 9.4, and a drawing of the middle-size loading rig is shown in Appendix 3, Figure A3.4, where the test control is also explained. Three loading frames were built, each suitable for testing two sizes: the smallest rig for the smallest specimens "A" and "B," the middle rig for "C" and "D," and the largest rig for "E" and "F." The size D in Figure 9.4 varied between 50 and 1,600 mm; the total panel length was always 1.5D, thus varying between 75 and 2,400 mm. The smallest section between the curved bays of the dog-bones was 0.6D; thus for the smallest specimen "A," the effective area where the crack would very likely develop was 30 × 100 mm, and for "F," the largest size, 960 × 100 mm. This implies that for the smallest specimens the thickness is larger than the width of the dog-bone, thus creating a plane–strain situation. For the largest specimens one can clearly expect a plane–stress state. This is a consequence of the decision to scale specimens only in two dimensions and keeping the thickness constant. As a matter of fact, until these tests were done nobody

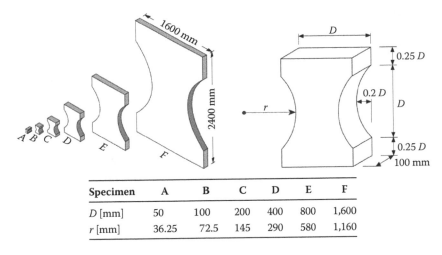

Specimen	A	B	C	D	E	F
D [mm]	50	100	200	400	800	1,600
r [mm]	36.25	72.5	145	290	580	1,160

FIGURE 9.4
Dog-bone-shaped specimens used in the size-effect experiments. The specimens were glued with a scaled eccentricity between the loading platens. The eccentricity was 1 mm for the smallest specimen labeled A and increased with a factor 32 for the largest specimen F. (From Van Vliet and Van Mier. 2000. *Eng. Fract. Mech.*, 65(2/3): 165–188. With permission from Elsevier.)

ever seemed to have realized that a transition from plane–strain to plane–stress would occur. This obviously must have consequences for the analysis of experimental data. Van Vliet (2000) shows in his doctoral thesis that the rotational behavior of the specimens during growth of the main crack depends on geometrical variations.

Due to the small size of the neck of the dog-bone size A, it was decided to use a concrete with $d_{max} = 8$ mm. The cement used was an ordinary Portland cement CEM I 42.5 R. The 28-day compressive strength (150-mm cubes) varied around 50 MPa for the various batches, and the splitting tensile strength (also using 150-mm cubes) varied around 3.5 MPa (see Van Vliet 2000) for all the details). Compared to the neck width of 30 mm for the A-size specimens, the W/d_{max}-ratio was 3.75, which is small compared to the usual RVE of $5d_{max}$. In the experiments of Carpinteri and Ferro (1994), the situation was even more critical: in that case the ratio W/d_{max} decreased to 1.875, barely enough to consider these results of any use. This is of course the dilemma: in order to extend the size range it is attractive to downsize, but if a specimen becomes smaller than the RVE the results become useless, and a high scatter must be expected. In all the discussions about the small-size asymptote, which we reviewed in the last section, there is no discussion whatsoever about this constraint. The situation should be clear by now: downscaling is useless, and only increasing the sample size will lead to very large size ranges that are needed to validate the various size-effect approaches. For very rough concretes, for example, using $d_{max} = 32$ mm, would lead to minimum width of "A"-size specimens $W = 120$ mm, and for the "F"-size, 32-times larger; that is,

$W = 3,840$ mm. The panel length would in the latter case be 9.6 m. This is too large to handle in many laboratories; the only option left would be to load such specimens horizontally, for example, floating on water, as was done in the largest known size-effect tests by Dempsey et al. (1999a,b). They tested floating ice sheets that were examined first on uniformity of their thickness. Due to problems encountered in hostile arctic conditions, these tests were conducted using flat-jacks, without closed-loop control.

The main results from the size-effect tests are shown in Figure 9.5 and 9.6. In Figure 9.5 the stress–deformation diagrams from the experiments are shown, labeled "A" (smallest) through "F" (largest). Results are shown in two diagrams to increase legibility: specimens "A"–"C" in Figure 9.5a, the other

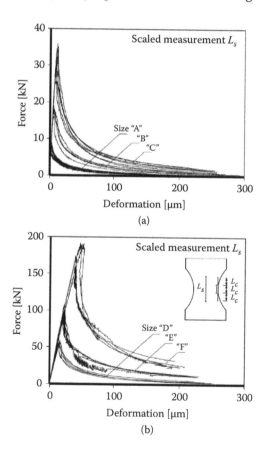

FIGURE 9.5

Tensile force versus deformation curves for the size-effect experiments on 8-mm concrete. The deformation is the average scaled deformation measured at the front and back sides of a specimen. The measurement length L_s is scaled along with the specimen size. (a) Results for the three smallest sizes are shown; (b) results for the three largest specimens. Each test has been replicated at least four times. (From Van Vliet and Van Mier. 2000. *Eng. Fract. Mech.*, 65(2/3): 165–188. With permission from Elsevier.)

three sizes in Figure 9.5b. Note the different scale along the force-axes (factor 5 difference). Because scaled deformations are shown in these diagrams we clearly see snap-back behavior for the "E" and "F"-sizes. The adopted control system can handle this easily. It is of utmost importance to conduct stable experiments because otherwise conclusions cannot be drawn and a lot of work is done without appreciable result. These results are likely the largest size/scale range ever done for concrete. At least four successful tests were demanded for each size. Exceptions were the smallest "A"-size specimen for which 10 replications were made, and for types "C," and "D." It is important to note here the large number of replications for the "A"-size, which relates to the discussion about the lower asymptote in the previous section. In order to continue the discussion it is important to show first the nominal strength versus size diagrams of these experiments; see Figure 9.6. Two diagrams are shown, for the "dry" series in Figure 9.6a and for the "wet" series in Figure 9.6b. These latter results are ignored for the moment, but we return to these as soon as we discuss small-scale asymptotic behavior.

In Figure 9.6a it can be seen that the "A"-size specimens showed a large scatter (average tensile strength for "A" was 2.54 MPa, the standard deviation was 0.41 MPa, i.e., 2–4 times as large as for the other sizes; see Van Vliet 2000 for all the details). The increasing scatter was expected because size "A" was below the RVE for the concrete used. Therefore it would make sense that the smallest size be discarded. This is normally not done because one would like to have results for the largest possible size/scale-range. In this case, however, it is shown to be meaningful to remove the results for the smallest size "A," and it is considered very important for each test series to make a thorough check. For example, the previously cited results by Carpinteri and Ferro (1994) concerned tests on five different size dog-bones. The smallest size

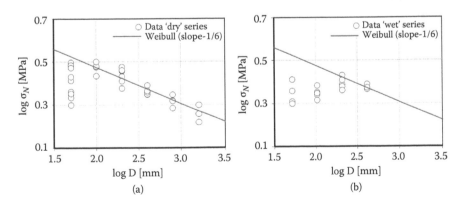

FIGURE 9.6
Effect of specimen size D on the tensile strength of 8-mm concrete. (a) Results from the "dry" tests ($P-w$ diagrams of Figure 9.5); (b) results from the "wet" series are shown (see main text). (From Van Vliet and van Mier. 2000. *Eng. Fract. Mech.*, 65(2/3): 165–188. With permission from Elsevier.)

specimen was just a factor 1.5625 larger than the largest aggregate particle in the concrete mixture used. This certainly makes these results doubtful. In addition, because the control system used by Carpinteri and Ferro (1994) was not capable to obtain stable load-deformation diagrams for the largest specimens (only one valid result was obtained for that size, which is clearly not sufficient). The remaining size/scale range in their experiments spans only three sizes with a scale range of factor 4, which is barely sufficient to make a good comparison to the various models.

In the experiments of Figure 9.6 the effective scale-range is 16 and not 32, as was intended. There is another important effect here, namely, the average tensile strength for the "A"-size decreases, which does not correspond to any of the assumptions made in the size-effect approaches discussed in Section 9.1. The available choices were constant strength for small sizes (Walsh and SEL), or an asymptote with negative slope (MFSL); see Figure 9.3. A notable effect during the experiments was the rotational behavior of the specimen when the main crack propagated in the softening regime. In Van Vliet and Van Mier (1999) it was shown that the rotational behavior caused by the combination of the chosen boundaries (free rotations) and the effects from nonuniform crack growth (a truly three-dimensional effect, that was discussed in Section 6.1.4.1) led to the decrease of apparent tensile strength for the "A"-size. Here it should be recalled that when keeping the specimen thickness constant, a transition from plane–strain to plane–stress occurs when the specimen is scaled in the other two directions. This effect is usually ignored, and one can doubt many comparisons of the various size-effect approaches in the small-scale range. Thus, two effects must be carefully considered when experimental results are compared to size-effect theories: (1) comparing data below the RVE is not very useful because the material sample is not representative, leading to an increased scatter, and (2) two-dimensional scaling does not imply that we can consider the problem in just two dimensions. Three-dimensional effects may appear when we go through the transition from plane–strain to plane–stress. The strength of the experiments shown in Figures 9.5 and 9.6 lies in the fact that they were very carefully conducted. It has taken more than five years to come to the first journal publications, which is for most unacceptable in the present-day research environment with its heavy publication pressure. Nevertheless, there is great need for carefully conducted experiments as well as a careful analysis of the obtained results.

The "dry" results from Figure 9.6a were fitted to Weibull, and it was found that a Weibull modulus $m = 12$ led to a favorable comparison with the five largest sizes. The negative slope of the Weibull theory is $-n/m = -2/12 = -1/6$ as shown in Figure 9.6a ($n = 2$, that is, two-dimensional scaling). This value lies in the range for tensile tests on concrete by Zech and Wittmann (1978). In an in-depth analysis Vořechovský (2007) derives a negative slope of $-1/7.91$ for these experiments based on a stochastic analysis with three different length scales.

Let us now return to the results of Figure 9.6, and investigate what happens when the environmental conditions change. In Figure 9.6 the log σ_N versus log D diagrams are shown for "dry" and "wet" specimens. Dry specimens were kept for a longer time (several months) in the lab, until equilibrium with the lab environment (specifically the relative humidity) was established (RH = 60%; T = 20°C). Wet specimens were stored in a fog room (RH = 90%, T = 20°C) until one day before testing. The wet specimens therefore suffered from significant moisture gradients, which may lead to drying induced microcracking as shown in Figure 9.7. Because the zone affected by micro-cracking from drying shrinkage appears to have constant depth (see also the examples of drying shrinkage cracking in Section A4.4), the effects from drying may have some influence on the size effect on strength and fracture energy. In particular toward smaller-sized specimens the fracture strength may be affected because the microcracked zone takes up a relatively larger part of the specimen's cross-section. The results shown in Figure 9.6 con-firm this: the difference between fracture strength is larger for the smallest specimens, and comes close to the results for dry specimens when log D > 2. Vořechovský (2007) also investigated the effect of damage zones on the size effect and concluded that the decrease for structure size "A" could be explained from just that.

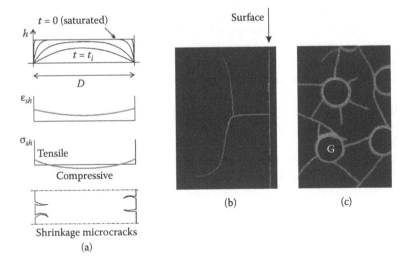

FIGURE 9.7
(a) Drying shrinkage of concrete may lead to differential deformations and differential stresses, eventually leading to microcracking in surface layers of the material; (b) typical drying shrink-age crack patterns in hardened cement paste (section perpendicular to drying surface); and (c) composite of cement and mono-sized glass beads (view parallel to drying surface after grind-ing the top-layer). Drying shrinkage cracking is explained in detail in Section A4.4 and Section 11.2. ((a) After Van Mier. 2004a. *Proceedings 5th International Conference on Fracture of Concrete and Concrete Structures (FraMCoS-V)*. (b),(c) From Shiotani, Bisschop, and Van Mier. 2003. *Engng. Fract. Mech.*, 70(12): 1509-1525. With permission from Elsevier.)

In Chapter 2 we discussed cohesive fracture models, which have been included in various finite element models for simulating fracture on the macroscale. Such models incorporate all the microscale behavior in a single cohesive fracture law. Basically, the behavior of a process zone is modeled as a nonlinear fracture phenomenon, which can be used in any type of structures loaded under any type of boundary condition. The analyses presented in Section 6.1.4.1, as well as experimental results shown in Section 2.5 showed that the boundary conditions have a significant influence on the shape of the softening curve in tension (see Figures 6.7, 6.8, and 2.11, respectively). The specimen size not only affects the maximum strength but the entire load-deformation response. The size effect must likely also be considered as a structural property, and can therefore also be analyzed by means of a micromechanical model. Thus, again, simply consider a specimen as a small-scale structure and include the boundary conditions from the experiment in the analysis. In the context of this book quite obviously the lattice model has been used. Particle models and other micromechanical models can be used too, however, as long as the intention is not to curve-fit the experimental results. Instead the model should be used as a tool to come to a deeper understanding of the size-effect phenomenon. Such an approach is quite different from those presented in Section 9.1 where in most cases ad hoc assumptions are made, in some cases lacking any sound physical basis. Before presenting the results from a number of lattice analyses, however, the size effect on fracture energy is shown here for the sake of completeness. In the previous section it was mentioned that the fracture energy increases with structural size. In Figure 9.8 the effect of size on fracture energy is shown in two different diagrams.

We do not know the width of the crack zone exactly: it is "assumed" that it is a either a line-crack (Fictitious Crack Model) or a crack-band with certain (usually, constant) width (Crack-Band Model), nor is it straightforward how the correction for the unloading of the elastic material inside the measurement length should be done. So, in Figure 9.8 we plot the raw data directly from two types of deformation measurements that were conducted: the so-called "'scaled" measurement (L_s), in which the measurement length is scaled with the specimen dimensions (see inset in Figure 9.8: two LVDTs are mounted at the front and back side of a dog-bone and the length L_s is scaled with D), and the "control" measurement length L_c. This latter measurement is related to test control: the LVDTs to the right side of the dog-bone were used in the test control system (see Appendix 3, Figure A3.4), but of course also deliver information about the deformations during the whole experiment. The fracture energy is defined as the area under the postpeak stress–deformation curve, following,

$$G_F = \int_0^{w_c} \sigma(w)\,dw \qquad (9.8)$$

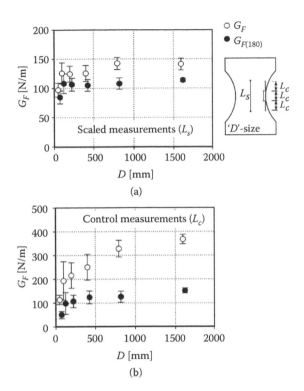

FIGURE 9.8

Effect of size D on the fracture energy G_F of 8-mm concrete: (a) results computed from the scaled deformation measurement (measurement base L_s, see right); (b) fracture energy computed from the control measurements (measurement base L_c, see inset). Solid symbols indicate the fracture energy G_F up to 180 μm; the open symbols until full separation. (After Van Vliet. 2000. *Size Effect in Tensile Fracture of Concrete and Rock*. With kind permission of Dr. Marcel van Vliet.)

Here we use the subscript F instead of the common f to indicate that a correction for elastic unloading has not been made. The area under the load postpeak deformation diagram has been calculated up to a 180-mm average crack opening and up till maximum crack-opening, when the load is zero and the specimen is separated into two parts. Both values are shown in Figure 9.8a for the scaled measurements and in Figure 9.8b for the control measurements. All diagrams show the same trend: steadily increasing fracture energy. It is not clear if the asymptotic value has been reached already for specimen size "D." The point of interest is that specimens of varying size give different values of the fracture energy, which, according to the two aforementioned theories (FCM and CBM) should be a material constant. This, again, implies that structural effects have an influence on the experiments; or that our interpretation of fracture energy as material constant is not correct. One way to find out what is happening is using mesomechanical analyses and studying the fracture process in detail, as we did in Chapter 6 for tensile fracture. In the next two sections we look to the size effect in tension and in 3-point bending.

9.3 Lattice Analysis of Size Effect: Uniaxial Tension

Van Vliet (2000) analyzed the results of Figure 9.6 with a two-dimensional lattice model. The restriction to two dimensions was made in view of the enormous computational demand, which was so large that even in two dimensions the largest specimen size of 2.40 m could not be handled within reasonable wall-clock time. One of the important issues that constantly returns is the way in which heterogeneity is handled and the selection of the type of lattice to be used (see Chapter 4). In Figure 9.9 the effect of randomness is shown. Variations are made in the type of lattice (regular triangular lattice) and the incorporated material structure (without or with particle structure). The material randomness increases from regular homogeneous, to regular with particle structure, to random homogeneous, and random lattice with particle structure. The load-displacement diagrams are becoming gradually less brittle in this order, and the crack patterns change from a linear crack (Figure 9.9a) to a highly irregular pattern (Figure 9.9b–d). The steps refer to the number of cracks until the end of the relative force–inelastic deformation diagram. Quite clearly there is a direct relation between the jaggedness of the force-displacement diagram, the tortuosity of the crack pattern, and the number of lattice elements removed. In considering size effects resulting from lattice analyses the number of cracks at peak load is of importance; the crack histories shown in Figure 9.9 actually represent the post-peak (unstable) crack process, which, in experiments is stabilized by using an appropriate servo-hydraulic test technique with a well-chosen feedback parameter (see Appendix 3). Because the crack histories in Figure 9.9 are so different from one another it may be suspected that the size effects from lattice analyses depend on the modeled heterogeneity. Indeed this is the case, as shown in Figure 9.10, although perhaps not that convincing from the chosen representation.

In Figure 9.10 the size effects from regular and random lattice analyses are shown, with and without particle overlay. The shaded gray band shows the range of the experimental results from Figure 9.6a. The analyses cover the range 1:16 and not the experimental range of 1:32, therefore only the first five specimen sizes are shown and size "F" has been omitted. The calculated range was smaller for the random lattice analyses (Figure 9.10b), that is, covering a size range of 1:8 (note the *D*-axes are different in Figures 9.10a and b). The specimen size "D" was explained in Figure 9.4. The horizontal dash-dotted line in Figure 9.10 makes a comparison between the two diagrams easier: the line shows the normalized stress for the fourth size type "D" specimen

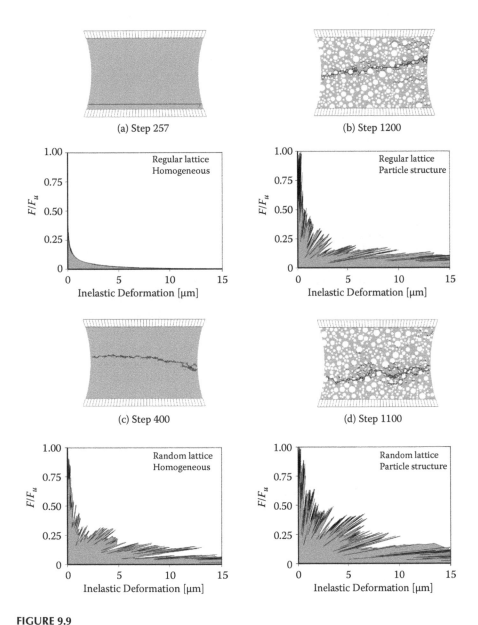

FIGURE 9.9
Effect of lattice type (regular triangular versus random beam length) and heterogeneity (i.e., homogeneous and with overlaid particle structure) for the third smallest dog-bone tensile test specimen (type "C") of the test series shown in Figure 9.4. The step numbers for each example refer to the number of removed lattice elements at the end of each analysis. (After Van Vliet. 2000. *Size Effect in Tensile Fracture of Concrete and Rock.* With kind permission of Dr. Marcel Van Vliet.)

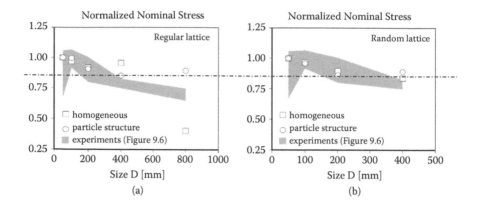

FIGURE 9.10

Size effect from (a) regular lattice analyses, and (b) random lattice analyses. (Reprinted from Van Mier and Van Vliet. 2003. *Eng. Fract. Mech.*, 70(16): 2281–2306. With permission of Elsevier.

(D = 400 mm). Comparison of the two figures shows that initially the nominal strength decreases faster in the random lattice analyses. Qualitatively the size effect appears to be correct for the smaller sizes; for larger sizes the comparison is not very favorable (regular lattice only, open squares in Figure 9.10a).

The results suggest that heterogeneity is the key factor that determines the size effect of concrete. Without particle structure, cracks are straight and the slope of the nominal stress-size diagram becomes steep. Generally it is assumed that when the effect of heterogeneity becomes less notable the size-effect curve would take a slope –1/2, which is the limiting value from linear elastic fracture mechanics, which is the large-scale asymptote in Walsh's model and in SEL (see Section 9.1, Figure 9.3). Drawing a straight line from the "B"-size specimen (200 mm) to the "E"-size specimen (800 mm) gives a slope of –0.52 in log–log representation, which is close to the LEFM limit. It should be mentioned that the analyses are too limited to lead to a final conclusion, and many more simulations are needed. Important in the results of Figure 9.10 is that heterogeneity has a vast effect on the (negative) slope of the size-effect diagram. The degree of heterogeneity is determined not only by the particle structure, but also by the randomness of the applied lattice. The randomness of the lattice can be varied (see Figure 4.12), but it is not very clear how a certain randomness would relate to, for example, the structure of the cement matrix between the coarse aggregate particles. The solution would of course be to model the material structure down to the smallest scale level, but this would result in an impossible lengthy computation. It is also no solution to change the fracture law of the lattice elements in, for instance, a softening fracture law as suggested by Ince Arslan and Karihaloo (2003), because this distracts from the real goal, namely understanding fracture in heterogeneous materials. There are simply too many ad hoc assumptions underlying cohesive models. Moreover, structural effects have a strong influence on the cohesive parameters. It is likely that an entirely different approach will be needed.

One final remark should be made with respect to the result shown in Figures 9.9a,b. The cracks in a regular triangular lattice tend to follow the mesh lines, and rather straight cracks develop (see, for instance, Figure 9.9a). Also when the particle overlay is used, small straight cracks connect the interfacial transition zone cracks, which can be observed from the crack pattern of Figure 9.9b. The use of a random lattice will, irrespective of the chosen degree of randomness, immediately cure this spurious crack mode. But the question remains, which beam-length distribution must be selected to have a correct representation for the material under consideration?

For the size effect on strength the crack distribution immediately before peak stress is reached is considered of importance. Actually it is considered *the* most important factor that may lead to a sound explanation of the size effect, because it defines the attainable strength of the material/specimen (structure). What happens beyond peak is merely (unstable) macrocrack propagation and localization, which is different for each structure. Again: softening cannot be considered a material property. Of course researchers (including me) were so pleased over the past years that it was possible to measure the softening regime, the stage of (unstable) crack propagation that much attention has gone to that part of the fracture process, that the most important part has been badly neglected, namely the prepeak crack distribution. In Figure 9.11 the crack distribution for the five considered specimen

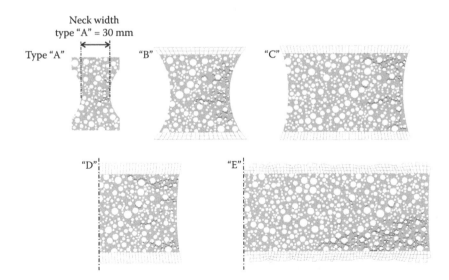

FIGURE 9.11
Crack distribution at peak-stress for the analyses with regular lattice and particle overlay. (After Van Vliet. 2000. *Size Effect in Tensile Fracture of Concrete and Rock*. With kind permission of Dr. Marcel Van Vliet.)

sizes with the regular triangular lattice with particle overlay are shown (size "A" through "E"). For sizes "A" through "C" the entire neck part of the specimens is shown; for sizes "D" and "E" only the right side of the specimen. Crack initiation at the right side was controlled by loading the specimens with a small-scaled eccentricity, that is, 1 mm for the smallest "A"-size, up to 32 mm for the largest "F"-size specimen. The main reason to apply a slightly eccentric load was that test control would become easier (see Appendix 3 for details). In the analyses of Figures 9.9–9.11 the particle structure was overlaid in such a way that the same particle geometry appeared at the right side, the most loaded side of the specimen (for details see Van Vliet 2000). Sizes "A" through "C" showed that crack initiation was near the same grains; deviations started to appear for the sizes "D" and "E." This is probably caused by the different curvatures in specimens of different size, which cause deviations in local stress distributions. At peak-load cracks have reached as far as half of the neck width in some specimens. Compared to the analyses without particle structure (i.e., the homogeneous samples), the number of lattice elements that failed before reaching peak is substantially larger.

In the homogeneous specimens, between one and three elements were removed before peak, which would be in agreement with the hypothesis made using Weibull's theory, namely the whole structure fails upon failure of the weakest link. In the analyses with particle overlay the prepeak damage increased significantly: from 66 elements removed for the "A"-size to 1,029 removed elements at peak stress for the "E"-size specimen. The removed elements are part of several cracks, as can be seen from Figure 9.11. Which of these microcracks will eventually develop into the critical macrocrack leading to localization of deformations in the postpeak regime depends on many factors. One of the main tasks at hand is to determine which of the prepeak cracks is critical. Van Vliet tried to establish the extent of microcracking in specimens of various size ("A," "C," and "E") by using the impregnation technique (see Section A4.1). One of the last examples he mentions on page 82 of his thesis (Van Vliet 2000) relates to the crack densities along the edge of an "E"-size specimen where the control LVDTs are attached. Indeed the thin sections show larger crack densities where the largest deformations are measured between two neighboring measuring points. This means that the deformation measurements are, as would be expected, a reliable measure for the amount of cracking in parts of a specimen. However, in this way we cannot answer the question of how long the critical microcrack is, and at what point it would start to propagate. The local strength distribution will be an important factor to decide this, but also possible crack interactions between neighboring cracks. As long as we cannot easily access these details of the fracture process, an averaging method will be needed. Progress in various crack detection techniques, in particular x-ray computed tomography (Section A4.2), will eventually help to unravel all the details. Also additional numerical simulations will prove to be helpful. In the next section some further examples are given, focusing in particular on the effect of the aggregate structure (particle shapes and particle content) on size/scale effects in the fracture of concrete.

9.4 Lattice Analysis of Size Effect: Bending

Many researchers revert to bending tests because, as we already mentioned, these are presumed simpler than uniaxial tension tests. Perhaps this is true as far as gripping a specimen is concerned, although nowadays strong epoxies are available that, given a reliable test protocol will lead to a plenitude of reliable results. Bending tests were already discussed in Section 6.2.2. The main objection against such tests is that strong stress and strain gradients are present in the projected failure plane from the very beginning of the experiment/ simulation. This has certainly an effect on the mechanical behavior, and the flexural strength of concrete (usually determined by applying linear elasticity to the result of a test; see Equation (6.2)) is always larger than the uniaxial tensile strength. In concrete, for example, under uniaxial tension (i.e., more or less uniform stress field) microcracks can—to some extent—be arrested in the heterogeneous material structure (see, e.g., Figure 6.3). In bending, in addition, cracks can be arrested in the stress and strain gradients, which allows reaching a higher failure load. The size of a specimen has influence on the absolute size of the part of the specimen involved in the prepeak microcrack process. Thus, the likelihood for a critical localized crack increases with specimen size, which may lead to the expectation that larger specimens will appear to be weaker. However, the fracture process may in detail be quite different from uniaxial tension where the entire cross-section is always approximately uniformly stressed (even in the tensile tests on dog-bone-shaped specimens with (scaled) load eccentricity; see Figure 9.4). Thus although the test may appear to be simpler, analyzing the obtained results is more tedious. Nevertheless, because bending is the most common type of loading in practical situations it is worthwhile to ponder a bit on such experiments.

Together with one of my most recent doctoral students we carried out a numerical study on the size effect of concrete prisms subjected to 3-point bending (see Man 2009). Following the simulations done by Van Vliet (see previous section) some improvements were essential to come closer to the needed answers. In the first place, full three-dimensional simulations would be needed. The aforementioned problem with two-dimensional scaling, where a transition from plane strain to plane stress is unavoidable can be circumvented. Next, the computer-generated particle structures used by Van Vliet would be replaced by realistic particle distributions from CT scans, as discussed in Section 4.4 (e.g., Figure 4.7). Also, as was shown in 2D simulations by Prado and Van Mier (2003; see also Section 6.1.2 and Figure 6.3), the particle density has a significant influence on the fracture process and may, as a consequence, also affect the size/scale effects. Finally, from a fully 3D analysis the crack-size distribution at various stages of the fracture process can be extracted, but most importantly at peak-load. This might give a clue about the critical crack size, which is an important parameter in the 4-stage fracture model explained in Chapter 10.

Prisms containing different types of aggregates were cast, and the aggregate structure (geometrical distribution of the particles) was determined by means of a medical CT scanner as explained in Section 4.4. Oval-shaped basalt particles were used in one concrete, and crushed basalt particles in another. Basalt has a higher density that the porous cement matrix, and the particle shapes are quite accessible in a simple medical scanner. The resolution turned out to be sufficient for our purpose. Three different aggregate densities were studied for the oval-shaped aggregate particles (P_k = 20, 30, and 40%), and just two densities for the crushed basalt concrete (P_k = 40 and 48%). Smaller parts were cut from the scanned prisms and overlaid with a regular triangular lattice to build the simulation models. All details can be found in Man (2009) and in Man and Van Mier (2011). Some preliminary results were published in Man and Van Mier (2008a).

In simulations selecting specimen sizes larger than the RVE is of less importance than in physical experiments. Not only can specimens containing very large aggregates at critical locations easily be identified, and if necessary for well-documented reasons be discarded, it is much easier to repeat some of the simulations than would be possible with experiments. In the latter case, casting new specimens would inevitably lead to more scatter, just owing to differences between batches. The simulations, therefore, give important additional insight to the fracture process, even from perspectives that would be impossible in experiments. One of these perspectives was, again, as was also used in the 2D dog-bone simulations of the previous section, to place the same aggregate structure along the most critical edge of the specimen/structure. In the case of bending we would have to take care that the aggregate distribution matches along the lower specimen surface for the various prism sizes. Figure 9.12 shows the applied procedure. The method allows having exactly the same aggregate structure at the midpoint of the front-lower edge of the prisms. It will of course be possible that crack initiation occurs away from this point. An alternative approach would just be to select random locations in the result from the CT scan for all prism sizes. Four different sizes were investigated; the smallest "A" size prisms were $L \times B \times H = 6.25 \times 2.38 \times 2.6$ mm³ (approx. ratio $2.6 \times 1 \times 1$), very small indeed compared to the size of the largest aggregate particles d_{max} = 15 mm. The specimen sizes "B," "C," and "D" were obtained by subsequent doubling of these dimensions.

Volume scaling has a large effect on the number of lattice elements needed in the analyses. The number of elements increased from 21,107 for size "A" to 9,269,081 for size "D." In view of this the number of replications of the analysis for each size differed: 10–12 replications for size "A," 8–10 for size "B," 5–6 for size "C," and just a single analysis for the size "D" prisms. They were simply at

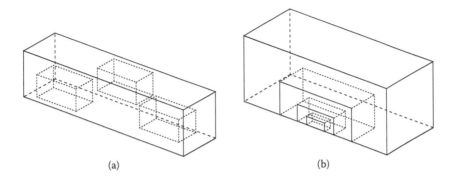

FIGURE 9.12
(a) Large prisms (size "D") were cut out at random locations from the CT scans of a real concrete prism; (b) subsequently the smaller specimens (sizes "A" through "C") were cut from the "D"-sizes. (From Man and Van Mier. 2011. *Cem. Conc. Comp.*, 33: 867–880. With permission from Elsevier.)

the brink of the possible given the available computing infrastructure (see also Appendix 1). The main difference between oval-shaped and crushed aggregates is in the crack patterns: depending on the location and shape, crushed aggregates may fracture, whereas for rounded aggregates the crack will always grow along the smooth interface; see Man (2009) and Man and Van Mier (2011). In load-displacement diagrams the differences in crack growth lead to slightly lower peak strength for smooth aggregates. The point of interest here lies of course in the differences in size effect. In Figure 9.13 crack growth in prisms of four different sizes are shown for oval-shaped aggregate particles ($P_k = 20\%$, i.e., a relatively low particle content). The size differences between the prisms

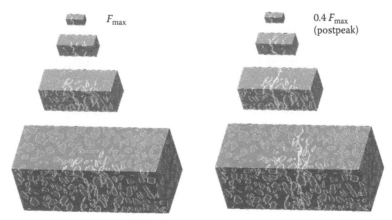

FIGURE 9.13
Crack growth in prisms subjected to 3-point bending. The concrete contains 20% oval-shaped basalt aggregates. At left the crack patterns are shown at maximum load (F_{max}), to the right the crack patterns at $0.4F_{max}$ in the post-peak regime. (After Man. 2009. *Analysis of 3D Scale and Size Effects in Numerical Concrete*. With kind permission of Dr. Hau-Kit Man.)

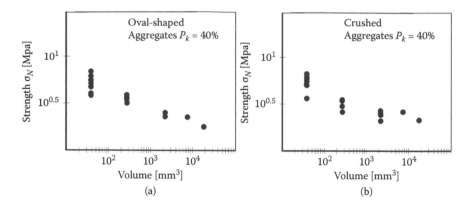

FIGURE 9.14

Size effect on nominal strength σ_N in 3-point bending, shown in bi-logarithmic representation for numerical concrete with (a) oval-shaped aggregates and (b) crushed aggregates. The particle content in both cases is $P_k = 40\%$. (From Man and Van Mier. 2011. *Cem. Conc. Comp.*, 33: 867–880. With permission from Elsevier.)

are clearly visible: the smallest specimens at the top are barely larger than the d_{max}, and are not suitable for a continuum representation of the concrete used. The cracks propagate predominantly along the interfaces between aggregate and matrix, which was modeled as the weakest elements. Some of the cracks, in particular in the postpeak regime and very well visible in the largest size "D" specimens at the bottom, show matrix cracking. Here large continuous cracks have developed, almost to the top of the beams. There is a remarkable amount of side cracking along the bottom part of all prisms, which all contributes to the fracture energy. Unless notched beams are used, bending tests will therefore lead to overestimating the fracture energy of concrete. Yet, notches have a detrimental effect as well because they force the main crack to develop at a specific location, which is not necessarily the weakest part of a prism. For an aggregate content of 40% oval-shaped or crushed particles, the bi-logarithmic size-effect plots are shown in Figure 9.14. Along the horizontal axis log V is plotted, along the vertical axis log σ_N, which has been calculated using Equation (6.2). The trend in the two diagrams corresponds to what we have seen for tension as well. The nominal strength decreases with increasing size and the scatter is largest for the smaller sizes because here the effect of disorder from the particle structure is dominant.

When it is assumed that Weibull theory applies, even in this case where substantial prepeak cracking is observed (see Figure 9.13, left column), the results can be approximated with the linear expression:

$$\log \sigma_N = a + b \log V \tag{9.9}$$

where the parameters a and b, as well as their confidence intervals (*CI*) and the regression coefficient R are shown in Table 9.1 for the various particle

TABLE 9.1

Slope b and Intersection a^a in Equation (9.9)

Aggregate Shape	P_k [%]	a	$CI(a)$	b	$CI(b)$	R
Crushed	20	2.16	0.152	−0.17	0.026	−0.90
	30	2.09	0.186	−0.15	0.033	−0.85
	40	2.27	0.166	−0.18	0.02	−0.90
Oval	40	2.36	0.148	−0.19	0.026	−0.94
	48	2.66	0.229	−0.24	0.04	−0.91

Source: Man and Van Mier. 2011, *Cem. Conc. Comp.*, 33: 867–880. With permission from Elsevier.

[a] From linear regression of the computational size-effect results for numerical concrete with various particle shapes and particle densities. The confidence intervals (*CI*) for parameters a and b are included, as well as the regression coefficient R.

densities. Parameter b is the slope, which appears to vary between −0.15 and −0.24 depending on the type of aggregate and the amount of particles included in the concrete. The translation into the Weibull modulus leads to m between 12.5 and 20 for $n = 3$ (using Equation (9.4) for three-dimensional scaling).The regression coefficients lie between 0.85 and 0.94, indicating that a linear fit is not all too bad. A few remarks should be made. First of all, it is debatable whether the results from the smallest size "A" should be included in this analysis. They should probably not be included considering the arguments given in the previous section. Secondly, the Weibull modulus varies with material compositions, which is of course expected. The approach becomes a bit more rational than applying SEL and MFSL, because the actual material composition is now taken into account. For a certain concrete composition a CT scan will deliver the input parameters for the lattice model, specifically the particle geometry and particle distribution, which may then be used to compute the size effect. The analyzed range is still too small; we realize that in full, but the approach seems promising. Given time and improvements of the model, specifically the solver, and progress in computing infrastructure, will ultimately allow for analyzing larger size ranges. There is no need for making assumptions about the small-scale and large-scale asymptotes, as is the case in SEL, for instance. Refinements are in order, however. The linear approximation of Equation (9.9) is just that, an approximation, and the final model is more complicated. The last word has also not been said about the fracture law in the lattice model, and it is expected that there is room for improvement.

The lattice approach for estimating the expected size effect is for two other reasons also quite useful. In the first place, the load-displacement diagrams are computed, which helps to judge brittleness. Secondly, the crack-size distribution becomes available, which may be used as a starting point in the proposed 4-stage fracture model; see Chapter 10. SEL is only capable of fitting the nominal strength at peak; MFSL also gives access to stress–deformation behavior, but the lattice model (or any other micromechanical model) will deliver the full content. This is exactly what was meant in the beginning: a good fracture model/theory has the size effect implicit and no separate rule is needed. In the next section we give an example of how the crack-size distribution may be retrieved from a lattice analysis.

9.5 Damage Distribution in Structures of Varying Size

The strength of a material or structure depends on the (micro-) crack growth and the crack-size distribution just before reaching the maximum. This is nothing new, although one might ponder whether the point is well taken within the concrete fracture community. As may have been clear from the macroscopic models presented in Section 2.4, the main attention has focused on softening, which is the regime of unstable macrocrack growth (see also Chapter 10). The macroscopic models are all based on incorporating softening under the assumption that microcracking occurs. This erroneous assumption (made, for example, by Bažant and Oh 1983), somehow keeps dominating fracture studies, in spite of an abundance of contradictory experiments; see Chapters 2 and 6. The softening regime represents the behavior beyond peak strength, and thus the material/structure has failed already, unless, as we are accustomed to in concrete fracture experiments, special measures are taken to stabilize macrocrack growth (see Appendix 3). We return to these matters in Chapter 10.

In the field of ceramics there has been interest in describing the microcrack population for several decades; see, for instance, the work by Jayatilaka and Trustum (1977), Danzer (2006), and Danzer et al. (2001, 2007). For example, Jayatilaka and Trustum provide a probability density function $f(a)$ of semi-crack length a $(a \geq 0)$:

$$f(a) = \frac{c^{n-1}}{(n-2)!} a^{-n} e^{-c/a} \tag{9.10}$$

In this equation c is a scaling parameter and n determines the length of the tail of the probability density function, where the largest cracks are found. Jayatilaka and Trustum (1977) suggest that only a crack of sufficiently large length may lead to global failure, and the probability for failure of a single crack follows from

$$F(\sigma) = \int_{\kappa}^{\infty} \left[1 - \frac{\kappa}{a} \right] \frac{c^{n-1} a^{-n} e^{-c/a}}{(n-2)!} da \tag{9.11}$$

with $\kappa = K_{Ic}^2 / (\pi \sigma^2)$; that is, LEFM principles are assumed to be valid and incorporated in the formulation. Danzer, Lube, and Supanic (2001) summarize the situation very clearly. For homogeneous materials under uniform tension, with a flaw population described by a negative power law similar to Equation (9.10),

$$g(a) = g_0 \left(\frac{a}{a_0} \right)^{-r} \tag{9.12}$$

a Weibull distribution according to Equation (9.1) is found. Two parameters are of importance in Equation (9.12), namely the power r and the coefficient $g_0 a_0^r$. The parameter a_0 is a normalizing length. It is assumed that cracks are not hindered by obstacles (such as aggregates in the case of concrete), and also that no crack interactions occur. Danzer et al. (2001, 2007) continue on the path set out by Jayatilaka and Trustum and extend the formulation to flaw populations with any size distribution (thus not just a population following Equation (9.12)) and for specimens with inhomogeneous flaw size distributions. Again, the Weibull hypothesis (weakest link) applies, and crack interactions are ignored. After including the LEFM criterion,

$$a_c = \frac{1}{\pi} \left(\frac{K_{Ic}}{Y \cdot \sigma} \right)^2 \tag{9.13}$$

using Equation (9.1), and the relation suggested by Freudenthal (1968) relating the probability for failure to the mean number of critical flaws $N_{c,S}(\sigma)$,

$$F_S(\sigma) = 1 - \exp[-N_{c,s}(\sigma)] \tag{9.14}$$

Equation (9.12) can be rewritten as follows (Danzer et al. 2007):

$$g(a) = \frac{m}{2V_0} \left(\frac{K_{Ic}}{Y \sigma_0 \sqrt{\pi}} \right) a^{-n} \tag{9.15}$$

where σ_0 is a characteristic strength and V_0 the related volume (as in Equation (9.1)), $m = 2r - 2$ and Y is the geometrical function as described in classical LEFM (see Section 2.2; here we use a different notation to circumvent confusion with the symbols used for the flaw-size distribution). Note that $N_{c,S}(\sigma)$ follows from integrating over the defect density from the lower bound $a_c(\sigma)$ to infinity, which in turn is related to $g(a)$. The crack density $g(a)$ is shown as a function of semi-crack length a in Figure 9.15. At a critical crack size $a_c(\sigma)$ failure will be inevitable for a given external stress σ. With increasing σ the critical crack size decreases, and the number of "destructive flaws" increases (shaded area). Because it is assumed that no crack interactions occur, damage has to be sparsely distributed, and for that case Danzer et al. (2007) show that

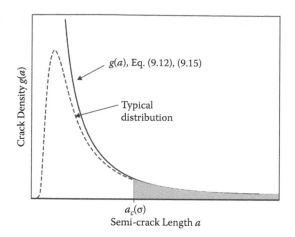

FIGURE 9.15
Probability density of crack sizes a. The function $g(a)$ shows the distribution necessary for a Weibull distribution. (From Danzer et al. 2007. *Eng. Fract. Mech.*, 74: 2919–2932. With permission from Elsevier.)

the probability of failure of ceramics can be described with a Weibull modulus $m = 15$. The above exposure is short here because the equations are not used in the remainder of this chapter. The reader interested in these matters is referred to the various cited publications by the group of Danzer.

The important question to be asked is of course if such an approach would apply to concrete fracture as well. The results from the lattice simulations in Chapter 6 (uniaxial tension), as well as to those presented in this chapter, reveal that the microcrack distribution certainly is not sparse, and crack interactions will be the rule rather than the exception. Moreover, the material "concrete" is far from homogeneous and crack-arrest and crack-deflection caused by large stiff aggregates appear anywhere. What can be done, however, is investigate how the actual crack-size distributions develop from lattice simulations. Because cracks are basically "removed elements" from the lattice, it takes some additional effort to compute the crack-size distribution. The final result is, as expected, quite similar to the density function plotted in Figure 9.15. In a lattice simulation all crack interactions are automatically incorporated: we simply compute the structure as it is and base our conclusions directly on the results obtained. A simple procedure to count the cracks from lattice simulations was recently developed by my student Man (2009). Figure 9.16 shows the procedure. In Figure 9.16a the cracked prisms are shown at a certain loading stage. The cracks can be isolated, as done in Figure 9.16b, where just the removed elements are plotted. The connectivity between these removed elements is hard to determine. Therefore a three-dimensional cubic grid is projected over the prism's volume. Cubes have a side length equal to the length of a single lattice beam. The cubes containing a crack are separated from intact cubes; see Figure 9.16d. In three dimensions

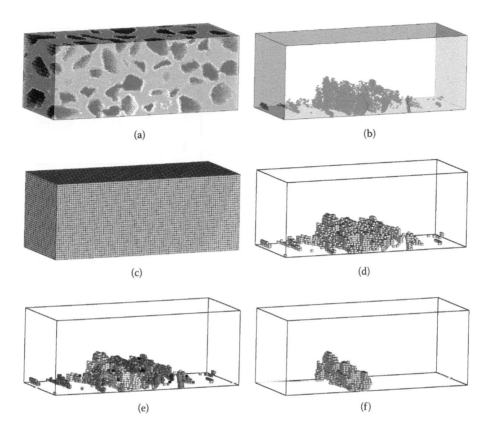

FIGURE 9.16
From a cracked structure (a) the removed elements representing cracks can be isolated (b). The connectivity of the elements is not clearly described; therefore, a 3D cubic grid is super-imposed on the structure, with base length equal to a single lattice beam length (c). The cubes containing a crack are plotted in (d). Considering the connectivity between neighboring cubes, crack clusters containing more than one cube can be identified and marked using different grayscales or colors (e), and it is possible to isolate the largest cluster (f). (After Man. 2009. *Analysis of 3D Scale and Size Effects in Numerical Concrete.* With permission of Dr. Hau-Kit Man.)

a cube has 26 neighbors, which reduces to 18 when the diagonal connectivity is excluded (or even less when edge cubes are considered). Subsequently all the clusters of connected cubes can be determined and visualized using a grayscale coding or color coding (see Figure 9.16e). From here it is straightforward to retrieve the largest crack cluster (Figure 9.16f). Note that these crack clusters are not necessarily curved planes; there may be branching, making the crack cluster a complex 3D object. There is much room for improvement at this point, but as a first approximation the results can be well used.

In Figure 9.17 results are shown from a series of analyses of lattice analyses on prisms of varying size, subjected to 3-point bending. The example relates to a crushed-aggregate density $P_k = 40\%$. We only show results for the "C" and "D"-size specimens here inasmuch as their volumes are larger than the RVE,

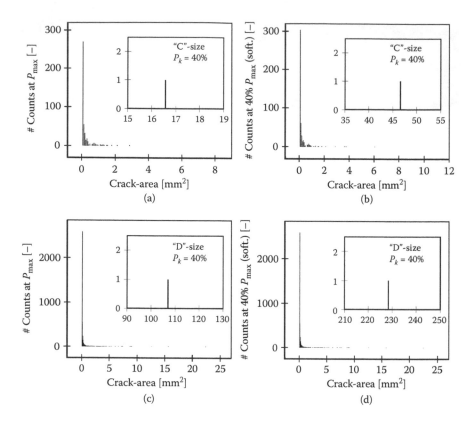

FIGURE 9.17

Distribution of crack-cluster areas for size "C" and "D" prisms containing crushed aggregates with $P_k = 40\%$. Shown are the results at P_{max} (a) and (c) and at 40% P_{max} in the softening regime (b) and (d). The top-row diagrams relate to the "C"-size prism (a), (b), the bottom two diagrams (c), (d) to the "D"-size prisms. The size of the largest crack-cluster can be read from the small inset diagrams. (After Man. 2009. *Analysis of 3D Scale and Size Effects in Numerical Concrete.* With permission of Dr. Hau-Kit Man.)

and variations due to accidental aggregate locations are reduced compared to the "A" and "B"-sizes (the complete range can be found in Man 2009 or in Man and Van Mier 2011).The area of the individual crack-clusters is shown along the horizontal axis, the counts for each size in the vertical direction. The crack-cluster distributions are shown at peak-load (Figures 9.17a and c), and in the descending portion of the load-deformation curve at 40% of peak-load (Figures 9.17b and d). The distributions are all quite similar: clearly the small clusters outnumber the larger ones. Due to the heterogeneity of the material structure, many isolated cracks develop. Only those under the most critical loading conditions (in combination with the local material strength) will grow to larger-sized clusters, and perhaps even develop into the last critical crack that leads to complete failure of the prism in the postpeak regime. Large differences start to appear when the largest crack-cluster is considered.

The largest clusters are shown in the inset of each figure. Going into the postpeak regime we see a negligible increase of small-sized crack-clusters, whereas the largest crack keeps growing: from 16.5 mm² to 46 mm² for the "C"-size prism and from 108 to 228 mm² for the "D"-size. Translated to the fraction of the total cross-sectional area for the type-"C" and type-"D" prisms we obtain the following results: at peak 19% and 25% cracked area for "C" and "D," respectively, and in the postpeak regime at 40% of P_{max} these values increase to 48% and 55%. Apparently the largest crack is the critical crack that will lead to separation of the prism in two parts at the end of the softening curve. Thus, again, as we have seen earlier for tensile loading (Chapter 6), the softening regime is dominated by the growth of a single large crack-cluster. Remember that this is probably not a single undulating crack but a complex 3D object complete with branches and possibly crack-face bridging (see Section A4.1). For the application in a simple macroscopic fracture model presented in Chapter 10, it is necessary to translate the above crack-area data to an effective crack length in two dimensions. This is attempted in Figure 9.18. For each crack-cluster the largest extension in the direction of the height of the prisms has been determined, and plotted for various loading stages for the variety of concretes investigated.

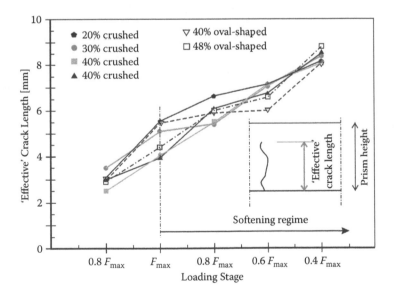

FIGURE 9.18
The effective crack lengths shown have been determined from a two-dimensional projection of the largest crack-cluster (defined according to Figure 9.16f) as indicated in the inset. In the main diagram the effective crack length is shown for the various concretes with different content of crushed or oval-shaped aggregates at five different loading stages. Results are for "C"-size prisms. (From Man and Van Mier. 2011. *Cem. Conc. Comp.,* 33: 867–880. With permission from Elsevier.)

The important issue at hand is how this crack-cluster should be interpreted and used in a global macroscopic model. In the spirit of the effective crack models developed in the past (see, e.g., Karihaloo and Nallathambi 1989), it seems a good idea to return to classical fracture mechanics. There are some differences, however. First of all, before reaching peak-load a whole family of cracks develops in the prisms. These cracks can be understood as a means to weaken the structure, that is, to prepare it for global failure. The critical stress intensity factor that would be a necessary input parameter in any classical fracture mechanics approach is therefore reminiscent of an R-curve model, where the stress-intensity factor keeps increasing until it reaches a maximum value. From the beginning of an analysis it is not straightforward to determine the first critical crack. Together with my former student Prado, I looked into the problem, trying to trace back the critical crack from the final crack map. Usually a small crack was found that did not seem very special compared to its neighbors, and it is very likely that the crack map has to be judged simultaneously with the distribution of local material strength. At this point we leave these results for what they are. In the next chapter they are of good use in the development of the 4-stage fracture model.

Perhaps a last remark is appropriate. Uniaxial tension and 3-point bending are among the simplest possible loading situations one may investigate. The subject matter (size effects) is rather complex, and so it is no surprise that the main attention goes to these cases for the moment. All cracks are more or less oriented perpendicular to the direction of the principal tensile stress. Under more complex loading situations, such as torsion (and multiaxial compression), the stress-direction may rotate affecting the orientation of microcracks. These can be measured using stereological principles (see, e.g., the applications in Stroeven 1973 and Van Mier 1985). The full picture has to include the crack orientations with respect to the loading direction; for the simple loading cases discussed here and as a first approximation we can refrain from that.

9.6 Concluding Remarks

Size effect is a major issue in fracture studies of concrete materials and structures. It may be obvious by now that separating the material effects from structural behavior is no simple task when it comes to describing fracture. This means that phenomenological models directly based on experimental results cannot be used. The best way is still deriving the size effect directly from a macroscopic fracture theory, and no special rule must be devised. This already eliminates SEL and MFSL as possible candidates. Another reason not to use either SEL or MFSL is that the assumed asymptotic behavior can never be verified, as discussed to some extent in this chapter. Micromechanical modeling is a great tool, but cannot be used either because of the lengthy computations needed, whereas it is still questionable if truly large-scale analyses can be performed at all considering the demands of the various length scales that must be included ([μm]- to [mm]-scales) and the sheer size of the structures ([m]-scale and larger) that need to be analyzed. At the moment the best bet would be using Weibull theory, which has the clear advantage that it is based on sensible assumptions. Expanding the statistical theories in a way comparable to developments in the field of ceramics seems a workable approach. Again, micromechanical modeling is quite useful here because one may obtain insight to how the crack population evolves throughout the loading history. In doing so, it is best not to trust any model used, but combining the analyses with sound experiments.

10

Four-Stage Fracture Model

Thus far we have concentrated on existing model approaches for the fracture of concrete. As an extension of the cohesive model developed by Barenblatt and Dugdale in the late 1950s/early 1960s, the Fictitious Crack Model was developed by Hillerborg and co-workers in 1976. The similarity was taken very loosely because the fracture process zone is very small for the metals that were considered initially, in comparison to the large (global) size of the process zone for concrete materials. By starting out from this similarity the original "local" model for the fracture of plastic metals is transformed into a "global" model where the process zone may even exceed the size of the entire structure or specimen. As a result, the softening parameters needed in the model are size-dependent (see Figures 9.5, 9.6, and 9.8) and also are heavily influenced by the boundaries (see Figure 2.11). It is obvious that, even though the model is used extensively for the moment, albeit sometimes with rather debatable outcomes, eventually a better and more reliable model needs to be developed. One might think that a micro- (or rather meso-) mechanical model may be the answer to many of the questions. Yet, one must accept lengthy analyses in that case (see, e.g., Appendix 1), and also at a smaller level of observation the same uncertainty will remain about the validity of the parameters used. We return to these matters in Chapter 11.

The intention here is to develop the framework for a possible successor of the fictitious crack model. This is done for tensile fracture (mode I in classical terminology), but at the end it is shown that the same approach may well be used for another important failure mode of concrete (and for related geomaterials), namely compressive fracture. To set the stage, in Section 10.1 we first review the fracture process of concrete subjected to uniaxial tension, as it emerges from all the experiments described thus far and the various lattice analyses done. In Section 10.2 the new framework is explained, and an example of its application is given. Next, in Section 10.3 the fracture process in compression is summarized, citing the most important evidence for the view given, and the similarities and differences with the approach proposed for uniaxial tension are outlined. The model is barely fit for practical applications. Yet it is considered as an important tool (or guideline) for future experimentation and a better understanding of the fracture of concrete. For other materials the model appears to apply as well. This is discussed in Section 10.4.

10.1 Fracture Process in Uniaxial Tension

In Chapters 2, 6, and 9 quite some attention was given to mode I fracture of concrete at different sizes/scales. Because of the low tensile strength of concrete, mode I fracture will usually prevail, unless high confinement is applied, fibers are added to the concrete, or very fast loading rates are applied. In most other situations global failure is always initiated by the nucleation and growth of mode I cracks. The crack orientation can change when the loading on the structure is redistributed during crack propagation (see, e.g., Chapter 7 dealing with mixed-mode fracture). The fracture process understood from many experiments and numerical simulations at the level of the particle structure of concrete (lattice model and related micromechanical models) conveniently can be subdivided into four stages:

(0)   Elastic stage
(A)   (Stable) Microcracking
(B)   (Unstable) Macrocracking
(C)   Bridging

As can be seen from Figure 10.1 the first linear-elastic stage (0) starts from the origin. Next, when the first microcracks nucleate and grow, stage (A) leads to a curved prepeak stress–deformation diagram. This continues up till peak stress, when, with the growth of the single most critical microcrack, unstable macrocrack growth (stage B) leads to the steep part of the softening curve. Finally, the macrocrack growth is stabilized, to some extent, by bridging (stage C). Stages (B) and (C) very likely overlap as indicated in Figure 10.1: bridging starts almost directly after the macrocrack propagates. In stage (B) the material specimen (or structure) is basically failed. During this stage the macrocrack is unstable, but may be stabilized depending on the composition of the material (tensile or compressive failure) or the presence of confinement (compressive failure only). Let us now examine the evidence for each of these four stages in the following subsections.

10.1.1 Stage (0): Elastic Behavior

Immediately after the beginning of loading the material or specimen may behave elastically: if no initial defects are present the loading and unloading behavior will be identical. Initial defects can be caused, for example, by drying or wetting, leading to differential shrinkage or swelling, respectively, or differential expansion caused by nonuniform cooling after the cement has heated due to the hydration process, and so on. Depending on the composition of the material, and the local properties of the constituents the initial

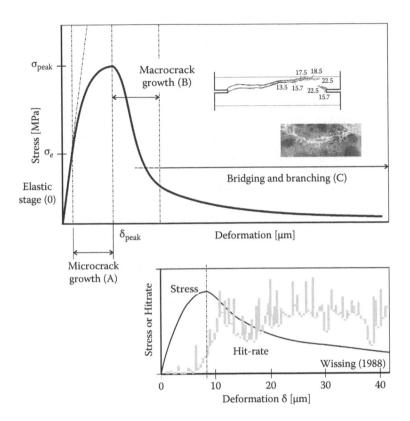

FIGURE 10.1

Schematic representation of the fracture process of fracture of concrete under uniaxial tension, along with the load-deformation diagram. The AE data shown below the diagram were obtained by Wissing (1988, *Acoustic Emission of Concrete*), the photoelastic results of macrocrack growth are from Van Mier and Nooru-Mohamed (1990, *Eng. Fract. Mech.*, 35(4/5): 617–628; see also Appendix 2, Figure A2.1), and the aggregate bridging is from Van Mier (1991. *Cem. Conc. Res.*, 21(1): 1–15; *Fracture Processes in Concrete, Rock and Ceramics*, 1991; see also Section A4.1).

response is either linear-elastic or nonlinear elastic. In general it is assumed that linear elasticity dominates stage (0).

10.1.2 Stage (A): (Stable) Microcracking

As has become clear from the meso-level analyses in Chapters 6 and 9 before peak stress is reached many small microcracks will develop along the interface between aggregates and the cement matrix; see, for instance, Figure 6.3: at peak load a considerable number of microcracks have developed, the amount being determined by the aggregate content. Of course the lattice is a rather simplified model, but the results (foremost the crack patterns and their temporal appearance during the loading process) are in agreement with experimental data: from optical microscopy with or without the use of fluorescent dye (e.g., Stroeven 1973), and from dyeing experiments with ink at various loading stages up to and just beyond peak (Hsu et al. 1963). In these experiments compressive loading was applied, which is just a bit different from tensile loading, but not in principle. Indirect evidence for pre-peak (micro-) cracking in tension can be derived from acoustic emission measurement carried out, for example, by Wissing (1988). He tested small prisms loaded with deformation control in uniaxial tension. Some results by Wissing are included in Figure 10.1. Before maximum stress is reached, considerable AE activity is monitored (number of counts, or "hit-rate"). Shah (1990) reports similar results on single-edge notched tensile specimens: the AE count is significant at peak stress. Using a technique called ESPI (electronic speckle pattern interferometry), Meda (2003) showed early microcracking on Serena Sandstone. Note that in his specimens notches were machined, which may have affected microcrack growth at earlier than normal loading (see, e.g., the uniaxial tests on granite by Labuz, Shah, and Dowding (1985), who showed a decrease of tensile strength and a larger prepeak curvature when notches were used). Many tests have been done under bending, but such results are affected by stress- and strain-gradients caused by the specimen geometry and the type of loading. Unfortunately the resolution of x-ray tomography is not sufficient at the moment to reveal the microcracks, which are expected to have submicron openings at the most. It should be mentioned here that when small cracks nucleate or grow, elastic energy is released as explained in Section 2.2, Equations (2.17) through (2.20). Summing the energy released by all microcracks leads to an increasing curvature of the stress–deformation curve as sketched in Figure 10.2. The increasing curvature in the prepeak regime can easily be identified from experiments; see for example the σ–δ diagram measured by Wissing shown in Figure 10.1.

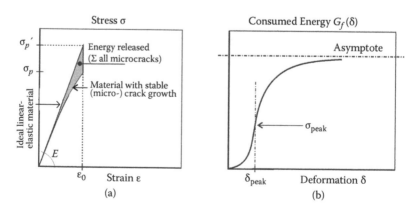

FIGURE 10.2
Curvilinear stress–deformation response in stage (A) caused by energy release from the nucleation and stable growth of microcracks (a), and total energy $G_f(\delta)$ consumed in the complete fracture process (b).

The energy release can be interpreted as the deviation from "ideal" linear-elastic behavior. The area below the stress–deformation diagram is the amount of energy needed to create the cracks, and equals the fracture energy G_f postulated by Hillerborg and coworkers in the Fictitious Crack Model. The prepeak energy is usually subtracted from the total energy as the definition in the FCM is limited to crack growth during the softening regime only. The complete prepeak curve is commonly linearized, which is equivalent to eliminating the crack initiation stage. If the area under the stress–deformation diagram is plotted against the axial deformation, an S-shaped curve is found, with the point of maximum stress located at the bend-over point, as shown in the inset of Figure 10.2. When the maximum energy is reached, the specimen is broken into two parts, at the end of the softening regime.

10.1.3 Stage (B): (Unstable) Macrocracking

Demonstrating macrocrack growth is much easier. A macrocrack has length and depth of the same order of magnitude as the dimensions of the specimen or structure that is considered. Beyond peak crack-widths are substantially larger, and after the steep part of the softening curve have become as large as 30–50 µm. Such cracks can be seen with the naked eye, but with tools such as dyeing, impregnation, photoelastic coatings, and ESPI the cracks are even more easily detected. In Figure 10.3 we show a result obtained by Labuz et al. (1985) on charcoal granite, which behaves almost identically to

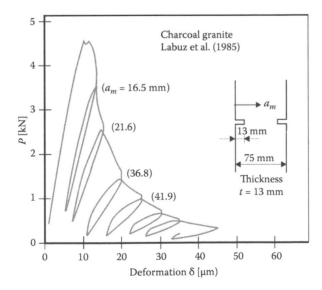

FIGURE 10.3
Measurements of optical crack length a_m in an uniaxial tension test on granite between freely rotating loading platens. (From Labuz, Shah, and Dowding. 1985. *Int. J. Rock Mech. Min. Sci. & Geomech. Abstr.*, 22(2): 85–98. With permission from Elsevier.)

concrete in tension. The test was done on a double-edge notched plate loaded in deformation-control between freely rotating loading platens. Several loading-cycles were applied in the postpeak regime showing that some quite irreversible deformation occurs. The optical crack length a_m, measured from the side of the specimen as indicated (thus including the initial notch depth of 13 mm) is shown between brackets along the softening curves. At a load of 1 kN the ligament between the two notches (49 mm width) is for the largest part broken: the crack length from the tip of the notch is $41.9 - 13 = 28.9$ mm. No secondary cracking was reported. Indeed, considering the results from the lattice analyses in Section 6.1.4.1 no secondary crack should develop when freely rotating boundaries are used.

In Figure A2.1 in Appendix 2 the results are shown of deformation controlled uniaxial tensile tests on a double-edge notched (DEN) specimen of size 200×200 mm^2 and a thickness of 25 mm; the notch depth was 25 mm. A thin (1 mm) photoelastic coating was glued to the front and the back side of the specimen in the area where the crack was expected to grow. The main crack started to propagate just beyond peak, and because (contrary to the tests in Figure 10.3) fixed boundaries were used; two overlapping crack branches were found. The stage at 22.5 μm average crack opening (measured with LVDTs attached to the sides of the specimen; measuring length 35 mm) is also included in the main diagram of Figure 10.1. At this crack opening, the steep part of the softening curve changes into a shallow tail. It appears that most of the cross section is cracked, although some doubt remains because with the available equipment only one side of the specimen could be viewed during the experiment. Impregnation experiments at subsequent crack openings in the softening regime reveal that the macrocrack starts from the notch and extends farther along the specimen surfaces, leaving something like an "uncracked" core in the specimen, as shown in Figure 10.4. Further explanations and in-depth analyses of these results can be found in Van Mier (1997).

From the results included here the growth of macroscopic cracks in the softening regime cannot be doubted. Assumptions made in the Fictitious Crack Model and in the crack-band model (see Section 2.4) are hereby refuted. The hypothesis of a microcrack cloud in the softening regime does not appear to be correct, and an alternative model is required. This seems the "fast" conclusion from these observations, but it is very likely correct. There remain a few loose ends, namely the tail of the softening curve is not explained, nor the uncracked areas in Figures 10.4b and 10.4d at the 50-μm crack opening. Bridging (and branching) is part of the explanation; the other parts derive from typical structural behavior caused by environmental effects and structural boundary conditions. We return to these matters in the next section.

10.1.4 Stage (C): Crack-Face Bridging

The crack overlap that is visible in Figure A2.1d (Appendix 2) is caused by structural effects; the reason was given in Section 6.1.4.1. When the crack starts

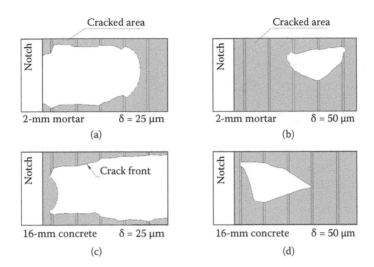

FIGURE 10.4
Macrocrack growth measured from impregnation experiments on single-edge notched specimens of 2-mm mortar (a) and (b), and of 16-mm concrete (c) and (d). The figures show the cross-section along the plane of the notch, perpendicular to the tensile loading direction. The gray-shaded area is cracked. (From Van Mier. 1991a. With permission from Elsevier.)

to propagate from the left notch, a gradually increasing load-eccentricity develops, which leads to a bending moment that hinders the crack from propagating (because nonrotating loading platens were used). Only when the tensile strength at the right notch is exceeded is the former symmetric situation restored, but as a consequence a second crack has grown into the specimen. The result is an increase of the fracture energy by as much as 30%, which can easily be explained from the increasing fracture area. At a smaller scale, that is, at the scale of the individual stiff and strong aggregate particles the same mechanism develops. The aggregates deflect cracks that originate at the interface between aggregate and matrix, and intact ligaments remain between overlapping crack-tips. The overlapping crack mechanism (sometimes also referred to as a *handshake crack*) is recognized in many materials and under a variety of structural conditions. In an interesting paper by Sempere and Macdonald (1986) crack overlaps are shown in a variety of materials at different length scales, such as at km-scale in the earth crust (e.g., the fault system in the African Rift Valley) to the μm-scale in ceramics. The reason for development of the crack overlaps is not always clear and sometimes subject to debate. The mid-Atlantic ridge is a place where two continents are drifting away from each other and magma wells up. The fracture zone that develops shows the same crack overlaps as those in Figure A4.2 for concrete and in Figure A4.6 for hardened cement paste. The impregnation tests that were the basis of Figure 10.4 yielded many examples of overlapping cracks, and an effect on the carrying capacity in

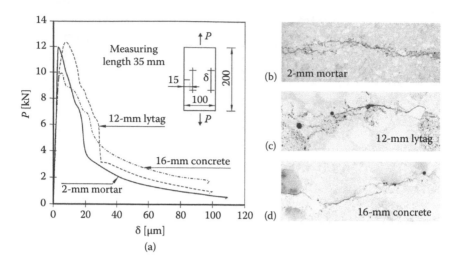

FIGURE 10.5
Load-deformation diagrams for three different concretes: 2-mm mortar (full line), 16-mm gravel concrete (dotted line), and 12-mm lytag concrete (dashed line). In (b)–(d) three examples of crack overlaps in these three materials are shown. In the 2-mm mortar and 16-mm concrete cracks are all located at the interface between aggregate and cement-matrix; in the lytag concrete cracks runs straight through the porous (and rather weak) lytag particles (speckled areas) but along the interface of the larger sand grains. (Adapted from Van Mier. 1991b. *Fracture Processes in Concrete, Rock and Ceramics.*)

the tail of the softening diagram could be established. The result is shown in Figure 10.5. Here we show load-deformation curves from uniaxial tensile tests on three different types of concrete: 2-mm mortar, 16-mm concrete, and 12-mm lytag concrete. The lytag concrete is a lightweight concrete containing soft and light aggregate particles in the size range between 4 and 8 mm. In addition, sand fraction is present up to 4-mm grains. As can be seen from Figure 10.5 the tail of the softening diagram increases with increasing aggregate size, and it seems that the size of the ligament between two overlapping crack-tips is a good explanation for the increasing carrying capacity in the tail.

In Figure 10.5 we also show clear examples of crack-overlaps in these concretes. In the lytag concrete (Figure 10.5c) the main crack propagates through the porous lytag particles, and forms crack-face bridges around the stiffer sand particles (see, e.g., the cracking around the white sand particle to the right in Figure 10.5c). Another example of crack-face bridging is included in the Appendix, Figure A4.2b. Finally, in Figure 10.6 the final stages of the failure in the area of two overlapping crack-tips are shown. Again, this is an example of cracking in lytag concrete; observations were made at the surface of the specimen using a long-distance optical microscope (Questar QM-100). The images were taken almost at the end of the tail, just before complete rupture. The stages are labeled along the tail of the softening branch in the

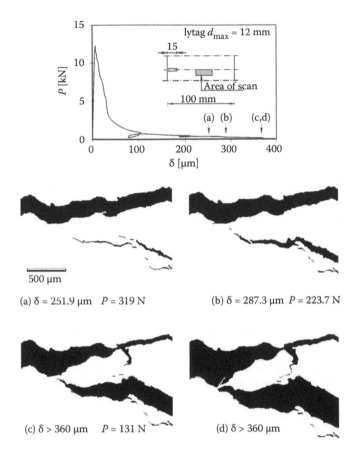

FIGURE 10.6
Four snapshots showing failure of the ligament between two overlapping crack-tips in a lytag concrete tensile specimen. The four stages (a) through (d) are indentified along the tail of the *P–δ* diagram. (After Van Mier. 1991b. *Fracture Processes in Concrete, Rock and Ceramics.*)

P–δ diagram. In the stage labeled (a), at a deformation of 251.9 μm we see the lower crack-tip approaching in the wake of the upper crack. The upper crack-tip is located to the right and not visible in the image. In stage (b) we clearly see a widening of the lower crack, and subsequently in stage (c) part of the ligament breaks off when a flexural crack develops from the top of the ligament. Instantaneously the upper crack-tip closes, and the path of the main crack is now the wide crack along the bottom of images (c) and (d). In the process it was observed that some debris fell down, which may be an explanation for the irreversible deformations that are observed when cyclic loading is applied as in Figure 10.3. It is interesting to note that final failure of the ligament is through bending (equivalent with mode I crack growth), which explains the continuation of AE activity along the entire softening curve as can be seen from Wissing's results in Figure 10.1.

The last part of the failure process is, in comparison with the steep portion of the softening curve just beyond peak-stress, extremely stable. A servo-controlled loading device is basically not required any longer; just a high stiffness of the loading frame would suffice. The bridging mechanism through the formation of a hand-shake crack follows quite naturally from lattice analyses, and other micromechanical models, as was shown in Figures 6.2d and 6.2h. Finally, the crack-overlaps are all shown here in two dimensions. In reality the overlaps are three-dimensional "flaps"; they do not necessarily have the same geometry over the specimen thickness. In microtomography experiments we recently observed bridging in hardened cement paste, and because the specimens were scanned in three dimensions the full geometry of the ligament connecting the two crack-faces could be established; see Figure A4.6c. The interested reader is referred to Van Mier (1997) for further details.

10.2 Four Fracture Stages, yet a Continuous Process

The four stages form the basis for a new macroscopic model for tensile fracture of concrete, which is shown schematically in Figure 10.7. The model is very simple in principle: it just follows the physical mechanisms elucidated in the previous section. It was recognized that the same four stages appear in practically all materials that fail in a quasi-brittle manner, through the development of one or more cracks, as explained in Van Mier (2004a).

In stages (0) and (A) in the prepeak regime the application of a continuum model seems the best approach. This is not different from any other current approach in engineering. In simplified form:

$$\sigma = \frac{F}{A} \quad \text{and} \quad \varepsilon = \frac{\Delta l}{l} \tag{10.1}$$

where A and l are the cross section and length of a tensile test-specimen, respectively. Assuming uniformity in the distribution of stresses and strains allows for this approach. In general one would simplify the prepeak behavior from a curvilinear function to a simple linear function up to peak stress (i.e., assume linear elasticity (LE)). Note that neither the linear elastic nor the nonlinear elastic (NLE) continuum approach for the prepeak regime delivers the parameters needed for the postpeak stage, namely the size of the (largest) critical microcrack. So, if LE or NLE is used for modeling prepeak behavior, the size of the critical flaw must be estimated using another methodology. Micromechanical modeling seems here the best choice, for instance, the

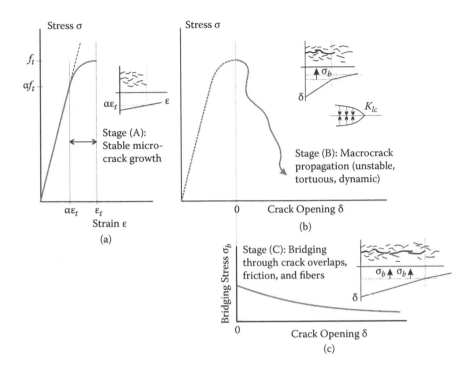

FIGURE 10.7
Universal four-stage fracture model for concrete and various other materials such as glass, metal, rock, fiber-reinforced composites, ice, ceramics, and so on. After a linear-elastic regime, between the origin and αf_t in (a), the prepeak microcracking in Stage (A) is modeled using a classical continuum approach. After the peak-stress has been reached, a phase-transformation occurs and a classical fracture mechanics approach is needed to model macrocrack growth (b). The macrocrack is not stress-free due to the development of crack-face bridges or friction and possibly the presence of fibers bridging the crack, which is incorporated via a bridging stress as shown in (c). Note that the bridging stress relates to the tail of the softening diagram only, and not the entire softening curve as in cohesive fracture models. (From Van Mier. 2008. *Engng. Fract. Mech.*, 75: 5072–5086. With permission from Elsevier.)

lattice-type analyses that led to estimates for the largest (and likely also the critical) flaw in 3-point bend beams shown in Figure 9.18.

As mentioned, a different approach is needed as soon as peak strength is reached. At this point a change occurs that can best be characterized as a phase-transition. Because at peak a macroscopic crack starts to propagate, applying continuum theory hardly seems appropriate. Due to the growth of a macrocrack, new boundaries are created in the specimen/structure, and trying to average the crack-width to retrain the classical state variable strain seems just impossible. Nor does it seem fruitful separating strains in different contributions as, for instance, attempted in higher-order continuum theories (see, e.g., Iacono 2007, who, working with a gradient plasticity model, concluded that for modeling the behavior of specimens of different size the same parameter set could not be used, including the length-scale parameter, which appears in such models). The length scales needed in higher-order continua are impossible to measure directly from experiments. A sound physical basis is lacking and one may wonder why advocates of continuum theory do not just accept the limitations of their approach and let nature decide which method to apply. Classical fracture mechanics, with some small amendments, appears to be the best option: it comes closest to the observations. Thus, directly beyond peak, the classical criterion from LEFM may be used

$$K_I = \sigma\sqrt{\pi a}\, f(a.\theta) = K_{Ic} \tag{10.2}$$

for describing crack propagation. Some clarifying remarks are needed. First of all, the crack propagates in a material with numerous smaller cracks, and a distinct heterogeneity. This means that the critical stress intensity factor must be determined under the same circumstance. It implies that somehow K_{Ic} must be derived from a test specimen at peak-load. At that moment a test is rather unstable, and it will be a challenge to carry out such measurements successfully. Moreover, as mentioned before, the main macrocrack is not a simple flat plane, but has numerous undulations, and branching and bridging occur frequently as a result of the existing microcracks and the heterogeneity of the material. Therefore it is essential to make an amendment to the classical formulation, and to include a bridging stress. The bridging stress is equal to the tail of the softening diagram exclusively. It is best measured under fixed boundary conditions where the crack-propagation stage (B) is clearly separated from the bridging stage (C). The specimen should be relatively wide, allowing for unhindered extension of the macrocrack, but not too large to create instabilities in test control when the second crack branch from the opposite notch develops. The bridging stress in the tail can be assumed as an almost uniformly distributed stress keeping the crack-faces somewhat together (see Figure 10.8). This is quite different from the cohesive model developed by Dugdale/Barenblatt and its counterpart for concrete developed by Hillerborg and coworkers; namely it is not assumed that stress-intensities from the farfield stress σ and the bridging stress σ_b cancel at the crack-tip. The bridging stress depends on the material composition: the size of the aggregate used, the strength- and stiffness contract of matrix-, aggregate- and ITZ-phases in the specific concrete considered, and fibers in the mixture. The latter addition may have a significant effect on the magnitude of the bridging stress, as we discuss in Section 10.4. Because the stress-intensities do not match,

$$K_I + K_I^b \geq 0 \tag{10.3}$$

with K_I the stress-intensity from the farfield stress and K_I^b the (counteracting) stress-intensity from the bridging stress, is still positive and a stress-singularity keeps controlling crack propagation. In Figure 10.8 the situation is sketched for two cases: in Figure 10.8a for normal concrete where bridging originates from (stiff and strong) aggregates only, and in Figure 10.8b for fiber-reinforced concrete where bridging comes not only from aggregates but also from fibers. In the latter case substantially higher bridging stresses may be measured (see Section 10.4 and Section A5.2).

With this, the three main components of the model for tensile fracture of concrete are defined: prepeak (non-) linear elasticity (continuum behavior may

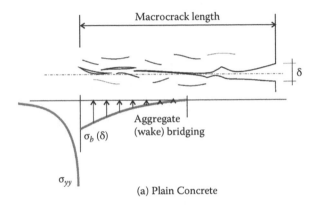

(a) Plain Concrete

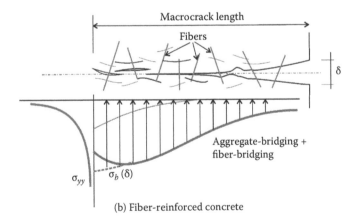

(b) Fiber-reinforced concrete

FIGURE 10.8

Concept of the bridging stress in the four-stage fracture model for plain gravel concrete (a), and for fiber-reinforced cement composites (b). Above each figure the geometry of the main crack and accompanying microcracks is sketched. The length of the macrocrack is measured from the stress-singularity to the free edge of the structure to the right. In the case of gravel concrete bridging stresses in the wake of the macrocrack are caused by aggregate bridging (see Section 10.1.4); in the case of fiber-reinforced cement composites, fiber bridging comes in addition to aggregate bridging. In each figure we see the stress-singularity at the tip from classical LEFM at the left; the diagrams to the right show the postulated bridging-stress distribution. (From Van Mier. 2008. *Engng. Fract. Mech.*, 75: 5072–5086. With permission from Elsevier.)

be assumed), postpeak macrocrack propagation (discrete model is required, viz. classical fracture mechanics) with bridging. The bridging stress is more or less uniformly distributed for ordinary concrete. For fiber-reinforced concrete (FRC) it may also be rather uniform until complete pull-out of the fibers. After pull-out the bridging stress rapidly decreases to zero, be it at rather large deformations (in average up to half the fiber length). Depending on the precise material composition the bridging stress may also gradually decrease for FRC; more about this in Section 10.4 and Section A5.2). The stress-singularity from the propagating main crack controls the process beyond peak.

As a simple example let us consider a beam loaded in 3-point bending. In Chapter 9 we computed the size of the largest crack in beams loaded in 3-point bending (Figure 9.18). The slenderness of these beams, expressed as $S/W \approx 3$ (S is the span, W is the height of the beam, and a is the notch depth), so an approximate choice may be to take the equations from Tada's handbook (Tada, Paris, and Irwin 1973) for $S/W = 4$:

$$K_I = \sigma\sqrt{\pi a}\, f\left(\frac{a}{W}\right)$$

$$f\left(\frac{a}{W}\right) = 1.090 - 1.735\left(\frac{a}{W}\right) + 8.20\left(\frac{a}{W}\right)^2 - 14.18\left(\frac{a}{W}\right)^3 + 14.57\left(\frac{a}{W}\right)^4 \tag{10.4}$$

This equation was derived by means of least square fitting and has an accuracy of 0.2% for $a/W \leq 0.6$. The crack-opening at the lower edge of the beam is given by:

$$\delta\left(\frac{a}{W}\right) = 0.76 - 2.28\left(\frac{a}{W}\right) + 3.87\left(\frac{a}{W}\right)^2 - 2.04\left(\frac{a}{W}\right)^3 + 0.66\left(1 - \frac{a}{W}\right)^{-2} \tag{10.5}$$

This expression has accuracy <1% for any a/W; see Tada et al. (1973). The computation is now relatively straightforward. From the onset of macrocrack growth at peak stress, the process is artificially stabilized by controlling the relative crack length a/W (this is essentially also what happens in deformation-controlled experiments, although strictly speaking in such experiments the crack width is controlled). The main assumption is that during crack propagation $K_I = K_{Ic} =$ constant. The remaining carrying capacity $\sigma_1 < \sigma_{peak}$ can be computed for each value of $a_1/W > a/W$ using Equation (10.4). The crack opening follows from Equation (10.5), and a point on the descending branch is found. It is rather straightforward to see that this will be a continuously decreasing function.

The lattice analyses of Chapter 9 suggested that for the various concretes with crushed or rounded aggregates the initial notch lies between $a_0/W = 0.39$ and 0.55. From Figure A2.2 in Appendix 2 it can be seen that the postpeak behavior of a SEN tensile specimen changes when the initial notch size varies. This is not different in this example: a shallower tail of the softening curve is found when the relative size of the initial notch increases. Or, in other words: beams with more pronounced prepeak cracking, leading to a larger relative crack at peak stress, will in general show a more shallow softening behavior. The increase of prepeak cracking will also result in a lower global strength. From experiments it is known that lower-quality concretes have a lower flexural strength and behave relatively more ductile. For any test that must result in stable softening behavior it is therefore recommended to cut a deeper notch. For example, in the bending test proposed by Hillerborg for determining the fracture energy in the fictitious crack model, a notch depth $a_0/W = 0.5$ is required, which is the right decision to guarantee test stability.

The bridging stress is not known for the beam analyses of Chapter 9. The only useful starting point is the increase of crack length in the softening regime as shown in Figure 9.18. One may use these results to determine the possible variation of the bridging stress via an inverse analysis. In the context of this chapter this result is not very important, and instead we discuss the advantages and disadvantages of the proposed approach, and see if it is possible to include compressive failure within the same simplified modeling strategy, as well as to ramify to different types of materials.

A final important remark should be made. It is obvious that the method is generally applicable for mode I crack growth. For each new structure, with its own specific size and shape, and boundary conditions the function $f(a/W)$ must be determined. Closed-form solutions for the geometric function are not always available and in many cases one has to use empirical formulations such as those given in Equations (10.4) and (10.5), or using approximations from numerically computed functions. As in the Fictitious Crack Model, the above model approach is empirical. The main question is whether we are now closer to physical reality, which fully depends on our interpretation of experiments. New results may force us to change to an alternative view.

10.3 Similarity between Tensile and Compressive Fracture

Describing (confined) compressive fracture in the same framework as in the previous section for uniaxial tension is quite straightforward. There are some important differences but the four stages can be recognized as well in the fracture process under (confined) compression. In Figure 10.9 we show the equivalent model for compression. It all starts with the linear elastic stage (0), assuming that damage from initial defects can be neglected. At a certain stress, onset of microcracking is observed. This may be between 30 and 50% of peak strength as classical experiments using either strain readings or acoustic emission measurements will tell.

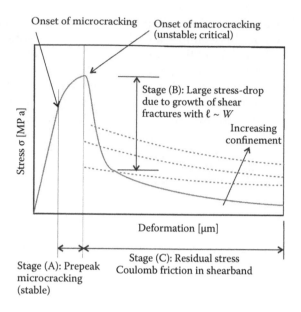

FIGURE 10.9

Four-stage fracture model for concrete subjected to (confined) compression. The main differences to the tensile model are that more stable microcracking leads to a more pronounced stage (A) and friction is the main component defining the residual stress level in stage (C).

Stage (A), stable microcracking, is larger than its counterpart in tension. This is obvious because the stability of a microcrack growing under (confined) compression is much improved: the farfield stress must be increased to increase the crack length, which is an important difference from tensile microcracking. We discussed these matters in Section 8.1, and do not repeat all the arguments here. At some stage peak-stress is reached, and now again, as in the model for tension, a phase transformation occurs. The most critical crack will propagate and form a mode II (in-plane shear) crack that will extend during the steep part of the softening curve. A transition occurs from a continuum-based approach to a discrete crack model. The discrete mode II crack(s) start(s), at least in laboratory-size specimens from one of the corners of a specimen, as was clearly shown in Figure 8.15. In Figure 8.15b several cracks have nucleated at different corners and the most critical crack will subsequently propagate to form a complete shear band. In the example of Figure 8.15 there are clearly two cracks, which again may be affected by the suppressed rotations of the loading platens. The shear band appears to propagate along a straight line, and is hardly affected by the heterogeneity of the material. In Figure 8.15e the shear band has formed a more-or-less straight path between two diagonally opposite corners of the specimen, but still considerable stress can be transferred. The reason is that the shear band is not a continuous fracture plane, but has a certain width. Rubble and grains in the wide shear band may develop a sliding/

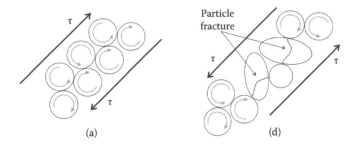

(a)   (d)

FIGURE 10.10
Rolling friction of (idealized) spherical aggregate particles in a shear band, that is, gear-box mechanism (a), and intraparticle fracture in grains that have become "jammed" in the shear-band (b). (From Van Mier. 2008. *Engng. Fract. Mech.*, 75: 5072–5086. With permission of Elsevier.)

rolling mechanism that can carry substantial load (gearbox mechanism, which is frequently considered in modeling granular assemblies; see, e.g., Bagi and Kuhn 2004).

When aggregates are "jammed" this may lead to rupture of such aggregates within the shear band (some examples of intrafracture of aggregates in shear bands were given by Van Geel 1998). In Figure 10.10 both mechanisms are illustrated. If lateral confinement is applied, depending on the magnitude of the confining stress, the residual stress level will increase as indicated in Figure 10.9, stage (C). Thus, comparing the framework for (confined) compression to the model for tension, two main differences must be considered: the extended stage (A) and the mechanisms in the bridging stage (C). Note that the effect of frictional restraint in the shear band (stage (C)) starts directly after peak stress, in the same way aggregate and fiber bridging was modeled in mode I in the previous section (see Figure 10.7).

An analysis will always start by calculating the inclination of the mode II crack. This qualification seems correct because it propagates in a straight line: no rotations are caused by the loading system which would turn the situation toward a mixed-mode crack. In a compressive test several factors will contribute to the direction of the shear crack: the compressive loading σ_a itself, the confining stress σ_c, and the frictional stresses τ_b along the boundaries of the specimen in contact with the loading platen. In Figure 10.11a the shear band, at an angle α to the axial loading direction, is shown together with the acting shear stress τ and the normal stress σ_n working parallel and normal to the shear-plane, respectively. The aforementioned external stresses are indicated as well. The normal and shear stresses along the inclined plane can be estimated by considering the local equilibrium:

$$\tau = \sigma_c \sin\alpha + \sigma_a \cos\alpha + \tau_b \sin\alpha \qquad (10.6a)$$

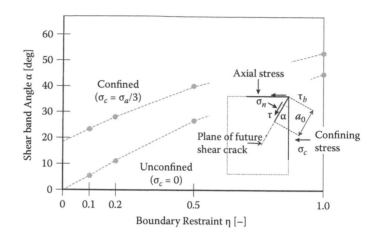

FIGURE 10.11
Effect of boundary restraint (defined through the coefficient $\eta = \tau_b/\sigma_a$) on the inclination angle α of a shear crack in uniaxial compression ($\sigma_c = 0$) and under confined conditions ($\sigma_c = \sigma_a/3$). The shear band with local stresses σ_n and τ is situated in a specimen subjected to axial stress σ_a, confining stresses σ_c, and boundary restraint τ_b as shown in the inset.

$$\sigma_n = \sigma_c \cos\alpha - \sigma_a \sin\alpha + \tau_b \cos\alpha \tag{10.6b}$$

These equations can be simplified by assuming a coupling between the boundary shear and the axial stress via $\tau_b = \eta\sigma_a$ ($\eta > 0$) as follows:

$$\tau = \sigma_c + \sigma_a(\cos\alpha + \eta \sin\alpha) \tag{10.7a}$$

$$\sigma_n = \sigma_c \cos\alpha + \sigma_a(-\sin\alpha + \eta\cos\alpha) \tag{10.7b}$$

If the shear crack propagates under pure mode II, it can be assumed that $\sigma_n = 0$. Substituting this in Equations (10.7a–b) the direction of the shear-band can be computed. The result is shown in Figure 10.11b, both for uniaxial compression ($\sigma_c = 0$), and for confined compression ($\sigma_c = \sigma_a/3$, assuming frictionless transfer of the confining stress to the concrete). When the boundary restraint, now indicated through the coefficient η, increases it can be seen that the inclination of the shear band increases too. This is in agreement with experimental observations; see, for instance, the results of Kotsovos (1983) shown in Figure 8.4. MGA pads and rubber as a friction-reducing medium between loading platen and specimen leads to inclination angles $\alpha \approx 0^0$; brushes with slightly more frictional restraint lead to $\alpha \approx 10^0$ and for steel platens $\alpha \approx 20^0$ was measured. Under confined conditions ($\sigma_c = \sigma_a/3$) the curve in Figure 10.11b moves upward: a higher inclination angle is predicted. Unfortunately, however, no systematic experimental results are available for comparison. The tests of Van Geel (1998) (Figure 8.15) suggest an average inclination of 22.5 degrees; tests by Van Mier (1984; Figures 8.12–8.14) indicate that α lies between 20 and 25 degrees. The confined results are likely limited to the lower-confinement regime only. When σ_c increases substantially, a transition from brittle to ductile behavior is observed. In this case it seems that stage (A) of the fracture process is extended and no macroscopic crack can develop. How specimens subjected to very high confinement would fail is unknown. No such tests have ever been conducted, but, judging from available experimental data it might well suffice to model the behavior at high confinement using an elastic-plastic model, that is, not going beyond stage (A) of our approach.

The solution shown here is likely not valid when α approaches zero. In that case a transition to vertical splitting is observed; see, for instance, the aforementioned results of Kotsovos when very low-friction systems are used, such as MGA-pads or even rubber. When rubber is used one should be very careful because this material has a Poisson ratio that is initially higher than that of concrete in the elastic stage and the boundary restraint changes sign and is directed outward, thus causing a tensile splitting stress at the specimen's far ends.

The four-stage model now proceeds as follows. The direction of the initial shear crack is known from the simplified analysis shown before. The length of the initial crack is a_0 (see Figure 10.11a), and is the result of pre-peak microcracking. The details are not simple, and there is need for a solid micromechanical model that can help to compute the crack-size distribution in compression, in particular in the regime just before peak stress is reached (see also Section 8.4). The growth of the macrocrack is controlled as long as deformation-control is available, and the residual carrying capacity σ_1 at crack length $a_1 > a_0$ can be calculated as before using:

$$\frac{\sigma_1}{\sigma_p} = \sqrt{\frac{a_1}{a_0}} \cdot \frac{f\left(\dfrac{a_1}{W}\right)}{f\left(\dfrac{a_0}{W}\right)} \tag{10.8}$$

where σ_p is the stress at peak, and $f(a/W)$ is the geometrical function for the considered problem. The propagation criterion is now $K_{II} = K_{II,c}$ = constant, thus reminding us that we are now dealing with a mode II problem. Bridging in stage (C) is now largely caused by (Coulomb) friction (or rolling friction; see Figure 10.10). Friction will delay macrocrack growth, and the net effect follows from summing the two contributions to the stress-intensity factor:

$$K_{II} = K_{II,loa} + K_{II,Cou.frict} \tag{10.9}$$

It should be noted that a stress-singularity still remains for low confining stress, but for higher confinement we may experience that the singularity vanishes as in the plastic crack-tip model. In that case we have a fully stable situation: a crack can propagate only when the loading (or deformation) is increased.

At this point we should be fully aware that the above describes just the framework or concept for a new model for compressive fracture. The strong point is that the physics of the fracture process is followed in great detail, and no debatable assumptions are needed. Of course there is a lot of work to be done. Not only should there be extensive experimentation to elucidate the microcracking process in stage (A), the role of friction on the stress intensity and the transition from brittle to ductile behavior are in need of further testing as well. The role of the model, or rather framework for a new model, is thus exactly what it should be: a guide to experimentation. Further simulations, either with a lattice model or a particle model may help to interpret the experimental results and assist in expanding our knowledge. For the moment we continue to discuss the possible ramifications to other materials.

10.4 Ramification to Other Materials

The above applications of the four-stage fracture model for tensile and (confined) compressive loading are both related to quasi-brittle fracture. In Figure 10.9 it was suggested that the increase of external confinement would lead to a more pronounced residual stress-level in stage (C). The four-stage fracture model may well be used for materials other than concrete when the relative effect of the four stages is varied. Behavior ranging from brittle to quasi-brittle to ductile behavior can be described by such a model, as illustrated in Figure 10.12. When the material/structure shows brittle behavior, the first microcrack likely will trigger global failure. It should be mentioned here that for very large structures brittle failure may occur as well; see, for instance, the size-effect tests of Figure 9.5, where the largest specimens showed a "snap-back" response, indicating that the energy release due to the developing macrocrack was already larger than the energy needed to create the new crack surface (see also the discussion on the brittleness number in Section 2.2). For the moment we ignore the structural effect and consider the response of the material independent of the structural conditions. Even though there is recent evidence that glass exhibits nanoscale fractures advancing the growth of a macrocrack (see Célarié et al. 2003) the behavior is generally classified as brittle. Generally one would expect finding, even for very small specimen sizes, unstable crack growth. Indeed, if one carefully studies the mirror–mist–hackle stages (see Figure 2.1) under an electron microscope, at the largest magnifications rather large roughness is still observed even in the mirror-zone, indicating that at least some effect from the heterogeneous material structure at very small scales must be expected,

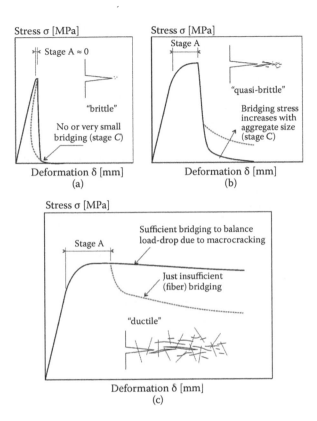

FIGURE 10.12

The four-stage fracture model for materials exhibiting (a) brittle, (b) quasi-brittle, and (c) ductile stress–deformation behavior; only tensile fracture is considered in these diagrams. (From Van Mier. 2008. *Engng. Fract. Mech.*, 75: 5072–5086. With permission from Elsevier.)

probably at the atomic/molecular level. All the lattice analyses presented in Chapter 6, for example, are independent of scale/size, therefore the heterogeneity arguments provided translate unrestrictedly to smaller size/scales, unless other physical aspects need to be considered, that is, other than included in the very simplistic lattice model.

Thus, in terms of the four-stage fracture model, glass, and other purely brittle materials will have an almost negligible stage (A); whereas crack growth during stage (B) dominates the behavior completely. If bridging in the main crack occurs it is believed to be insignificant, and will not help to stabilize the macrocrack; see Figure 10.12a. Quasi-brittle behavior was discussed at length in the two previous sections. In Figure 10.12b the case for tensile fracture is shown again: stage (A) has some significance, unstable macrocrack growth in stage (B) is partly balanced by bridging in stage (C); the bridging stress increases for coarser-grained materials. Then, in Figure 10.12c the possible application to more ductile materials is shown. Fiber-reinforced concrete may show ductile behavior depending on the mixture used. Examples of the mechanical behavior of hybrid-fiber concrete (HFC) are shown in Section A5.2; see Figure A5.7. The challenge here is to combine a high strength in tension with large deformability before and beyond peak; see, for example, Markovic, Walraven, and Van Mier (2003). This is in contrast to, for example, ECC, *engineered cementitious composites* (a strange term considering that all concretes are the result of engineering) where the tensile strength always lies around 4 MPa, but an enormous strain capacity up to 0.03 is reached; see, for instance, in Li (2010).

In the case of HFC, a strong matrix and large carrying capacity of fibers after fracture is asked for; for ECC the strength demands for the matrix are not very high, but large deformations of the fibers are striven for. This then leads to the choice of PVA as fiber material. Anyway, these details are not very important here. The fibers, in the case of HFC, arrest and deflect microcracks and may lead to a substantial increase of stage (A). The behavior in stage (A) can be characterized as "multiple cracking," as shown, for example, in Figure A4.3 in Section A4.1. The dense network of cracks with approximately the same width has developed before and around peak-stress; the main localized crack which is visible in this figure has developed after peak; see Stähli (2008) for details. The strength, even under uniaxial tension, is higher than the tensile strength of plain concrete. After the main localized crack starts to grow (again, see Figure A4.3b) substantial carrying capacity is derived from fiber bridging. An example of fiber bridging observed in an experiment on microfiber-reinforced cement is shown in Figure 10.13. At the right side of the image the crack width has become so large that some fibers have been completely pulled out, whereas closer to the tip fibers can be identified that still seem to connect the two crack-faces. The bonding properties of the fibers to the cement matrix, as well as the fiber geometry are decisive parameters in the pull-out process, which is essential for a large deformability beyond peak. Alternatively, long fibers

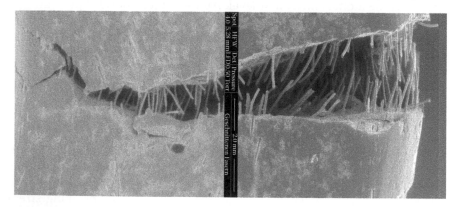

FIGURE 10.13
Bridging of a crack in microfiber-reinforced cement. Two images made with ESEM are stitched together. (After Rieger and Van Mier. 2009. *Advances in Cement-Based Materials.*)

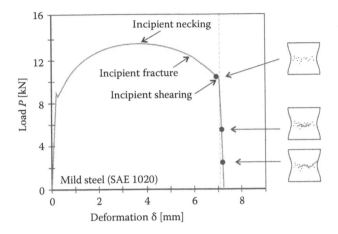

FIGURE 10.14
Tensile tests on mild steel. The crack-patterns have been traced from photographs shown in the original paper. At incipient shearing small voids were observed in the neck region; in the falling branch the growth of inclined cracks was reported. (Adapted from Bluhm and Morrisey. 1965. *Proc. ICF-1.*)

with low modulus will lead to large deformations; the aforementioned ECC is a good example for such response.

Figure 10.12c summarizes the behavior of ductile materials. When the carrying capacity of the fibers can match the peak-stress, large deformations occur without any appreciable stress-jump when localization of deformations in a main crack occurs. When the fibers cannot match the peak strength, a sudden drop of stress is measured, which accompanies the growth of the main crack. The four-stage model is capable of capturing this behavior as well. In brittle metals crack localization is observed too, for example, in the interesting experiments by Bluhm and Morrisey (1965). An example from their results is shown in Figure 10.14. In uniaxial tensile tests on copper and mild steel it was observed that after a long deformation trajectory starting from peak-stress, necking occurs, followed by the growth of an inclined macrocrack in the steep softening branch of the diagram. Even under uniaxial tension these metals fail through the formation of inclined cracks (shear fracture), which is the easier mode. The four stages that are discriminated in the model presented here can again be easily recognized. Now, however, the mechanisms up till localization are different from the microcrack process discussed for cement and concrete. In the case of metals dislocation movement and void formation at the onset of macrocrack growth are the important mechanisms. Thus, even though the micromechanisms deviate the four-stage fracture model can be applied.

For each case, brittle, quasi-brittle, or ductile, the importance lies in identifying the prepeak deterioration mechanisms, and possibly describing the behavior by means of an as-simple-as-possible micromechanical model.

Possible bridging mechanisms in stage C must be identified, as well as the closing stress distribution these mechanisms may exert on the faces of the macroscopic crack. Thus, the micromechanisms leading to peak may take different forms: dislocation movement and void growth in metals, nanoscale cracking in glass, microcracking in plain concrete, and multiple cracking in fiber-reinforced concrete. The mechanism that affects the main (in principle unstable) crack may vary from crack branching and bridging near aggregates and fibers in the case of tension, and frictional restraint for (confined) compression.

11

Multiscale Modeling and Testing

Fracture properties of concrete are hard to separate from their structural environment. This is in short the message conveyed in the previous chapters. In Chapter 10, with the four-stage fracture model, the influence of the structural boundary conditions and the geometry (shape and size) of the considered structure are incorporated via the geometrical factor, known from classical LEFM. Estimates for the size of the critical crack that leads to softening must be determined by means of an alternative model, for instance, the lattice model. Doing so seems unavoidable. The cohesive crack models will not work for concrete because the size of the cohesive zone is larger than the considered structure, and the original "local" plastic crack-tip model is changed into a "global" approach. It is quite essential that the structural component is included in the formulation. The 4-stage fracture model is an approach that might work. At the same time it should be mentioned that estimates for the geometrical factors are not always available, and it generally requires quite some effort to carry out the required analyses or experiments. Another approach that may be used is to return to a different form of lattice, which we refer to as "structural lattice." The equivalence between lattice and particle models can be used conveniently in developing this new approach. Again, as in the last chapter, the material presented here has not yet led to a fully operational model. Of importance is developing the framework of the model first, and showing its potential. Again, as in the 4-stage fracture model, the multiscale approach presented here is a good guideline for new experiments. After all, new insights are usually derived from carefully conducted and original experiments. Our modeling efforts are needed to summarize our knowledge in a uniform framework. The resulting model is no more than an approximation of physical reality. A critical assessment of the limitations and shortcoming of any modeling approach is therefore considered of the utmost importance.

At present there is enormous interest in multiscale approaches, where the behavior of materials and structures is analyzed simultaneously at several different length-scales. A stepwise upscaling methodology is used as shown in Figure 11.1. The most important assumption is that by following this sequence computational costs will decrease and the reliability of the model outcome will improve. It is obvious that the lattice model, which can be described as "upscaling from a predefined size/scale" leads to significant computational costs, as may be obvious from the short overview provided in Appendix 1. Whether computational effort is reduced by applying the multiscale methodology

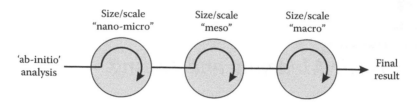

FIGURE 11.1

Multiscale modeling: principle. At the smallest size/scale, here the nano-/micro-size/scale, an ab initio analysis is carried out. The obtained result is used at the higher size/scale, the intermediate or meso-level. Finally this result is carried over to the next size/scale: the macro-level, where the final result is obtained.

depends on many factors. It is hard to say if it is really true. More experience with this new modeling approach is necessary. In this chapter we take a closer look to the multiscale approach applied to cement and concrete.

As mentioned, the basic principle of multiscale modeling is shown in Figure 11.1. At the smallest size/scale, for example, the nano- or micrometer scale, the behavior of hydrating and hardening cement paste is considered. A view of the structure of hardened cement paste was provided in Figure 4.1b (left and middle photos). Models are developed and experiments are carried out to feed the models with the necessary parameters. It is quite essential that these parameters be "pure" material parameters, and are not influenced by structural conditions in the same way as we have seen in other parts of this book, for instance, the size-dependency of softening parameters as well as their dependence on boundary conditions. The results from the efforts at the smallest scale are used in a model at the mesoscale, the intermediate level in Figure 11.1. If the results from the micro- or nano-level analyses are dependent on boundary conditions, one should take care that the same boundary conditions apply when the results are applied at the mesoscale. We have already seen a meso-level model, namely the lattice model, and shown several results that are obtained from it in Chapters 3, 6, 7, and 9. In Figure 4.1b (right photo) a view of the meso-level material structure of concrete was shown. A mesoscopic model operates at the [mm]-size/scale to the [cm]-size/scale, that is, the size/scale where the aggregate structure of concrete is considered.* Results from the smallest size/scale-analyses are needed for determining the properties of the cement matrix and the interfacial transition zone (ITZ). The results from the meso-level analyses and experiments are then used at a larger size-scale, the macroscale. Quite often the model used at this scale, that is, the scale of building structures ([m]-size/scale to [km]-size/scale) will be either a (modified) continuum model, or a discrete crack model in the spirit of classical fracture mechanics.

In this chapter, in Section 11.1, we first address the structure of hardened Portland cement and in relation to that say a few words on the mechanical

---

* Square brackets are used to clearly emphasize the dimension of the various size/scales used.

and fracture properties of this extremely heterogeneous, but interesting material. Next in Section 11.2 we discuss the role of water in the cement structure. Much of the water that is mixed initially with cement will react, but a substantial part of the water can move rather freely, even after hydration, and may have an effect on eigenstresses, eventually even resulting in cracking. Capillary forces resulting from "water-bridges" may, to some extent, add to the strength of the cement. This is a variable factor because, as mentioned, the water may move around depending on the environmental temperature and relative humidity. Spanning the size scales, *F–r* (Force–separation) potentials can be a useful tool to capture the size dependency of the fracture properties. As shown in Section 11.3, the potentials can be constructed from the atomic level to larger scales using relatively simple and straightforward formulations. The potentials can be applied directly in a "structural lattice model." The constitutive equations are the most troublesome factor in many, if not all of the models that have been debated to this point. Instead of solving the kinematic, equilibrium, and constitutive equations as done in a classical mechanics approach, it might be simpler to analyze structures directly at the force and deformation (separation) level. This may perhaps lead to complications because boundary effects must in some way be incorporated. On the other hand the unsolvable problem of the boundary and size effects on the constitutive level of commonly used fracture models may be overcome in this way.

11.1 Structure of Cement at the [μm]-Scale and Its Properties

Portland cement is the binding agent in concrete. The material has been known since 1824 when a patent was applied for by Joseph Aspdin. The raw material is marl, a mixture of lime and clay, from which four clinkers are produced at relatively high temperatures (around 1450°C) in a rotary kiln. If they are cooled down rapidly, the four clinkers can react with water, the amount added being one of the main parameters defining the porosity of the hardened cement paste, and with that the strength of the material. Also the ability of the material to transport fluids depends directly on connected porosity, or permeability. The main products from the reaction of the clinkers (C_3A, C_2S, C_3S, and C_4AF) with water are calcium silicate hydrates (CSH in short, in the notation used in cement chemistry; C stands for CaO, S for SiO_2, A is Al_2O_3, and H is used for H_2O) and calcium hydroxide (CH). The reactions develop from the surface of the cement grains (that are in contact with water in a fresh mixture) to their center.

Depending on the amount of available water, commonly expressed through the w/c-ratio (by weight), not all material will hydrate but unhydrated material will remain in the core of the cement particles. It is obvious that hydration is faster when small cement particles are used because the relative surface is much larger in comparison to coarser mixtures, and the resulting amount of unhydrated material at any time of the hydration process may be substantially smaller. The main hydrates responsible for strength are the calcium silicate hydrates. The CH is not particularly important for strength; instead it provides an alkali environment that protects the steel reinforcement against corrosion. Reinforcing steel in the form of ribbed bars or as fibers is commonly supplied to carry the tensile loads in structural applications of cement and concrete. CSH is found in two different forms, namely high-density and low-density CSH; see, for example, Jennings (2000) and Tennis and Jennings (2000). The structure of hardened Portland cement was shown in Figure 4.1b (middle image). The lightest shade of gray in this image is unhydrated cement. The darker grays are low- (lighter) and high-density CSH (darkest gray). Small pores are visible as well: they are the irregularly shaped black areas: their form develops as a consequence of the nonuniformity of the hydration process of the rather irregularly shaped cement particles. In Figure 11.2 another view of hardened cement paste is shown, before and after an indentation test using a Berkovich diamond tip. We return to the indentation later, but Figure 11.2a shows at higher magnification the structure of hardened cement paste (hcp). The large unhydrated particle in the center clearly has internal structure, and it seems important to consider that when modeling the material. Also here, darker and lighter gray are the high- and low-density CSH, respectively; porosity is black.

Thus, at the observed size/scale hardened cement is perhaps even more heterogeneous than plain concrete at the [mm]-size/scale. The methodology developed with the lattice model (i.e., projecting the measured material structure over the mechanical lattice) is as simple and straightforward for hardened cement paste as it is described in Chapter 4 for concrete. The size/scale is the difference, and the lattice elements will be one or two orders of magnitude smaller. When making the projection, it is important to decide which material phases should have clearly distinct elastic and fracture properties, and at what locations interfaces appear. For plain concrete this has been quite clear for several decades (the inherent weakness of the aggregate–cement bond was recognized in the early 1960s; see, e.g., the papers by Alexander and Wardlaw 1960, Hsu et al. 1963, and Alexander, Wardlaw, and Gilbert 1965), and for hardened cement paste these decisions must be made, based, for example, on local measurements of the material properties, such as by means of indentation tests (as shown in Figure 11.2b), or by scratching the surface (see, e.g., Akono, Reis, and Ulm 2011). Before going into this matter in more detail, a few words on the role of the amount of water mixed in the cement and the ensuing porosity are needed.

For complete hydration a quantity of water equal to approximately 40% of the weight of cement is needed: 25% for the chemical reactions and 15% physically absorbed water. When cement, sand, aggregate, and water are mixed, the water is absorbed at the surface of the particles. When the reactions start and CSH is formed, the amount of free water decreases and the particles

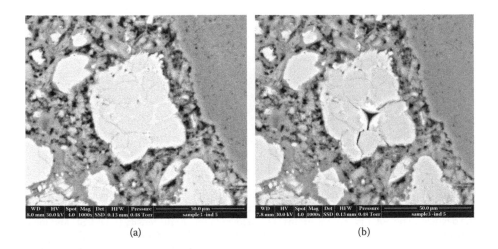

(a)　　　　　　　　　　　　　　　　(b)

FIGURE 11.2
(a) Polished surface of hardened cement paste. The almost white particles are the remaining unhydrated cores of partially hydrated cement grains. Light gray and dark gray are the low- and high-density CSH; black is porosity; (b) same area after an indentation has been made with a Berkovich diamond tip. (From Van Mier. 2007. *Int. J. Fract.*, 143(1): 41–78. With permission from Springer.)

move to a more compact configuration. With increasing degree of hydration the CSH becomes denser and gains strength. The reactions between cement and water are more violent when more cement surface is in contact with water; that is, finely grained cements react faster. If more water is added than needed, excess water will remain between the hydrates, and after evaporation open pore-space is created. This porosity increases with increasing w/c-ratio, and is detrimental for the cement's strength (this is simply caused by stress concentrations around pores; see also Section 2.1). The connectivity between adjacent hydrating cement particles is a point of concern. In Figure 11.3 the hydration process between two neighboring cement particles is shown schematically. Initially, as mentioned, water is absorbed on the surface of the grains, and capillary water bridges may develop in between (we return to these in the next section).

When the chemical reactions between water and cement clinkers start, high-density (HD-) CSH and low-density (LD-) CSH are formed as indicated in Figure 11.3b. LD-CSH forms along the grain's perimeter, and HD-CSH more toward the core. Accessibility of water must be guaranteed to keep the reactions going. The core may remain unhydrated for some duration, which typically will depend on the amount of available water. Note that if concrete dries out at some stage hydration will come to a halt, but may start again upon re-wetting. Seasonal drying and wetting of exposed concrete may show the development of "year-rings" in the hydrated cement structure, quite similar to year-rings in wood. The main question for a model, similar to the meso-level lattice model is if, and where, interfaces must be

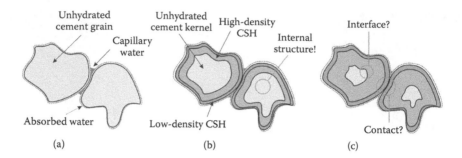

FIGURE 11.3
Schematic drawing showing the hydration process between two neighboring cement particles. Going from (a) to (b) to (c) we see a gradual decrease of the unhydrated material in the core of the particles. The outer part of the grains is transformed into high-density (HD)-CSH and low-density (LD)-CSH as indicated. The contact between two particles appears to be the result of contact/entanglement of the LD-CSH zones enveloping the grains, whereas the capillary water that is collected at the nearest point may further affect strength (see Section 11.2). One of the questions to be answered is if the clear separation among the respective material phases (viz. the unhydrated cement core, the LD-CSH, and the HD-CSH) and between adjacent cement particles, must be treated as interfaces, similar to the interfacial transition zone between aggregate particles and hardened cement-paste at the meso-level. (From Van Mier. 2007. *Int. J. Fract.*, 143(1): 41–78. With permission from Springer.)

assumed. Is this between the HD-CSH and the unhydrated core, or between the HD-CSH and the LD-CSH? Or is the latter phase transition just a gradual density shift of the hydrates going from the inside to the outside of the grains? And what to do with the interface between the LD-CSH of two adjacent particles: must this also be seen as an interfacial transition zone? Moreover, the ESEM-image of Figure 11.2a shows the internal structure in the unhydrated core of the cement grain: when an indentation test is carried out cracks seem to run along the visible particles inside the unhydrated core. Are these genuine interfaces as well? In all, more questions than can be answered at this stage, and rather demanding experiments are needed to provide the answers. To pose a few additional questions: what are the elasticity constants of the various material phases depicted in Figure 11.3, and what is their fracture strength? Do these materials exhibit softening? Or are they behaving elastic-plastic or just purely elastic brittle?

Over the years several numerical models have been developed for simulating the ever-changing structure of cement in time. The reactions of the four clinkers with water from which CSH and CH are formed are complicated, and most models are limited to describe the hydration of C_3S only. The reaction of this clinker with water leads to CSH, the main product defining the strength of hardened cement. Other reactions develop in part to control the entire process, in part because burning the aforementioned raw materials simply leads to these clinkers. The other limitation often imposed in the simulation models is that round (in 2D) or spherical (in 3D) grains are assumed. Under this assumption one should be careful drawing conclusions on the geometry of the pore-space, in particular the capillary porosity that is left between the hydrating particles. The divergence between real (see Figure 11.2) and assumed grain shapes will directly cause a misfit between the simulation and the real process. Unfortunately insufficient attention is usually given to the importance of having the particle geometry right. It simply directly affects the rate of the reactions and the geometry of the ensuing hardened cement structure. Models based on C_3S reactions only are, for example, the Hydrasim model by Berlage (1987), and more recent developments by Koenders (1997) and Ye (2003) of the same model, now under the name Hymostuc. In these models round spherical particles are always assumed, which, as mentioned, consist of C_3S only. Other hydration models include those by Maekawa, Chaube, and Kishi (1999), Pignat, Navi, and Scrivener (2005), Bishnoi and Scrivener (2009), and the NIST-model by Bentz et al. (1994) and Bentz (1997). The last model is a cellular automaton, which includes the most complete description of the clinkers and chemical reactions with water. Grain shapes can be as irregular as in real cements because, as in the particle overlay method in a lattice (see Section 4.5), in this model the cement structure (clinker distribution in the cement grains and the grain's geometry) is also directly mapped onto the model. The great advantage should not be underestimated, in particular when the simulator's goal is to make a direct comparison between model outcome and results from physical experiments.

We include just one result here, the simulation of the structure development in the ITZ between cement matrix and aggregate. The simulations were done several years ago by Garboczi and Bentz (1991), and are a marvelous illustration of how these simulation tools should be put to good use. In Figure 11.4 the two starting conditions and the calculated porosity as a function of the distance to the interface are shown. In Figure 11.4a the situation is as we would find it in any practical concrete, except that the shape of the grains does not match, which, in this particular example does not matter. The initial porosity along the aggregate–cement interface is rather high in this case, which is simply caused by the wall effect. A dense packing of cement particles is prohibited along the interface simply because of geometric constraints. In the structure of Figure 11.4b the aggregate particle has been placed over a structure of cement grains and water, having the same initial w/c-ratio as the case in Figure 11.4a. The average water-filled pore volume before hydration is about 60% in both analyses; in the case shown in Figure 11.4a the interfacial porosity increases to 100% next to the surface of the aggregate particle. The wall effect does not occur in Figure 11.4b: cement particles along the interface are simply cut into parts; all the cement-grains "under" the aggregate particle are excluded from further analysis. The initial porosity measured at a distance from the interface is in this second case almost constant.

Now hydration starts. This process is modeled in a very simple and straightforward manner. In a nutshell it proceeds as follows. Just consider the type of clinker in the cement grains, in particular those in contact with the water. Following the basic hydration equations (see Bentz et al. 1994, or also

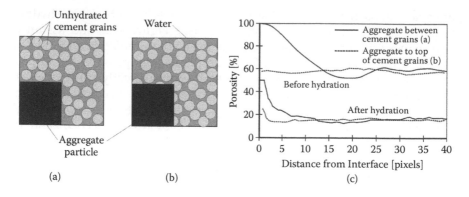

(a)   (b)   (c)

FIGURE 11.4

Two variations of an idealized particle structure of cement (light gray circles) near a square aggregate particle (black) are shown here in (a) and (b). Water is assumed between the cement particles. The water is consumed in the hydration reactions. A neighboring cement-pixel will be replaced by pixels of the reaction products (stoichiometry should be correct). In (a) the situation resembles that of a practical concrete where cement, water, and aggregates are mixed together. In (b) the cement grains were positioned first over the entire area, using the same w/c ratio as in the first example, and next the black aggregate particle is simply overlaid on top of the cement particles. The difference is in the wall effect. In (a) a normal wall effect develops: the grains can only be positioned until a point at their circumference touches the aggregate particle. This will lead to enhanced porosity near the surface of the aggregate particle. In (b) no wall effect occurs: the initial porosity is the same everywhere; (c) shows a diagram of the initial porosity of the two cases and the porosity distribution after complete hydration. (From Garboczi and Bentz. 1991. *J. Mat. Res.*, 6: 196–201. With permission.)

summarized in Van Mier 1997), pixels of clinker and water are transformed to the respective quantities of hydration products (CSH, CH, and open space). As a result a new material structure emerges; the step size is a measure of the elapsed time in the hydration process. The results in Figure 11.4c indicate that the porosity along the interface has increased substantially after full hydration. The increase has happened in both cases: where a wall effect was allowed, and in the case (Figure 11.4b) where it was suppressed. The conclusion is that hydration occurs in a direction away from the solid aggregate particle, and is part of the reason why enhanced porosity always occurs along the cement–aggregate (or cement–steel) interface; see, for example, the experiments by Scrivener (1989). The high porosity of the ITZ implies weakness of the material. Indeed, measurements, for example, by Zimbelmann (1985), show that the strength of the ITZ is very low in comparison to the strength of bulk cement. In the meso-level analyses in the Chapters 6, 7, and 9 a low ITZ-strength was always assumed when the analyses were to resemble normal gravel concrete.

It is amazing to see how much can be explained from simple geometrical considerations. Ignoring these may lead to erroneous conclusions, which are often happening, in particular in those model simulations where from the onset large deviations from the real situation are assumed and one keeps insisting on a perfect match between simulation and experiment. Surely this has to be attributed to sheer coincidence. I do not mention specific examples here; they are plentiful and are easy to detect with the above arguments in mind. In those simulations one should not be tempted to make a quantitative comparison; rather the simulations must be used in a qualitative way as in the example of Figure 11.4. They are a tool to increase our understanding of the complex subject matter at hand. The simulations are thus not a goal in itself and cannot be an attempt to rule out experiments altogether. Such a goal would be totally unrealistic because new ideas generally will originate from experiment.

Now let us return to the main topic: the structure and properties of hardened cement paste. The basis for a micromechanical model for hardened cement paste could look like the simple spring and dashpot model of Figure 11.5, which distinguishes three different material phases (unhydrated cement, LD-CSH, and HD-CSH), and three types of interfacial elements as indicated. The input parameters are more-or-less similar to what we have seen in the mesoscopic lattice model for plain concrete. The elastic parameters, Young's modulus, and Poisson ratio, and the strength for each of the three solid phases, as well as the spring stiffness of the interfacial elements (both normal and shear components) must all be estimated. The "easy solution" is to work with relative differences and use the results from simulations only in a qualitative way. In order to establish realistic values of the elastic properties and the strength of the three material phases rather demanding experiments have to be carried out. In the first place these experiments are very difficult, in part owing to the small size of the cement grains. Several types of experiments are needed. Not only those revealing the properties of the abovementioned material phases, including fracture properties, but also experiments on hardened cement paste at different size-scales to determine the overall properties. These latter results can be used for a comparison to model simulations at the nano- or micro-size-scale. In line with the previous paragraph: only the trend needs to be the same; no exact fit of the simulation results should be aimed at inasmuch as our knowledge is very likely incomplete. Next to these two types of experiments, one may also be interested to find out how two hydrated cement grains are bonded together, that is, the properties of the central interface spring in Figure 11.5.

So, let us show a few rather preliminary results from the three types of experiments mentioned. First we discuss the tests for measuring local properties of unhydrated cement, and the two types of CSH. Indentation and scratching are two tests that could be useful. Indentation was already shown in Figure 11.2. In that example a Berkovich-shaped diamond indenter was pushed in the surface of a large unhydrated cement grain. During indentation the contact area increases, plastic deformation occurs in the material close to the diamond tip, and the external load increases. There is a choice to drive the indenter until substantial damage is made in the substrate, or to stop at a prescribed depth of the indentor. Schärer (2005) carried out many

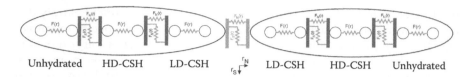

Unhydrated   HD-CSH   LD-CSH   LD-CSH   HD-CSH   Unhydrated

FIGURE 11.5

Principle for a micromechanical model for hardened cement paste. Included are three material phases, namely unhydrated cement, low-density CSH, and high-density CSH, as well as three different types of interfacial springs: between the unhydrated cement and LD-CSH, between LD-CSH and HD-CSH, and between two adjacent cement grains, indicated by the central spring. All interface elements shown here have a normal and a shear component (in 2D). A more realistic model would incorporate three dimensions, which implies that a second shear component must be added to the interface elements. (From Van Mier. 2007. *Int. J. Fract.*, 143(1): 41–78. With permission from Springer.)

indentation tests in order to see if the results published by Constantinides and Ulm (2004) could be reproduced. These latter results are shown in Figure 11.6, and suggest that the modulus of elasticity of the two types of CSH is indeed quite distinct and that the respective frequency distributions can be easily recognized. Schärer carried out indentation tests in an Environmental Scanning Electron Microscope (ESEM) in which the environmental conditions can be controlled to some extent. The indenter was fixed in the chamber, as can be seen in Figure 11.7a. This setup has the obvious advantage that the indent can be viewed at large magnification without having to move the material from the well-defined environment in ESEM. By moving the specimen underneath the indenter at regular spatial intervals indents were made to a depth of 500 nm, both with a Berkovich tip (shown here) and with a corner-of-cube tip. In Figure 11.7b three indents can be seen. The distance between the indents was large enough, and no mutual effects

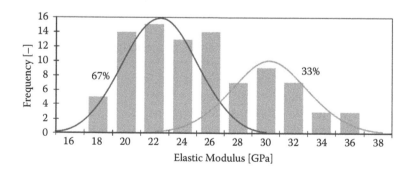

FIGURE 11.6

Young's modulus from nanoindentation tests on nondegraded plain cement. Two distinct frequency distributions mark the appearance of two distinct CSH phases, the low-density and high-density CSH, with lower and higher elastic modulus, respectively. The percentages indicate the area proportion under the curves. (From Constantinides and Ulm. 2004, *Cem. Conc. Res.*, 34(1): 67–70. With permission from Elsevier.)

between indents were expected. After the indents were made the underlying material was characterized: unhydrated, CSH, and at the boundary between these two phases. The results by Schärer showed a quite irregular frequency distribution for the Young's modulus; see Figure 11.7c for tests at a chamber pressure of 20 torr (RH = 90%). With some imagination one can see two "peaks" that would correspond to the LD-CSH and HD-CSH, at 18 GPa and 29 GPa. However, perhaps the number of tests is still too limited; in the example of Figure 11.7c results from just 70 measurements are included.

There are many disturbing factors in indentation tests that may have affected the outcome. First of all, the diamond tip exerts a high local force on the surface of a (usually polished) surface, and the resistance experienced will depend on the material under the surface, as much as on what we can actually see on the surface. Indentation is a truly three-dimensional experiment. Even if the surface shows that an indent is made in, let's say HD-CSH, this might be a sliver of just 10-nm thick with a larger unhydrated cement particle underneath with much higher stiffness. In this case a higher apparent modulus will be measured than in a test where the entire indent is made in the HD-CSH phase. Calcium-hydroxide crystals (CH) will be abundantly present as well, and here the orientation of the CH-crystal may affect the indentation result. In order to get a closer view of the three-dimensional damage exerted by an indenter on a cement sample a number of experiments were carried out in the tomography beamline of the synchrotron at the Paul Scherrer Institute in Villigen (Switzerland); see also Section A4.2 where the method is briefly described. Indeed the damage exerted on the specimen is not limited to cracking at the surface (see, e.g., Figure 11.2b), but under the tip of the indenter a vertical crack may propagate along some distance in the substrate. In Figure 11.8 two views from the synchrotron experiment are shown, a 2D-section along the deepest point of the indent, as well as a three-dimensional view. The appearance of the vertical crack is quite clear. Recently, using the lattice model described in Chapters 3–6, Chiaia (2001) and Carpinteri, Chiaia, and Invernizzi (2004) published results from simulations of a hard indenter pressed in a brittle material and a heterogeneous material resembling concrete, respectively. The simulations showed a similar result as the experiment in Figure 11.8: abundant surface cracking and a vertical splitting crack growing from the tip of the indenter into the substrate. It therefore seems that the indentation test is foremost a splitting experiment; the homogeneity of the material beneath the indenter will certainly affect the outcome of the experiments, and should be part of any analysis based on such results. It is quite obvious that much additional work is needed. Just doing the indents and deriving conclusions from that is "quick and dirty"; substantial additional effort is needed to characterize the material in which the indents are made more precisely, also in the third dimension.

An alternative to the indentation test is a scratch-test, in which a sharp object is dragged along the surface of a material, keeping the depth of the cutter-blade constant; see, for instance, Akono, Reis, and Ulm (2011). The

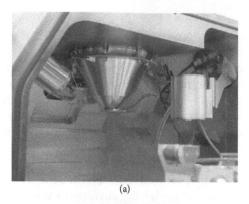

(a)

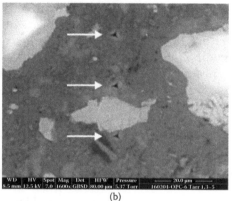

(b)

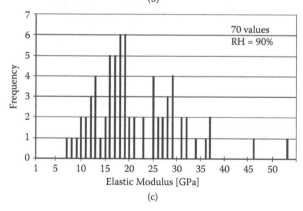

(c)

FIGURE 11.7

(a) View of the nano-indenter in the chamber of an environmental scanning electron microscope (ESEM). In the center the ESEM detector is visible; the cylinder to the right in the nanoin-
denter and the light gray element in the foreground are part of the specimen table. (b) Three
Berkovich indents at the surface of a polished cement sample, in a row, each with a depth of 500
nm, and (c) the frequency spectrum of the modulus of elasticity at RH = 90% is shown. (After
Schärer. 2005. *Micromechanical Properties of Portland Cements.* (b) and (c) with kind permission
of Mr. Reto Schärer.)

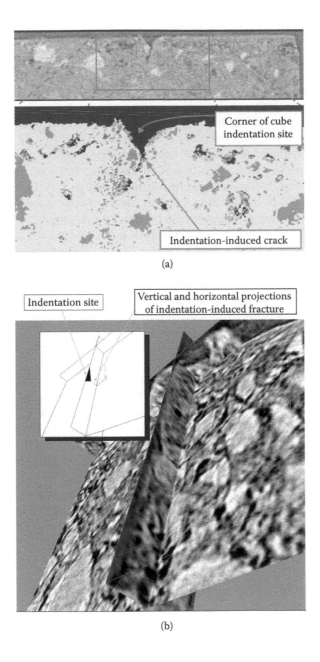

FIGURE 11.8
Two views of an indentation in hardened cement paste viewed in the tomography beam-line of the synchrotron at Villigen (CH). (a) Two-dimensional section along the deepest point of the indent shows a vertical crack running into the substrate. (b) Three-dimensional view reconstructed from the tomography experiment and the extent of cracking in the third dimension is clearly visible. (After Trtik et al. 2005.)

"scratch" develops over a certain length, on the order of [mm] to [cm]. Akono and colleagues concluded that scratching is a genuine fracture experiment, and for hardened cement a critical stress intensity factor in the range of values from quite different tests was reported. This conclusion certainly requires independent confirmation. It has to be clarified whether the method could apply at the very small scale of the hydration products.

The second type of experiment needed is a test that provides detailed information on the global behavior of hardened cement paste. The obvious way to go is to repeat the tensile tests mentioned in Chapter 6—uniaxial tension, 3-point bending, or a Brazilian splitting test—on small samples of hardened cement paste. A good example, which may be recalled here, are the classical 3-point bending tests on notched beams made of ordinary Portland cement and a material called "macro-defect-free" (MDF) cement by Birchall, Howard, and Kendall (1981). Varying the notch depth showed that the MDF samples were very sensitive to the smallest notch depth of 0.1 mm, whereas the normal Portland cement samples were hardly as affected by the notch depth until it reached 1.2 mm. The difference was attributed to the large porosity of ordinary Portland cement compared to the almost negligible porosity of the MDF cement, in which all possible measures were taken to prevent pores as much as possible. The MDF cement can be seen as a forerunner to the modern-day high-strength concrete. Indeed, a maximum flexural strength larger than 60 MPa was reached for MDF cement, which is very high compared to the maximum of 10 MPa for Portland cement. Birchall et al. concluded that in the regime where the notch depth affected strength, both materials followed the Griffith criterion. It is obvious that, although these tests are very interesting, we would need quite a bit of additional information for comparison to the outcome of simulation models. Not only the full load-deformation diagrams would be needed, but also some insight in the fracture process.

Like at the meso- and macro-size/scales the best way to proceed is to perform a uniaxial tension test and try to perform experiments where the stress-distribution over the specimen's cross-section is as uniform as possible during a significant part of the experiment, in particular before the localized critical crack starts to propagate. Loading a specimen and at the same time monitoring the fracture process would be the ideal setup. Recently we made some preliminary attempts in this direction; see Trtik et al. (2007). Small cylindrical specimens (diameter 130 μm, length 250 μm) were loaded in a newly developed miniature tensile loading device. In this loading apparatus the load was applied by varying the voltage over a piezo-crystal, as explained in Section A4.2 (see also Figure A4.5). The complete loading device was small enough to fit inside the tomography beam-line of the synchrotron at the Paul Scherrer Institute in Villigen (CH), which allowed monitoring crack growth while the specimen was under load. The gamble was if the test would be stable, because at this size/scale deformation control was difficult to realize. Unfortunately the experiments were unstable in the postpeak regime, but

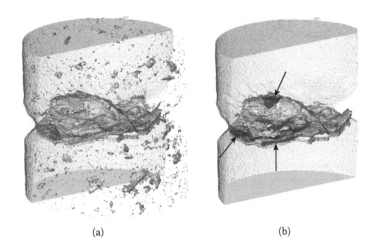

(a)   (b)

FIGURE 11.9
Three-dimensional reconstruction of the voids in a cylindrical specimen loaded in the µ-tensile testing device. (After Trik et al. 2007. *Proc. 6th Int'l. Conf. on Fracture Mechanics of Concrete and Concrete Structures (FraMCoS-VI)*. (a) All voids are shown, which included not only the cracks, but also the porosity in the hardened cement paste. The main crack is shown in (b), where the darker patches indicated with arrows are the locations where the main crack intersects unhydrated cement particles.

revealed some interesting results regarding the fracture process nonetheless. In Figure 11.9 the void distribution in a specimen after applying the loading is shown. In Figure 11.9a all detected voids are shown, in Figure 11.9b only the largest void, which is in fact the main crack. The main crack formed in the plane where a notch was machined (using femto-second (fs) laser pulses). The crack-plane is highly undulated and remnants of crack-face bridging were found at closer scrutiny. The view in Figure 11.9 is a reconstruction of 401 angular projections; see Trtik et al. (2007) for full details of the test procedure. Figure A4.6a shows how the crack touches the unhydrated and hydrated cement phases, which gives a clue about the relative strengths of the material phases in partially hydrated Portland cement. As mentioned, an interesting observation from these experiments is that bridging seems to occur, much like the crack-face bridging in mortar and concrete, which was mentioned in Chapter 6 (Figures 6.2d and 6.2h), in Chapter 10 (Figures 10.5 and 10.6) and in Section A4.1, Figure A4.2). The miniscule bridges are the white spots visible on the fracture plane of Figure 11.9; a three-dimensional reconstruction of a bridging event in hardened cement paste is shown in Figure A4.6c in Section A4.2. The conclusion is that hardened cement paste will show softening behavior in small specimens like those tested here, which is of interest for the model developed in Section 11.3.

The challenge in the tomography experiments includes the manufacturing of geometrically correct specimens, the test control, and the data handling. At the same time only a handful of experiments can be conducted per year, due to the aforementioned challenges, but also because only limited access to the tomography beam-line can be obtained. Nevertheless, reliable information on the fracture process of hardened cement, at a size/scale which is three orders of magnitude smaller than, for instance, the size effect tests on the "F"-size specimens shown in Figure 9.5. It will be obvious that in coming years substantial research effort is needed to reveal the properties of hardened cement paste at this size/scale, including variations of many parameters such as type of cement, water–cement ratio, degree of hydration, and the effect of admixtures.

The third experiment has to reveal how the contact between two neighboring hydrating cement grains is established during and after hydration. This test is thus related to the properties of the central interface spring in Figure 11.5. The experimental difficulties have been partially overcome at this point, and at least a few more years of tedious work are needed to fully master it. The idea is sketched in Figure 11.10. Two cement grains, selected for their shape and size, are brought together on the cooling/heating stage in ESEM. This can be done by using micro-manipulators in the chamber of the microscope. In Figure 11.10a we see the grains lying side by side, touching

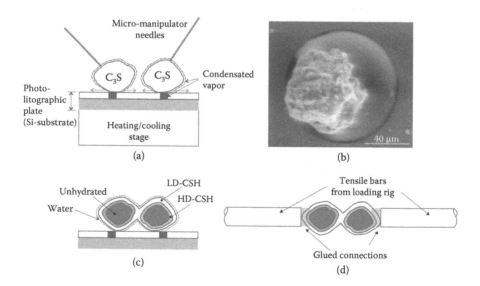

FIGURE 11.10

Three stages of development in an advanced experiment for studying the contact zone between neighboring hydrating cement grains. (a) Condensation of water at the surface of the cement grains can be achieved by properly cooling the grains down, just a bit more than the immediate surrounding. Localized condensation (b) will bring the water to those locations where the next step, hydration, should commence (c). Finally, after the hydration process has stopped, a tensile test is carried out by fixing the two hydrated cement grains in a tensile loading stage (d). In the ideal situation this all should take place in a controlled environment (T and RH). The chamber of ESEM would be quite well suited for that, but it means keeping the apparatus in use for quite extended periods of time.

perhaps at their nearest point. Next, by cooling down the grains it is possible to condensate water vapor on their surface. It is quite essential that only the grain's temperature is changed, at least more than the temperature change of their immediate vicinity. By placing an appropriate isolator on the surface of the cooling/heating table, with holes at the location of the grains this can be achieved. It is not straightforward to select the right isolation material. Figure 11.10b shows a cement grain with water condensed at its surface. The next step is to initiate the hydration process, and to try gluing the grains together at the nearest contact point; see Figure 11.10c. This step is most cumbersome. The contact has to be strong enough to carry out a uniaxial tensile experiment. For this to be successful the two grains must be fixed in a loading stage; see Figure 11.10d. The experience gained in the μ-tensile experiment described in the previous paragraphs is a good start for accomplishing this part of the experiment. An alternative to tension would be splitting: simply driving knives between the two hydrated particles. The load and the separation of the grains should be measured. Where in the first method, uniaxial tension, perhaps deformation-control could be performed, in the latter method this is much more difficult to achieve (compare the splitting tests of Section 6.2.1, which are a much larger size/scale, however). The challenges are enormous, and substantial support is needed before they can be overcome. In the present-day funding system continuous support for experiments of this degree of complexity is rather difficult to obtain. The number of publications produced is too small, progress will be at a snail's rate, but, when successful, the results are very valuable because they may help to improve our understanding of the origin of the strength of cement.

It can be concluded that the initial steps for determining the properties of hardened cement have just been made. It is hoped that the coming years will show increased activity in this field. A few more experiments than those mentioned here can probably lead to useful results; however, we refrain from extending this section too much, and leave it to the imagination of the reader. As mentioned, support for an extended period of time is needed to conclude these experiments; it is hoped that the funding system would allow for such developments.

11.2 The Role of Water at the [μm]-Scale

By now it may be obvious that water plays a large role in the structure of cement. The initial cement–water mixture gradually hardens during the hydration process. The chemical reactions consume the water for a large part, but some of it remains in the cement structure, either adsorbed to the walls of the hydrates or as free water in the larger pores. Particle size has a large influence on the adsorption of water on the surface. In a relatively simple

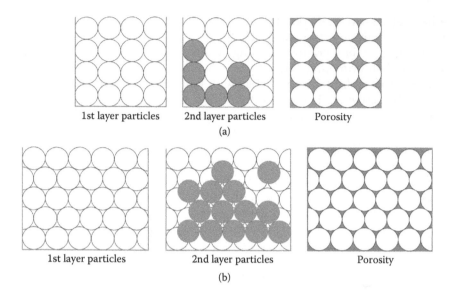

FIGURE 11.11
Regular packing of mono-sized spheres and resulting porosity (a) and hexagonal close-packing (b).

analysis the importance can be demonstrated. Mono-size particles (spheres) can be placed in varying geometrical packing. Figure 11.11 shows two quite distinct possibilities: regular packing (abbreviated rp; see Figure 11.11a), and hexagonal close-packing (or hcp; see Figure 11.11b). Two boxes of given size are filled with particles using these two packing arrangements: the first one, regular packing, will contain fewer particles per unit of volume than the second one, especially when incomplete spheres are allowed, that is, when it is assumed that from an infinite arrangement simply a cubical box is cut, thereby splitting some of the particles in segments as shown in Figure 11.11b.

In Figure 11.11 just the principle of the packing is shown: the second layer of particles lies directly on top of the first layer in the 'rp' scheme, whereas with 'hcp,' particles in the second layer are shifted to fall in the 'valleys' left by the first layer particles as shown. The density of these two types of particle packing differs significantly:

$$\varphi_{rp} = 0.5236 \ll 0.7405 = \varphi_{hcp} \tag{11.1}$$

Porosity is $1 - \varphi$; the pore volume decreases for 'hcp.' Even larger density difference would be observed when particles of different sizes are packed in a box. For the sake of simplicity we do not consider particle size distributions, other than saying that the Fuller distribution we introduced in Section 4.3, Equation (4.3) is the densest packing of spherical particles. The total area of the particles in the two situations of Figure 11.11 varies with the number of particles in the box, and is thus larger for 'hcp.' If a box of unit volume is considered, and one decides to reduce the particle size by a factor of 10, the number of spheres will increase in both cases by a factor of 1,000. However, although the total sphere volume remains constant, the total area of all spheres increases by a factor of 10. It is thus quite obvious that denser packing of smaller particles will lead to an increase of specific surface. The amount of physically adsorbed water is in that case larger if not the total pore volume is filled.

A variety of interaction forces can be active depending on the size/scale of the particles in question. One can distinguish, at extremely small scales, hydrogen bonds, primary chemical bonds, double-layer forces, and Van der Waals/London attraction. At a somewhat larger size/scale capillary water may be present, forming liquid bridges such as the one shown in Figure 11.12b, causing attractive forces between neighboring particles, for example, cement grains or sand. The simple fact that a granular medium is wet causes it to have some basic strength, which may lead to problems, for example, in the case when particles flow in a silo, or when caking of fine powders occurs. For very dry concretes used in the production of concrete bricks the capillary forces between the grains in the wet state allow us to remove molds relatively quickly after casting, which of course increases the production rate. In the end the interaction potential between two neighboring particles will be the result of the sum of all forces, and may take the shape of Figure 11.12a. When the particles are very close together, they may expel one another; at larger spacing the attraction force may reach a maximum, after which, with increasing distance, it will decrease again.

Now let us return to the capillary forces between adjacent particles. In this example we assume that only capillary forces make up the potential. When the particle size decreases, and the total amount of water in the porous particle stack is kept constant, the thickness of the water layers adsorbed on the surface of the spheres will decrease. As a consequence the radii of all menisci defining the liquid bridges will decrease, and with that the interaction force p_d, which is described with the Laplace equation,

$$p_d = \frac{-2\gamma}{r} \tag{11.2}$$

will increase; see Figure 11.12c. In Equation (11.2), γ is the surface energy for the air–water interface and r is the radius of the meniscus. The result is quite well known: with decreasing water content the porous material will become stronger. In three dimensions both the contributions from the internal pressure p_d and the surface tension γ along the water–air contact must be added to obtain the attraction force F following

$$F = \pi r_0^2 p_d + 2\pi r_0 \gamma \tag{11.3}$$

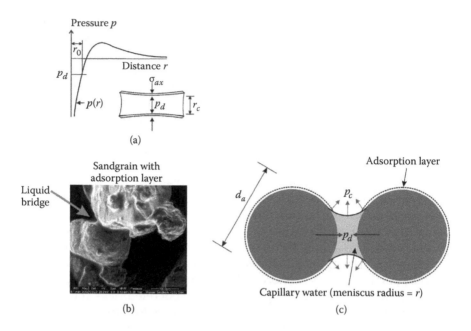

FIGURE 11.12
(a) Interaction potential between two spherical particles has a form as shown. At the smallest scale, the atomic scale, the atomic potential is retrieved; at larger scales other interaction forces contribute and blur the picture. Liquid bridges caused, for example, by condensation of water on the surface of sand grains (b) showing an image of sand grains viewed in ESEM where condensation is achieved by cooling the particles down in a vapor-saturated environment and are capable of building up capillary forces, which to some extent may keep particles together. The capillary attraction forces in a liquid bridge depend on the radii of the particles, the total amount of water, and the relative vapor pressure, which all determine the radii of the water menisci. (From Van Mier. 2007. *Int. J. Fract.*, 143(1): 41–78. With permission from Springer.)

In this equation, p_d is given by Equation (11.2), with $r = r_0 + r_1$, where r_0 and r_1 are the radius of the smallest water column between the two particles and the radius of the meniscus, respectively. This result is well known; see, for example, the application in the discrete element model by Muguruma, Tanaka, and Tsuji (2000). An interesting contribution to the problem is the recent paper by Rabinovich, Esayanur, and Moudgil (2005). The point made here is that the material-structure geometry has a significant influence on the behavior of the considered material. This has, as a matter of fact, also been demonstrated in some of the fracture analyses in Chapter 6, more specifically the influence of the particle density in concrete on tensile strength and ductility (Figure 6.3).

The above example is valid only for materials built up from equal-sized particles, which is far from reality, in particular for cement and concrete, where usually a certain particle distribution is used to fill up a volume as densely as possible, without too large porosity since that is detrimental to strength and durability. The distribution of water in heterogeneous particle composites such as concrete is far from uniform. When a porous material is brought into an environment of 100% RH smaller pores will be filled first; in larger pores water will condensate on the pore walls, and it will take much longer to completely saturate them. The reason is not just the difference in volume, but differences in water vapor pressure in the pores, which is related to the pore radius. The Kelvin equation,

$$\ln\left(\frac{p}{p_0}\right) = -\left(\frac{2\gamma V_m}{RT}\right)\cdot\frac{1}{r}$$

(11.4)

shows that the relative pore pressure p/p_0 increases with decreasing pore radius r. In this equation γ is the specific surface energy (0.072 J/m^2 for water), and V_m is the molar volume of the liquid (18.10^{-6} m^3/mol). Thus, smaller pores fill at much lower relative humidity, and much later the larger pores will be saturated. In heterogeneous particle composites smaller pores will occur closer to places where smaller particles meet, which will, according to the Kelvin equation, be saturated first.

The mutual attraction between the small particles is larger, and these places may act as attraction kernels to which all other particles will "move." When particles move under mutual capillary attraction forces, the water will be redistributed, which continues until a new equilibrium is reached. In some places between regions where particles are attracted to different kernels, voids may appear which can be interpreted as shrinkage cracks. This process may occur in material like clay, or in hardening cement paste, where often cellular-type crack patterns are observed at the surface; see, for instance, Figures 11.13a and A4.9b,c in Section A4.4. In Figure 11.13 the mechanism is clarified. The attraction kernel-development is shown in a birds-eye view of the surface in Figure 11.13b, and in a cross-section view in Figure 11.13c. Thus, the cell-like pattern of shrinkage cracks is caused by the geometrical arrangement of the attraction kernels, which is purely probabilistic. The relative strength of the material at various spatial locations and at a certain moisture distribution is mimicked in the crack patterns. In the vertical direction it can be shown that shrinkage cracks in materials like hardened cement develop almost instantaneously, in the first few minutes after the surface has been exposed to a drier environment; see Shiotani, Bisschop, and Van Mier (2003). In the formation of the vertical cracks capillary suction may play a limited role (see Wittmann 1978, where the mechanism is elucidated). Curiously, horizontal branching cracks are often observed, at a relatively shallow depth. The shrinkage cracks do not seem to extend beyond a depth of 4–5 mm in hardened cement paste; see Figure A4.12a. The reason may be that the moisture distribution is more uniform in deeper parts of the material, and the attraction kernel mechanism cannot work there. The horizontal cracks mark the depth where the moisture distribution is sufficiently uniform to prevent further growth of shrinkage

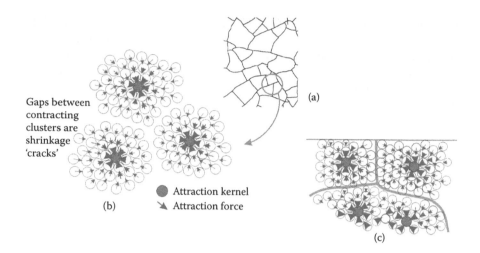

FIGURE 11.13

The relative movement of particles connected through liquid bridges, which create capillary forces between the particles, will ultimately lead to drying shrinkage crack patterns as shown. In a heterogeneous material the particles will have varying size, leading to preferential places called attraction kernels in (b), where the material is relatively stronger than its immediate environment. The capillary forces between the particles draw them nearer to the nearest attraction kernel, leaving larger "gaps" between "clusters of particles" surrounding the respective attraction kernels. These "gaps" are interpreted as being shrinkage cracks. (a) Top-view of a shrinkage crack pattern in hardened cement paste, with a typical cell size of 55–90 mm. In (b) the clusters of particles around the attraction kernels leave gaps, dubbed shrinkage cracks. In the third dimension the shrinkage cracks penetrate into the material in a vertical direction. Often, as in hardened cement paste and clay it is observed that horizontal crack branches develop (c), which may be explained from the formation of clusters of particles around attraction kernels. Additional drying through cracks and capillary suction are believed to contribute significantly to crack branching. It is pointed out that in the drying situation sketched here water evaporated through the top surface, and a strong moisture gradient developed as the interior of the material dried more slowly; see also Figure A4.11 in Section A4.4. Thus, deeper inside the material the attraction kernels will not develop because water is distributed more uniformly. The horizontal crack branch marks the location where the moisture distribution is sufficiently uniform. The drying shrinkage experiments on hardened cement shown in Section A4.4 suggest that the horizontal branches develop at a depth of 4–5 mm. (From Van Mier. 2007. *Int. J. Fract.*, 143(1): 41–78. With permission from Springer.)

cracks. The situation changes markedly when large aggregates are present, as explained in Section A4.4.

Shrinkage of hardened cement paste and concrete is a very important problem. Much damage can be caused, which can easily be prevented by controlling the climate just outside the fresh concrete until strength development is sufficiently large, or by adding fibers; see also Appendices 4.4 and 5.2. The crack growth caused by drying shrinkage can be modeled using lattice-type models comparable to the model described in Chapters 3 and 4. Notable models are those by Meakin (1991) and Leung and Néda (2000). A frequent problem remains that the moisture transport during the shrinkage deformations is not handled correctly; usually this factor is completely neglected due to the fact that these flow problems in porous media are not completely understood, and coupling to mechanical problems is not really straightforward either. During shrinkage the size of liquid bridges changes constantly as does the degree of saturation of pores. The contact forces are thus directly affected.

In a defined volume more small spherical particles can be placed than larger particles, for example, of one magnitude larger diameter. The increase of number of particles is by a factor of 10^3. The number of contacts between particles increases by a factor of 10^3 as well. If the same water volume is added to a porous particulate material with small or tenfold larger particles, the first material will react stronger. In a deforming partly saturated porous particulate material a constant redistribution of the relative volumes of liquid bridges leads to a constant redistribution of the interaction forces, and the outcome is not straightforward. When the particle shape changes, from spherical to, for example, ellipsoids, like M&M candies, the number of contact points between particles increases significantly; see, for instance, Donev et al. (2004). Again, hydration models based on realistic particle shapes, such as the NIST model by Bentz et al. (1994) are quite essential to obtain valid results. Results from models based on spherical particles must be regarded with some skepticism unless they are used in a qualitative way to elucidate certain principles such as the formation of the ITZ shown in Figure 11.4.

Cement has an incredibly complicated structure; see, for instance, Tennis and Jennings (2000), Jennings (2000), Pellenq and Van Damme (2004), and Gatty et al. (2001), among many others. Before hydration the situation is still relatively simple: the distribution of the four clinkers and other residues forming during the hydration process can be determined by means of optical microcopy, and incorporated in a numerical simulation model such as the NIST model. During hydration and at complete hydration the situation becomes more tedious because the length-scale of the CSH is much smaller than that of the particles before hydration. Although controlling humidity during viewing is of extreme importance, high-resolution TEM is probably the only realistic tool to be used (Pellenq and Van Damme 2004 and Gatty et al. 2001). According to these authors the nanometer-scale structure of cement can be described as a combination of nanocrystalline regions, microscale ordered regions, and amorphous matrix, that is, substantially more complicated than the simple model of Figure 11.5. For developing a model for the mechanical behavior at the moment simplicity is called for, and to start with the principle of Figure 11.5 seems quite workable.

Capturing the porosity of a sample of cement may prove quite difficult. In Figure 4.8a the pores in a 130-μm diameter cement cylinder of hardened Portland cement are shown. Using the advanced tomography beamline at the Swiss Light Source in Villigen (Stampanoni et al. 2002), or equivalent equipment at other places (for instance, the European Synchrotron Research facilities in Grenoble, which was used by Bentz et al. 2000) is helpful for getting a better insight to the porosity of hardened cement paste. Another tool is milling with a focused ion beam (FIB), which can be done in combination with high-resolution electron microscopy; see, for instance, Holzer et al. (2004). In this technique, thin layers of material (10–100 nm thickness) are removed step by step; between steps a high resolution SEM image is captured. The three-dimensional structure of a material can be captured in this way, but a stack of 100 images will easily take three to four hours. Thus, for porosity this seems OK; for fracture the technique is too slow to follow crack growth, even under quasi-static loading. Holzer et al. (2004) found deviations between the porosity measured by means of conventional mercury intrusion porosimetry (MIP) and their FIB-technique. Of course MIP is a partly destructive technique: when the mercury is pressed into a porous sample, pore walls may break. In addition, one has in some way to account for the "ink-bottle" effect: a large pore connected to the other porosity via a very narrow opening will be counted as small porosity. Recently Diamond (2000) argued against the use of MIP, and I can fully agree with his view. Of course tomography and FIB-milling are very time-consuming methods, and the necessary facilities are only sparsely available. Yet, the accuracy seems much improved, which is essential for comparing to the outcome of simulation models. Here one has to be careful: if the hydration model is based on spherical particles, the outcome may substantially deviate from the experimental measurements. This is not always recognized, and claims of excellent fits, as in the work by Ye (2003), must be viewed with skepticism. Hydration and the resulting material structure is largely driven by the shape and size distribution of the original cement particles.

It is obvious from the above exposure that the development of a fracture model at the cement-level is far from complete. As a matter of fact just some very humble initial steps have been made, and much research is needed before a workable model will be available. The assumption that cement can be regarded as a particulate material seems perhaps farfetched in view of the above description of the hardened cement paste structure and the role of water therein. Yet, as a first step this assumption may be good enough as simplicity is warranted. Some initial steps in the development of a particle model and the interaction potentials that are needed in such an approach are formulated in the next section. An important conclusion from the aforementioned μ-tensile experiment is that hardened cement paste shows softening behavior, which is part of the model development in the next section.

11.3 *F-r* Potentials: From Atomistic Scale to Larger Scales

The main parameters needed in the simplified model of Figure 11.5 are the interaction potentials, the properties of the interface springs between the

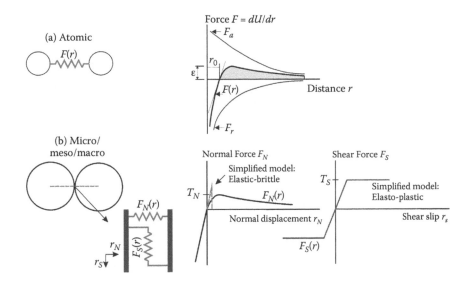

FIGURE 11.14
(a) Potential for the interaction force F between two neighboring atoms as a function of the separation distance r, (b) and an extended potential for interactions at the micro-, meso-, or macro-level consisting of a normal force-separation law ($F_N - r_N$) and a shear force-slip relationship ($F_S - r_S$). (Reprinted from Van Mier. 2007. *Int. J. Fract.*, 143(1): 41–78. With permission from Springer.)

solid unhydrated cement grains and the grains consisting of the two types of hydrates, LD-CSH and HD-CSH. The type of model can actually be seen as a lattice model or a particle model, which we discussed in Chapters 3 and 4. The main differences are, firstly, that we have to deal with many more material phases and interfaces, and, secondly, that the size/scale is significantly smaller, about three orders of magnitude smaller.

In view of the complexity of the physical interactions in hardened cement paste, which includes the role of water, it seems most appropriate to start with an interaction potential that lumps all effects into one single equation. The notion that the same type of lattice or particle model can be used at any size/scale-level suggests that starting from a very fundamental level may lead to a useful result. The potentials describing the interaction between two atomic particles is a well-defined starting point, where we should recognize that the proposed model here is active at a much larger size/scale-level, the nano-/micro-size/scale. In Figure 11.14a an atomic potential is shown. The attraction force or energy between two neighboring atoms depends on the separation distance. At a distance r_0 the system is at rest: attractive and repulsive forces are in balance. When $r < r_0$ the repulsion overtakes and particles are driven apart, when $r > r_0$ the attractive force or energy binds the particles together.

A well-known atomic potential is the Lennard–Jones (LJ) potential, which can be written as

$$\frac{V_{LJ}(r)}{\varepsilon} = -4\left[\left(\frac{\sigma}{r}\right)^{12} - \left(\frac{\sigma}{r}\right)^{6}\right] \qquad (11.5)$$

where σ and ε are units of length and energy, respectively. When the distance between two interacting particles is reduced below the equilibrium distance r_0, a substantial amount of potential energy (or a large force) is required. When the confinement of the system is strong enough, particles may merge to form a new element, or they may break down and a number of subatomic particles be created. The very steep slope of the potential indicates the enormous energy that is required to bring particles at very close distance, something that is aimed at in fusion or splitting of atomic particles. Many interaction potentials are known today. At this point we may refer to the beginning of Chapter 2, where a simplified version of an atomic potential, namely a sine function expressing the variation of the interaction force with separation distance was used to estimate the ideal strength of materials. For the interactions between Si-atoms potentials were proposed by Stillinger and Weber (1985), Bazant and Kaxiras (1996), and Bazant, Kaxiras, and Justo (1997). Some of the proposed potentials include interactions between more than two atoms; see, for instance, Stillinger and Weber (1985) for silicon; we briefly return to multi-particle interactions at the end of this section. Atomic potentials can be used in molecular dynamics simulations of the behavior of solids; see, for instance, Holland and Marder (1999).

The attraction part of the atomic potential ($r > r_0$) very much resembles a softening stress-deformation curve for concrete at the macroscopic size/ scale level; compare, for example, Figure 11.14a with Figure 9.5, which shows force-deformation diagrams for concrete specimens of different size/scale (a factor 32 difference in size/scale was applied in these tests). After an initial rise, a decreasing interaction force is observed with increasing separation distance. The analogy could be as follows: a concrete plate is separated into two parts through the nucleation and growth of a crack at the weakest location in a specimen. In the dog-bone-shaped specimens used for the measurement of the curves in Figure 9.5 this is obviously in the neck region, but not necessarily at the smallest cross-section due to the heterogeneity of the concrete. Note that the compression regime is not shown in the diagrams of Figure 9.5. Compression was discussed at length in Chapter 8 and we further elaborate on that at the end of this section. Unless extremely large confinement is applied (which is very hard to achieve for large concrete specimens; see also Appendix 5), the huge energies are not measured as compared to the very small-distance energies resulting from Equation (11.5). This is just a matter of practical limitations; at very high confinement; after an initial pore-collapse stresses up to 1,500 MPa are easily reached; see, for instance, the tests by Schickert and Danssmann (1984). Beyond that stress-level not much is known about concrete, simply because too large specimens must be used, and devices such as a diamond anvil cell, which allow for the extreme high pressures that are needed, unfortunately cannot be used. But let us not deviate too much from the description of a model based on potentials, and return to the similarity observed for the attraction part of the stress–deformation curves. The variation in shape, from snap-back behavior for very large sizes, to stable softening for very small sizes may conveniently be captured by varying the exponents m and n in the LJ-potential, which is rewritten as

$$\frac{F}{F_u} = \alpha\left[\left(\frac{\sigma}{r}\right)^m - \left(\frac{\sigma}{r}\right)^n\right] \tag{11.6}$$

Note that the state variables are force F and separation distance (or displacement) r. A convenient choice of the parameter α and the two powers m and n allows constructing an accurate fit to physical experiments like those shown in Figure 9.5. An example of a set of curves obtained with $\sigma = 1$ (separation distance scaling), $n = 6$ and $\alpha = -4$ leads to the family of curves of Figure 11.15 when m is varied between 2 and 20. When $n = m$, the trivial result $F = F_u = 0$ is obtained. Thus, at any given size/scale-level defined by L the maximum load $F_u(L)$ and the shape of the interaction potential should be fitted to the outcome of the experiments by properly selected values for the free parameters. In line with the foregoing discussions in the previous chapters we are here not interested so much in arriving at a close fit; rather we would be interested to learn what the underlying reasons are for the specific shape of the potentials and the implications of using them. At the meso-level some of them were already recognized (see Chapters 6 and 10). In contrast, much

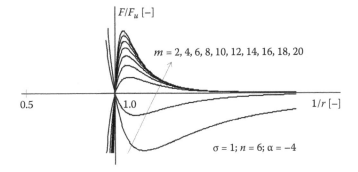

FIGURE 11.15
Set of interaction potentials through variation of the exponent m in Equation (11.6), which can be used to model the size-dependency of softening curves (attraction part of these curves only). The other parameters are constant at the values indicated in the graph. Different types of behavior are observed; the main problem is understanding the underlying physical mechanisms. (From Van Mier. 2007. *Int. J. Fract.*, 143(1): 41–78. With permission from Springer.)

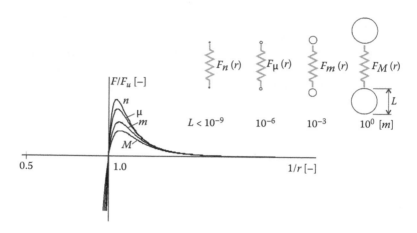

FIGURE 11.16
Simple particle (or lattice) model for the analysis of the mechanical behavior of solids based on pair-potentials. Note that the same mechanism appears for any size/scale-level L. At each level a different interaction potential $F(r)$ must be used, for instance, characterized by the size-dependent potential Equation (11.6), visualized in Figure 11.15. The subscripts n, μ, m, and M are used to distinguish between the atomic (nano- and smaller), micro-, meso- (or intermediate-), and macro-size/scale-levels, respectively. Note that the separation distance, which is here shown by a spring with clear physical length, has actually zero length. Therefore, at the atomic level we are dealing with an almost ideal point-contact, whereas at the other three levels the physical dimensions of the particles increases to a size that cannot be ignored in any analysis.

knowledge is still lacking at the size/scale of the hardened Portland cement and smaller.

The important aspect of the approach proposed here is that at any size/scale-level the material is thought to be composed of (rigid) particles or material points. Thus not only at the atomic-level, but also at the macroscopic size/scale-level we lump the entire F–r behavior in the contact law between two neighboring particles as shown in Figure 11.16. In this figure four distinct size/scale-levels have been included: the atomic-/nano-level (n), the micro-level (μ), the meso- (or intermediate-) level (m) and the macro-level (M). The size of the particles in contact increases from smaller than 10^{-9} [m] to 10^0 [m] at the macro-level, as indicated. The spring indicating the interaction force has in reality zero dimension. The pair-potential can be the LJ-potential or any other suitable function. The importance is to recognize that the entire attraction curve (from the equilibrium state to full separation; at both points the interaction force is zero) must be included for describing the interaction between two neighboring rigid particles (spheres may initially be assumed to simplify the problem). In softening models for concrete fracture (see Section 2.4, Equations (2.25)–(2.27)) the part leading to the maximum cohesive stress is usually ignored and linear-elastic behavior is assumed to describe the pre-peak behavior. The main difficulty, however, lies in the fact that two distinct

TABLE 11.1

Identified Mechanisms in the Fracture of Cement and Concrete at Size/Scale Level (L-1) Underlying the Interaction Potential at the Higher Size/Scale Level L

Size/Scale Level (L)	Size/Scale Level (L-1)	Identified Mechanisms at (L-1)
Macro (*M*)	Meso (*m*)	(1) Prepeak microcracking defines maximum tensile force (Figure 6.3)
		(2) Rapid load-decrease postpeak signifies decreasing contact area (Figure A2.1)
		(3) Aggregate- and fiber-bridging (in FRC) cause a stable tail postpeak (Figures 10.5, 10.6)
Meso (*m*)	Micro (μ)	(1) Crack-face bridging appears in the tail of the diagram (Figure A4.6c)
		(2) Capillary forces in water bridges affect the tensile strength (Figure 11.12)
Micro (μ)	Atomic (*n*)	(1) Capillary forces in water bridges affect tensile strength (Figure 11.12)
		(2) Equilibrium between attractive and repulsive atomic forces defines the shape of the diagram (Figure 11.14a)

regimes are distinguished, each relying on different state variables: prepeak behavior is described in terms of stress and strain, the postpeak behavior in terms of stress and displacement. The potentials suggested here are all based on force and displacement, in the entire diagram, and in this way a discontinuity in the model is prevented. This is certainly an improvement, yet it has far-reaching consequences for the mechanics and the type of experiments needed for determining the requisite model parameters. We return to these issues in Section 11.4. In order to clarify the hierarchy between the size-/scale-levels it helps to identify the main mechanisms leading to the specific shape of the interaction potentials at each level. Table 11.1 shows which mechanisms at size-/scale-level L-1 are responsible for the behavior observed at size/scale L, that is, one level higher. At the smallest size-/scale-levels physical forces are active, and one may wonder to what extent purely mechanical models still apply. Instead a full-fledged materials science approach is the only suitable approach at the very small size-/scale-levels.

Now, with the above description of the interaction potential at different size-/scale-levels the model is not yet complete. One of the extensions needed concerns multiple particle interactions. Thus, not only the pair-potential discussed thus far, but also triplet-potentials and higher-order interactions involving several interacting particles simultaneously must be considered. The general expression for the potential energy function Φ describing the interactions between N identical particles can be written as (Stillinger and Weber 1985):

$$\Phi(1,\ldots,N) = \sum_i v_1(i) + \sum_{\substack{i,j \\ i<j}} v_2(i,j) + \sum_{\substack{i,j,k \\ i<j<k}} v_3(i,j,k) + \ldots + v_n(1,\ldots N)$$

(11.7)

The single particle potential v_1 describes external forces and wall effects. The pair- and triple-potentials v_2 and v_3 can be written as:

$$v_2(r_{ij}) = \varepsilon f_2(r_{ij}/\sigma)$$

$$v_3(\mathbf{r}_i, \mathbf{r}_j, \mathbf{r}_k) = \varepsilon f_3(\mathbf{r}_1/\sigma, \mathbf{r}_2/\sigma, \mathbf{r}_3/\sigma)$$

(11.8)

where ε and σ are again units of energy and length, respectively, as introduced before in Equation (11.5). What can be seen here is that f_2 is a function of particle distance only, whereas f_3 is function of all translations and rotations between the three considered particles. The problem thus becomes notably more complicated. The LJ-potential that we showed in Equation (11.5) is mostly used for describing the interaction between atoms in noble gases including argon, krypton, or xenon. For silicon, Stillinger and Weber (1985) proposed a different function for f_2, and added one for f_3, which were used in molecular dynamics simulations. Introducing rotations in the formulation is equivalent to incorporating "structural" effects. As a matter of fact, using a beam-lattice model, such as the one presented before in Chapters 3 and 4, with various applications in later chapters, is just that: rotations between the elements have become part of the formulation, and are even used in defining fracture in the lattice (viz. by using a fracture law based on normal force and bending such as Equation (3.36)). One important consideration in using these potentials is that the particles are considered rigid and cannot be separated into smaller elements. If that is expected to occur, as in the unhydrated cement grain of Figure 11.2, one simply has to decrease one size-/scale-level, namely, the size/scale of the constituting particles of the unhydrated cement.

We have indicated the similarity between lattice and particle models in Section 3.4. Particle models have the obvious advantage that friction between the particles can be included, which is an important aspect of the behavior of materials such as concrete and other particulate geomaterials (rock, sand, ice) subjected to compressive loadings. We speculated about that in Chapter 8. Just considering the pair-potential (i.e., including the normal interaction between particles only) would not suffice; instead also shear force-slip relations must be included as sketched in Figure 11.14b. This is common practice in particle models; see, for instance, Cundall and Strack (1979). In a lattice model this could be mimicked by selecting the correct higher-order interactions that describe the nodal rotations.

Thus by carefully selecting pair- and higher-order potentials the response of multi-particle systems may be extended to compressive fracture. In order to finalize the discussion of Chapter 8, let us have a closer look at Figure 11.17. In Figure 11.17a the structure of sandstone with equally sized sand grains that are bound together by $CaCO_3$ is shown. The calcium-carbonate layers between neighboring sand-grains are very thin, leaving a large porosity in the material. The enormous porosity of such sandstones, for instance, for Felser sandstone where sand grains are bound with clay, Hettema (1996) reported porosity values up to 21.1%, which is the reason for the relatively low compressive strength. In concrete the particle structure and the cement content are selected in such a way that the large voids between the rough aggregates are filled completely, thereby reducing the porosity and increasing the compressive strength.

Another factor defining the strength of granular media such as sandstone is the quality of the "glue" between the particles. In the aforementioned Felser sandstone, particles are bound together by means of clay, and water may have a significant influence on the strength of the sandstone; see, for instance, Visser (1998). But, let us return to cement and concrete. Before hydration a complete hierarchy of small and large grains is found in concrete, as

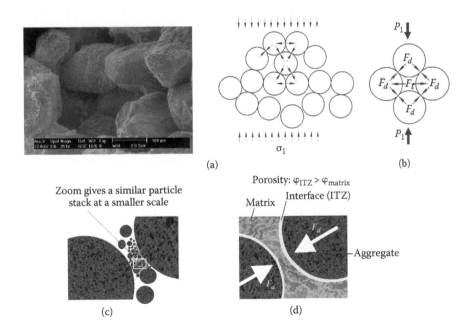

(a)

(b)

Zoom gives a similar particle stack at a smaller scale

(c)

Porosity: $\varphi_{ITZ} > \varphi_{matrix}$

Matrix Interface (ITZ)

Aggregate

(d)

FIGURE 11.17
Structure of sandstone: mono-sized sand grains are glued together by means of thin layers of lime, clay, or other adhesive material (a); mechanical response of the grain skeleton under external compression (b); in contrast to sandstone the particle structure of concrete contains many small particles (silica, sand, and cement grains) (c); after hydration the matrix forms a more-or-less solid material where the particle structure is less pronounced, at least at the meso-size/scale considered here (d). (From Van Mier. 2007. *Int. J. Fract.*, 143(1): 41–78. With permission from Springer.)

shown in Figure 11.17c. The application of the interaction potentials would require many different parameters to capture interactions between particles of different size. Considering that in modern concrete very small filler materials are added to close the smallest voids in an effort to increase the strength, the size-scale range of the particles varies between 10^{-7} and 10^{-2} [m], about 5 orders of magnitude. Working with an "average" potential could be useful at the beginning, but likely this would not suffice. After the cement hydrates, the matrix material between the rough aggregate particles forms a more-or-less continuous solid as shown in Figure 11.17d. This means that the larger aggregates with a size-/scale-range between 10^{-3} and 10^{-2} [m] can be interpreted as approximately mono-sized, and a model based on a single "averaged" potential may well work. This example may show that one has to be careful in selecting the right interaction potential, at the size/scale levels important for the material at a given moment of its history (e.g., unhydrated, partly hydrated, or fully mature).

11.4 Structural Lattice Approach

In the classical framework for structural analysis, three sets of equations must be solved: the equilibrium equations describing the relation between internal stresses $[\sigma]$ and external loads $[F]$, the kinematic equations that describe the relationship between the overall displacements $[u]$ and strains $[\varepsilon]$, and thirdly, the constitutive equations, describing the stress $[\sigma]$ –strain $[\varepsilon]$ relations. Constitutive behavior is what we mostly discussed thus far. There appears to be a certain indeterminacy when it comes to defining appropriate parameters for fracture of concrete. When it comes to softening, the use of strain is not realistic any longer, but considering 3D localization, the notion of stress is no longer very useful either. The above-mentioned approach in structural analysis implicitly assumes that the material can be seen as a continuum; it does not apply when deformations localize in a crack or in a shear band. The discussion in the previous section points toward an alternative approach for fracture where the constitutive equations are replaced by a direct relation between force and displacement for an element in, for instance, a lattice or particle model. This means that the interaction potential $(F–r)$, describing the relation between the applied normal force F and the crack opening r in tension, as we have named these relationships in the previous section is regarded as a structural property, and the notion of "material parameter" is abandoned altogether. In view of the size and boundary condition effects on the softening diagram used in cohesive models, just going one step back and negating the existence of appropriate material parameters for fracture seems the best possible approach that needs to be investigated in coming years. Doing so has enormous implications, not only for the

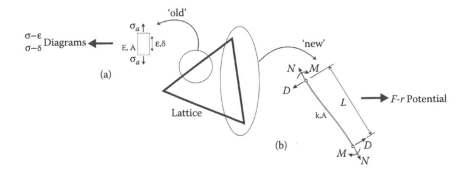

FIGURE 11.18
(a) In the classical approach, a material element is isolated from a structural lattice, and the properties of the material are determined in a separate test in which uniform stress is applied to the material specimen. The material specimen needs to be an order of magnitude larger than the defining material structural elements in the RVE for the material. (b) In the structural approach the properties of a complete lattice element are determined, of given size L, and under the governing boundary conditions of its position in a global lattice. The resulting interaction potential $F-r$ may have any shape, for example, the one discussed in Section 11.3.

theoretical framework, but also for the type of experiments needed to derive the structural properties of lattice elements. Figure 11.18 shows the difference between the old situation where one still would rely on constitutive equations and the newly proposed approach where one directly focuses on the relation between interaction potentials for a complete structural element, implying that structural size and boundary conditions form an integral part of the potential. It takes some time to get used to this weird idea, but in view of the problems around the definition of cohesive fracture properties it certainly is a workable hypothesis.

The structural approach requires that for a lattice with given element size, exactly for that element size and the governing boundary conditions for the particular place in the lattice, the structural properties must be determined. Here just for the sake of clarifying matters a bit better we return to Equation (3.11):

$$k = T^T C^T \bar{S} CTv = Sv \qquad (11.9)$$

which describes the relation between forces and displacements in a linear-elastic element. The matrix $S = T^T C^T \bar{S} CT$ describes the structural behavior of the element, and no further detail is considered. We may actually consider the product of the combination matrix and the local stiffness matrix as one, which leads to $S = T^T KT$. The constitutive equations are simply discarded; what matters are just the structural properties of the entire element via the structural matrix K. In Figure 11.18b the "new" approach is shown. The essential property for the analysis of the global lattice is the interaction potential $F-r$. Note that the interaction potential will, in general, be a highly nonlinear function in which all effects caused by element size and boundary conditions are included. For practical purposes it seems appropriate to distinguish two or three basic lattice element types in a global lattice. The structural element of Figure 11.18b is just an example: a beam element in a 2D-lattice with nodal forces and moments; the support conditions are set by the flexural constraints of the other lattice beams connected in the same node. Thus, lattice elements located near the free surfaces of the global lattice will have a different connectivity, which will affect the rotational stiffness in the two nodes of the beam.

Finally, it seems necessary to emphasize once more the role of heterogeneity. In the lattice model as it was applied in Chapters 6 through 9 the coarse heterogeneity observed at the meso-size/scale of concrete was incorporated directly in the model. In Figure 11.19a we show a small lattice consisting of just two elements. The two elements are placed symmetrically with respect to the loading axis: at the top a displacement δ is applied which leads to tensile loading in both elements. For the argument it is not important to specify whether the elements are beams or trusses. If all properties of the two elements are identical (EA, EI, f_t, ν, etc.) the structure will deform symmetrically. This is all right considering the mathematics, but it will never correspond to physical reality in general. Concrete is perhaps, next to rock, one of the most extreme examples. The variability of the material properties is so large, caused by the rough meso-scale heterogeneity, that symmetry in the behavior of a structure such as that in Figure 11.19a is hard to imagine; if symmetry occurs that would be sheer coincidence. In Chapter 4 we discussed various ways of including heterogeneity in a lattice model. When the principle of interaction potentials is used, the only variability would come from differences in the connectivity from the lattice elements, unless, as in Section 4.5 the material structure is projected on top of the lattice structure and different properties

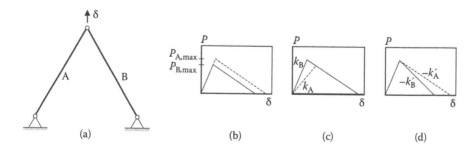

(a)   (b)   (c)   (d)

FIGURE 11.19

(a) Small two-dimensional lattice consisting of just two identical elements A and B placed at an angle with the loading direction. The structure is loaded at the top by a displacement δ. When the two elements have exactly the same interaction potential $F–r$, the deformation of the global structure will remain symmetric. When variations are made like those depicted in (b)–(d), the global structure will deform asymmetrically and flex to one side. This is exactly what experiments on cement and concrete will always show (also see, e.g., Figure 2.11).

are assigned to the various elements according to the local properties of the material phases in which they would fall. If variations in (simplified bilinear) interaction potentials are made, as shown in Figures 11.19b–d, we have made full circle, and it would seem we return to the FCM, in particular Figure 2.8 where various forms of softening curves are shown. Of course this is just a superficial correspondence. Here we obtain asymmetric behavior, and mechanisms such as the nonuniform opening in tension between fixed platen loading (Figures 6.9–6.10) emerge naturally from the analysis of the structure of Figure 11.19a. Bringing in a variation of the maximum load an element can carry (Figure 11.19b), the initial stiffness (Figure 11.19b), or the softening slope (Figure 11.19d), in all cases, nonsymmetric global structural behavior would be calculated. Some researchers have claimed to find the asymmetric mode without including variability of local properties, for example, De Borst (1986). Close scrutiny of these results shows, however, that extremely small load steps were made, small enough to take advantage of a tiny round-off error in corner elements where localization would start. It is amazing that some researchers do feel it necessary not to reveal such "tricks." When next comparisons to experimental results are made, one must take care to choose the boundary conditions in complete agreement with the experiment. For example, the erroneous claim that boundary rotations in tensile experiments would not affect the global softening behavior by Rots and De Borst (1989), can simply be traced back to fundamental errors made in the interpretation of the boundary conditions used in their simulations (see also Appendix 2.2).

Thus, in the end we have returned to considering the interaction potentials as the structural property of a lattice element. Because of the similarity between lattice and particle models, this can equally well be seen as a model of the contact laws between the particles. In a beam lattice, the flexural stiffness of the elements could be used to simulate friction; other possibilities that can be translated directly from the lattice model can be found in Section 3.5.

For modeling fracture the suggested fundamental modification of mechanics theory seems quite essential. Ignoring the constitutive equations is a simple step. At the experimental front no further discussion is needed as to whether an element would correspond to the demands for an RVE; these matters have now become totally superfluous.

12

Conclusions and Outlook

Science is what you know, philosophy is what you don't know.

Bertrand Russell (1872–1970)

I noticed the above line during a recent visit to a museum in The Netherlands. It was printed on a wall in an exhibition gallery. How true this expression is: it was a good reminder of how we fare in science and technology. It also expresses in a wonderful way the contents of this book: from the basic mathematical theory of linear elasticity to the more elusive debates on applying multiscale approaches, which stand at center stage in current fracture research. Although this book is about fracture of concrete, the way of approaching the topic proceeds in a similar way to that of many other branches of engineering. Engineers are interested in models. Models describe and summarize our understanding about a certain area of interest in terms of mathematics, in almost all cases interlaced with empirical content. Including empiricism to some extent is due to the fact that an area can be so complex, containing many influence factors that are barely understood, that it is impossible to even start thinking about an appropriate mathematical model that would capture all these factors. In that respect, the mechanical behavior of materials is an easy subject area, although here we already encounter substantial difficulties. The book enfolds very much as summarized in this powerful expression of Bertrand Russell. In linear elasticity, as presented in Chapter 3 one simply follows the rules laid out in the theory and one comes to the (exact) solution. A mathematical mind is what is required; understanding the language and knowledge of the logical rules are the essential tools. Linear elasticity is a well-defined mathematical construct. As soon as a comparison is made to real-world behavior, as we see it and interpret it from physical experiments, complications arise. Suddenly it is recognized that the world is not as black and white as the mathematics suggest, and some of the elasticity parameters may perhaps be understood in a different way. Do our interpretations have to develop along the lines set out by mechanics, or are alternatives possible, or even necessary?

Dealing with fracture is inherently more complicated than applying linear elasticity. In fracture one is confronted with a limit state: how do materials and structures break (fail)? How can we capture the observed behavior in a useful and correct model? I am hesitant to use the term "theory" because this seems to imply that the construct stands as a rock, and can only be turned over by extremely new insights. A theory is more than a model. Somehow

257

the word implies that the insights are well developed and stand for eternity. Over the years I have learned that we have to be careful with this terminology, and if we are to qualify a model with words such as "theory" or "law" (which is another qualification that one should use sparsely) it is very dependent on our ability to prove that the parameters used, and which provide the link to real-world behavior are properly defined, and can be determined from well-defined and independent experiments. The limitations of any approach should be known in as much detail as the phenomena that can be described. This attitude is quite lacking in modern science and engineering, to the point where a certain arrogance starts to affect progress in the field.

Exclamations such as "My theory (or law) is better than your model," are true killers, in particular when the considered theory or law, like the competing model relies heavily on curve-fitting. Not willing to admit that the parameters used in a theory, a law, or a model, are not as well defined as would be required, is a way of falling short of conducting science and engineering in a responsible manner. It is very damaging, and may lead again to delaying progress in the field. Perhaps this all has to do with the extreme competition to which scientists and engineers are driven today. The situation seems to depend to a large extent on the way research has been funded over the last decades, where the best strategy appears to call out, "My theory is the best." The person with the loudest voice may (temporarily) win, but in the end, deepening our insight over time, as well as the accumulation of further experimental data may point to quite another winner.

Knowledge and insight in the fracture behavior of materials and structures is what this book is all about. The models, and for a few decades, also numerical simulations, are just tools to arrive at a better understanding of the phenomena at hand. They should not, as in 95% of the publications in the field of fracture mechanics of concrete, be aimed at deriving the closest possible fit to experimental data. The reasons are manifold, but it is obvious that the empiricism brought in to make the close fit is not helping very much to improve our understanding of fracture mechanisms in concrete. It is a golden rule in fracture mechanics that in order to carry out the mechanics, the mechanisms need to be known first. How can we summarize our knowledge about fracture mechanics today in a coherent, not necessarily mathematical, framework? Chapters 6 through 9 set the framework. They define our current state of knowledge about tensile, mixed-mode, and compressive fracture, as well as the effect of size/scale on fracture behavior. A limitation had to be made: only quasi-static loading is considered; no rate effects (thus, very fast loading (impact) and slow loading rates (creep) are not discussed), which, however, in the end should become an integral part of these musings. In setting the framework, it directly becomes clear that classical fracture theories and models, developed for other materials such as glass and metal, are not always are the most appropriate choice for concrete. The material has a very rough, distinct particle structure at the intermediate level, the meso-level ([mm]-size/scale), where, when it comes to understanding some of the important details, one cannot simply ignore the complexities of the hardened (and also, hardening) cement structure at the micro-level ([μm]-size/scale). In particular for understanding the structure and properties of the interfacial transition zone between cement matrix and aggregates (or fibers) the behavior of cement is an important factor. Considerations about the material's microstructure are relatively new in mechanical models aimed at describing the fracture behavior of concrete, only since the development of the finite element method, including the structure of the material in the model has become feasible. One of my own first endeavors was published 25 years ago (Van Mier 1986b); an example that has not been previously published has become part of a somewhat historical overview in Van Mier and Man (2009). In Chapters 4 and 11 we discussed in detail the structure of cement and concrete and the implications for the mechanical and fracture behavior. The heterogeneity of these materials is very much the source of deviations from classical mathematical models for fracture, in particular those models based on presumed homogeneity or uniformity of the material. These erroneous ideas lead to proposing continuum-based models where the effects of the heterogeneity of the material are assumed to be effectively captured by using nonlinear fracture criteria.

Models expire after a certain date; theories and laws are for a (limited) eternity. New insights, mostly from experiments, will eventually lead to overturning a certain approach for a new one. This means that models, theories, and laws, when applied in the gray area between true mathematics and physical reality, are constantly revised. They stand for second-hand knowledge, where the true knowledge is the insight directly obtained from physical experiments. Experimental results stand the flow of time much better. I am always amazed to read that much of our modern-day insight in shear banding dates back for almost a whole century. Carefully conducted experiments carried out in those days such as Nádai (1924) and Richardt et al. (1929) still contain many insights that haven't been replaced today by repeating the same tests using modern computer-controlled devices. The modern devices are excellent tools that help to get a more detailed view of the mechanical behavior of materials and structures; yet, one should not overestimate their role. Interpretation of experimental results is still a human task, where it is often quite essential to combine knowledge from different areas to come to truly new insights and understanding. But let us not get too "philosophical" and return to the heart of the matter: understanding fracture of concrete and cement.

12.1 Fracture Mechanisms

First let us review our knowledge about fracture mechanisms. Seemingly impulsive conclusions, such as those made in conjunction with the development of the Fictitious Crack Model (FCM, proposed by Hillerborg, Modeér, and Peterson (1976) and its continuum counterpart, the crack band model (CBM), namely, that softening can be explained from a "cloud of microcracks advancing in front of a stress-free macrocrack" bring confusion and appear to be false afterwards. Under global uniaxial tension the fracture process can be divided in four distinct stages, namely (0) elastic stage, (A) stable microcracking, (B) unstable macrocracking, and (C) bridging. Stage (A) is related to the prepeak stress–deformation curve in tension, and not the softening curve as assumed in FCM and CBM. Distributed microcracking is found in the prepeak regime. The microcracking is "relatively stable" under global tension, but the situation is very much improved under global compression. In the latter case the external loading must increase to propagate the microcracks that tend to align in the direction of (compressive) loading. This explains the much higher compressive strength of concrete. Controlling microcracking, for instance, by means of fibers added to the concrete may also enhance the tensile strength of concrete.

The softening part of the stress–deformation curve is related to the growth of an unstable macrocrack (under global tension) or shear band (under global (confined) compression), which causes a gradual decrease of the cross-sectional area that can carry load. The softening regime can be referred to as the stage where localization of deformations occurs, both in tension and in compression. This means that the deformations mostly occur in the main crack separating a test specimen in two parts. This is indeed, the most important contribution made in the Fictitious Crack Model in 1976. By recognizing this important phenomenon during softening, the picture is not complete because the growth of the major crack is to some extent stabilized or slowed down by means of crack-face bridges, that is, intact material ligaments that connect the parts of the test specimen in the wake of the growing macrocrack. In compression the shear band is slowed down by frictional restraint in the band, which may eventually lead to considerable residual carrying capacity.

It will not go unnoticed that global tension and global (confined) compression are treated here as if they were identical phenomena. Indeed they are: there is a large resemblance between failing concrete in tension and in compression. Differences appear where the stability of the prepeak microcracking is concerned, and the modus of the main stable macrocrack leading to localization of deformations. Also under compression localization of deformations has been demonstrated, initially by me in 1984; subsequently confirmed many times. Differences with global tension during the softening stage relate to the crack mode (mode I in tension, and mode II in compression, to use the conventional notation from classical fracture mechanics), and to the way the main crack is stabilized (crack-face bridging versus frictional restraint).

Details about the fracture process under global tension, mixed-mode tension and shear, and compression are revealed by means of meso-scale models of concrete. Such simulations can be considered as the first multiscale analyses of the mechanical behavior of concrete. The first to try this were

Roelfstra, Sadouki, and Wittmann (1985) and was named "numerical concrete." The lattice analyses shown in Chapters 6 and 7 show that a local mode I criterion suffices to simulate crack mechanisms under mode I and mixed-mode I and II correctly. Mode III (torsion or out-of-plane shear) is captured realistically as well, as long as the confinement in the longitudinal direction is limited. The analyses show that, although the fracture mechanisms are simulated correctly, getting the overall load-deformation behavior right is less straightforward. Using a local elastic purely brittle fracture law leads in many of the examples to underestimating the global strength and the pre-peak brittleness is often overestimated. Using a softening model at the meso-level might seem the solution, yet there are strong arguments against such an approach, as discussed in the next section.

The prepeak microcracking can be attributed to the weak strength of the interfacial transition zone between aggregates and cement matrix. Because the structure of this transition zone very much depends on how the water moves in the fresh mixture during the hydration process, and because of its flimsy dimensions, it is mostly the properties of the hardened cement paste itself that decide how the properties of the ITZ develop over time. For understanding in detail the formation of the ITZ as well as estimating its properties it is necessary to repeat the meso-level analyses mentioned before at a smaller size/scale: with this step the multiscale approach is complete. An interesting observation is that many features of the fracture mechanisms observed at the meso- and macro-level reappear at the micro-size/scale level. For example, bridging in the final stages of tensile fracture has been identified, but needs further confirmation. However, there are several complications at the cement level, namely the structure of the material is inherently more complicated and moisture plays an important role. Except for the fact that several phases are recognized in the hydrated cement structure, and that they can be identified by means of indentation tests, detailed knowledge about the local cement properties is very much lacking. A wide field of experimentation lies open for the near future. A number of potentially useful experimental ideas have been included in Chapter 11.

12.2 Theoretical Models

For explaining fracture of concrete and cement several types of models have been used throughout the book, but foremost criticized. The main criticism relates to the unrestricted use of softening (sometimes even referred to as strain-softening, which of course is false simply because deformations localize during the softening stage) in many macroscopic fracture models for concrete. In the first place the material is usually considered as an isotropic continuum; in the second place handling negative material stiffnesses during softening leads to many complications and errors that are avoided by considering fracture over several size/scales, and through the inclusion of the heterogeneous structure of the material. The importance of material heterogeneity was shown in Section 11.4. For a material like concrete it seems just not possible to neglect its coarse heterogeneity because it will always lead to deviations from the exact mathematics of continuum theory. For the derivation of FCM (or CBM) usually a single element is considered, and over its length at some point the element will break; that is, a crack will appear that separates the element in two parts. It is not important where this happens; it is important that the element stress σ will decrease with increasing deformation δ. The point that is missed, and which is caused by assuming that an average state variable such as stress can still be used to describe the loading on the element, is that the localization will not spread over the full cross-section of the element instantaneously, but it will take time for the crack to traverse the cross-section. This is just the same simple idea as that of localization in the direction of loading: the heterogeneity of the material (and with that the nonuniformity of the internal stress distribution) will simply decide where cracking starts first, and how the subsequent crack growth will unfold. Now, one may argue that for large structures, well beyond the representative volume element, average state variables including stress and strain can well be used. Yet, as soon as cracks start to develop, causing a discontinuous deformation field, the continuum theory will simply not relate to the physical reality. A theory must be used that can capture the displacement jumps. In higher-order continuum theories this is just what is tried to be accomplished. By separating the total deformation in various contributions such as elasticity outside the cracked zone, strains within the cracked zone (where the notion of strain can only be maintained by introducing an internal length scale) it is to some extent possible to construct a displacement jump in the theory. The costs for this construct is an increase of complexity, and a new parameter must be introduced, the internal length scale, which has no clear physical basis. Indeed, as shown by Iacono (2007), when trying to model the behavior of specimens of different size by means of a gradient-enhanced continuum damage theory, it appeared that for each specimen size another value for the length parameter was needed.

So, to make a long story short: the original idea in FCM was the notion of localization of deformations. What is missed is that the localization is three-dimensional, and thus not only develops in the direction of the applied load. The error that persists to date is that softening is assumed to be a material property. It clearly is not correct, and easy to demonstrate in experiments through a variation of boundary conditions (BC) and specimen size (see Figures 2.11 and 9.5). Through the provision of this experimental falsification (Popper in action!) the theory should have been abandoned two decades ago. As long as continuum state variables are used in a fracture problem, and the softening curve is seen as a material property, there will be no improvement in the predictive qualities of such models, no matter how many internal length scales are introduced; they will remain an artificial way to capture the effects from material heterogeneity in a continuum model.

Putting matters in an historical perspective may show that it was a necessary step to go through the development of the Fictitious Crack Model. Now that it has become clear that not only the crack but also some of the model assumptions are fictitious, it seems a better strategy to develop a new model based on the physics of fracture. The fracture process as sketched in the previous section may serve as a guideline.

In this book several models have been discussed and/or applied; they are the following:

- Linear elastic fracture mechanics (LEFM, in Chapters 2 and 10 and Appendix 2)
- Cohesive crack models (FCM and CBM for concrete, in Chapter 2)
- Micromechanical models incorporating the structure of the material (Lattice model in Chapters 3–9)
- Multiscale models, extending over three or more size/scales (Chapter 11).

Normal procedure in many publications is to show the advantages of a certain model approach only. This is no surprise because future funding may depend on (presumed) success. Yet, it is unfair to newcomers in a field, who usually show no preference to any approach, but would like to hear about the advantages as well as the disadvantages of certain approaches. So, let us examine the four abovementioned approaches a little more closely.

- *Linear elastic fracture mechanics.* This theory is based on the stress-intensity at the tip of a slitlike crack. The theory can be applied at any size/scale. The determination of the critical stress-intensity factor is the crucial problem. Boundary and size effects are incorporated by means of a geometrical function, which, in many cases can only be determined in an approximate manner. The elegance of this theory lies in the separation between material (critical stress-intensity factor) and size and BC-related influences (geometrical function). The disadvantage is that for every new problem the geometrical function must be determined. For three-dimensional crack geometries this may be a tedious job. Of course the same disadvantage applies to the 4-stage fracture model that was discussed in Chapter 10.

- *Cohesive crack models. These* were developed with the idea of incorporating local nonlinear effects from crack-tip plasticity in a discrete fracture model. The original model was derived for metals, in which the size of the plastic zone is small compared to the structure size. Equivalent models for concrete have been proposed (FCM and CBM), but because of the extreme low tensile strength of this material, the plastic zone, or rather process zone, generally extends beyond the size of the considered structure. As in LEFM it is therefore important to incorporate BC and size effects. The models are preferentially used in conjunction with the finite element method. This latter method is assumed to take care of BC and size effects, but this is generally not correct. Mesh sensitivity is often a problem, mainly because the crack is not given sufficient freedom to grow in any desired direction. Numerical models based on LEFM generally fare much better in this respect (e.g., Ingraffea and Saouma 1984), where instead of having cracks growing through a predefined element mesh, the mesh is constantly adjusted depending on the growth direction of a crack. Softening is certainly not a material property, but is dependent on BC and size of the considered structure. It is by no means clear which softening relation must be used under certain structural conditions; it will vary from case to case.

- *Micromechanical models.* In these models the macroscopic behavior is calculated by incorporating the structure of the material directly. This is mostly done using a numerical model, such as the distinct element model, the lattice model, the finite element method, or another. The major disadvantage is that these approaches are computationally demanding, as shown in Appendix 1 (note that this disadvantage also applies to higher-order continuum models, where internal iteration loops tend to lengthen an analysis; this is, however, rarely mentioned). An advantage is that effects from material heterogeneity on the macroscopic fracture behavior can be explained. Size and BC effects can be analyzed in a simple and straightforward manner. The determination of the local fracture properties is possible only through inverse analysis, which does not guarantee uniqueness of a found parameter set.

- *Multiscale models. These* are similar to lattice or particle models, but, in comparison to the micromechanical models mentioned extend over more than two size/scale levels, for example, from micro- via meso- to macro-size/scale. The original idea is that results from the lowest size/scale are used as input in the meso-size/scale, and so on to the macroscopic level. The main issue here is to determine the interaction potential at the different size/scales. Moreover, it is not clear how the boundary conditions are included in the transition from one size/scale-level to the next one. It is common to use periodic boundaries, but this is a way of ignoring the real problem. The multiscale approach proposed in Chapter 11 returns to the fundamentals of mechanics. Constitutive equations are omitted completely, and the properties of the elements in a lattice model are directly determined as a structural property, thus including effects from the element size and the governing BC.

In all the aforementioned model approaches determination of the fracture parameters is of major concern. For LEFM applications, the process zone must be infinitely small, which means that the critical stress-intensity factor for a material such as concrete can only be determined by means of extremely large specimens (estimated size/scale range 1–10 [m], perhaps even larger). Cohesive crack models for concrete suffer from the fact that the required softening relation depends on specimen size and governing BC. For micromechanical models, which are basically a multiscale model at two size/scale-levels, the determination of the model parameters is also considered a major problem. Here inverse analysis is the common way out, but uniqueness of the parameters is not guaranteed. The same is true for higher-order continuum models in which an internal length-scale must be determined. In a more general way it can be stated that the "potentials" that must be used in any fracture model always presents the largest obstacle. Over the years it has become very clear that an intensive and fundamental search for the physical processes leading to fracture must be carried out, which would need to extend over all size/scale-levels, from nano- to macro-level. There is great need for new original experiments; less for additional simulations using models that suffer from the aforementioned fundamental problem of parameter identification.

Finally, there is a natural tendency for engineers to favor continuum-based theories. This is no surprise: it evolves from the way mechanics, materials, and structures are taught in our curricula. For fracture models it is difficult, if not impossible to measure material properties independent of size and BC effects. If this problem cannot be solved, and the outlook for a solution is certainly not bright, it seems better avoiding it altogether. For this reason, lattice models, or particle models might be used at any size/scale-level, in which the constitutive equations are simply omitted, which means a fundamental change in the way we apply mechanics. This implies that the properties of structural elements need to be determined. Such structural elements have the actual size of the individual elements in the global lattice, and the boundary conditions as they appear in the connectivity to adjacent elements in the global lattice. Expectations that something like a property of a material might exist that can be used in a fracture model are then simply denied ab initio. All we see are structural properties, and in particular in limit-state situations such as we encounter when studying fracture the full consequences become visible.

Appendix 1: Some Notes on Computational Efficiency

The first analyses with the lattice model were conducted in 1990–1991 using the DIANA Finite Element Package, and were published in Schlangen and Van Mier (1991, 1992a). Those first computations used meshes of up to a few thousand elements: for example 4,812 elements were used in the SEN tensile tests reported in Schlangen and Van Mier (1992a), and were carried out on a SUN workstation. By modeling only the part of the specimen where cracks were expected to grow as a lattice, and the remainder of the structure by means of continuum elements (typically 8-noded isoparametric plane-stress elements that were available in DIANA), realistic specimen dimensions could be handled. With the increase of workstation capacity, performance increased and larger lattices could be analyzed. In 1993 the largest lattice of beam elements measured up to about 35,000 elements: a detailed fracture analysis of anchor pull-out. This latter analysis was performed at a CRAY system available at TU Delft (see Vervuurt et al. 1993b). It was considered essential to look for different solvers such as a conjugate gradient solver (see Schlangen and Garboczi 1996). In 2000 the lattice algorithm was implemented in ScaFiep (Scalable Finite Element Package) developed by Lingen (2000). This allowed the use of parallel computing systems, which proved quite useful for three-dimensional analyses; see Lilliu and Van Mier (2003). These latter analyses contained up to 449,179 lattice beam elements and were done at the SARA-NSF computer center in Amsterdam using SGI-Origin 3000 TERAS. From this moment on the use of continuum elements was abolished because in many cases it was observed that small but very energy-consuming cracks also developed outside the region of interest.

At present two different systems are used: an in-house Silicon Graphics system with 16 Itanium 2 processors (single core) and a CRAY XT3 computer system at CSCS in Manno, Switzerland. The latter system has 1,656 AMD Opteron processors (1,100 single core and 556 dual core), where maximum 1,024 can be used. The result of a benchmark simulation of a 3D lattice structure (974,403 beam elements) with 20 fracture steps using 1 to 1,024 processors is shown in Figure A1.1. The parallel speedup is almost perfect up to 256 processors. For most analyses that were, for example, published in Man and Van Mier (2008a), 64 processors proved to be quite optimal with regard to wall-clock time. A problem of 7,448,383 elements (3D beam lattice) would take about 60,000 CPU hours. This number may show a substantial variation depending on the number of fracture steps that must be made. Further improvements in software and hardware are needed for

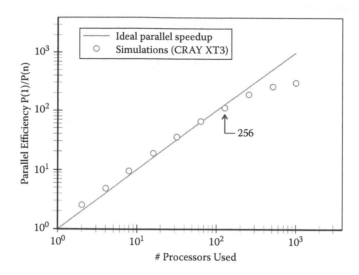

FIGURE A1.1

Parallel speedup in a benchmark analysis of a 3D lattice; 20 fracture steps were made. (From Man and Van Mier. 2008a. *Mech. Mater.*, 40(6): 470–486. With permission from Elsevier.)

including more detail in 3D analyses as they are carried out today. For further information see the most recent PhD thesis completed on the lattice model by Man (2009).

Appendix 2: Simple Results from Linear Elastic Fracture Mechanics

Ever since the pioneering work of Kaplan (1961), most researchers would agree that LEFM cannot be applied to materials like concrete and rock. The reason is that LEFM principles apply to single sharp cracks, and the presence of the microcracks can generally not be accounted for in this framework. Nevertheless, LEFM analyses can help to understand a few basic observations, and explain a few limit cases. In this appendix two cases are discussed: (1) the limit size effect on strength for very large samples, and (2) the effect of allowable rotations on the shape of the softening curve in tension.

A2.1 Size Effect

According to LEFM the stresses near the tip of a sharp crack in a thin plate (plane stress) can be written using Equation (2.11a) from Chapter 2:

$$\sigma = \frac{K_I}{\sqrt{\pi a}} f(\theta) \tag{A2.1}$$

where K_I is the mode I (opening mode) stress intensity factor, r and θ are the polar coordinates of a point close to the tip of the crack, and the function $f(\theta)$ is the geometrical factor that contains information on specimen geometry and boundary conditions. The half-length of the crack is equal to a. For a finite size structure $f(\theta)$ is often replaced by $f(a/W)$ for convenience. Now let us assume that for a notch length a_0 exactly the critical stress intensity K_{Ic} is reached and the structure starts to fail. The failure stress of the structure, with width W_1 can then be formulated as

$$\sigma_{p,1} = \frac{K_{Ic}}{f\left(\dfrac{a_0}{W_1}\right)\sqrt{\pi a_0}} \tag{A2.2}$$

For a second structure with size $W_2 > W_1$ and an initial notch size at which the maximum stress is reached such that $(a_0/W_1) = (a_0/W_2) = \xi$ leads to a strength ratio

$$\frac{\sigma_{p,2}}{\sigma_{p,1}} = \frac{\dfrac{K_{Ic}}{f(\xi)\sqrt{\xi}\sqrt{W_2}}}{\dfrac{K_{Ic}}{f(\xi)\sqrt{\xi}\sqrt{W_1}}} = \frac{\sqrt{W_1}}{\sqrt{W_2}} \tag{A2.3}$$

The strength ratio is thus proportional to the square root of the size ratio. For example for a structure having a size $W_2 = 4.W_1$ the peak stress is half of the peak stress for the smaller structure $\sigma_{p,1}$. On a log-log scale this leads to a slope $-1/2$ of the size-effect diagram.

A2.2 Boundary Rotations in Uniaxial Tension

In a similar way it is possible to calculate the shape of the softening branch, for example in a uniaxial tension test (see also Van Mier 1991a). The assumption made is that the main crack that fails a specimen starts to grow at the maximum stress in the stress-crack opening diagram. There is quite some experimental evidence to support this assumption. In the past this was shown, for example, using a photoelastic coating technique by Van Mier and Nooru-Mohamed (1990); see Figure A2.1. Using a contact-free method, for example, moiré interferometry (see, e.g., Raiss, Dougill, and Newman 1989) or digital image correlation (Sutton et al. 1983) the growth of macrocracks in the postpeak regime can also be visualized. The latter technique is nowadays commercially available, also in a three-dimensional version, and the resolution steadily increases (see Appendix 4).

Assuming that the macrocrack starts at peak stress σ_p with a small size a_0, the stress σ_1 after elongation of the macrocrack to length a_1 can be written as

$$\frac{\sigma_1}{\sigma_p} = \frac{f(\xi_0)}{f(\xi_1)}\left(\frac{\xi_0}{\xi_1}\right)^{0.5} \tag{A2.4}$$

The assumption is made that during macrocrack growth the critical stress intensity factor K_{Ic} governs the growth process and is constant. Moreover, the crack propagates in a straight path, perpendicular to the tensile loading. The critical stress intensity for the initial crack a_0 is exactly reached at the peak of the stress-crack opening diagram. In Equation (A2.4) the parameter $\xi_i = a_i/W$, where W is the width of the considered test specimen and a_i is the length of the macrocrack ($a_0 < a_i < W$). The key to the solution of Equation (A2.4) is in the geometrical factor $f(\xi)$. For uniaxial tension between freely rotating loading platens the approximate numerical solution given in the stress intensity handbook by Tada, Paris, and Irwin (1973) can be used, for the case between

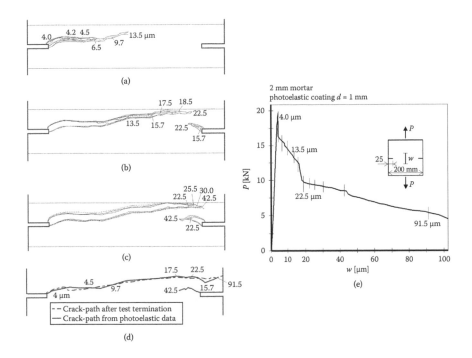

FIGURE A2.1

Crack propagation observed by means of photoelastic coating technique in a single-edge notched specimen subjected to uniaxial tension between fixed (nonrotating) loading platens. The main crack first starts at the left notch (a), jumps over to the right notch at 22.5 μm (b). Next the two cracks shield and propagate in a very slow and stable manner (c,d). (From Van Mier and Nooru-Mohamed. 1990. *Fracture Toughness and Fracture Energy: Test Methods for Concrete and Rock.* With permission from Elsevier.)

nonrotating (fixed) loading platens a more complicated numerical solution by Marchand, Parks, and Pelloux (1986) is used (see Equations (2.13)–(2.16) in Chapter 2). The result is plotted in Figure A2.2. Depending on the initial notch length a_0, the smooth shape of the softening curve between freely rotating platens (Figure A2.2a) follows a steeper or smoother descent after peak-stress. For the case of nonrotating platens the propagating crack is after a relative length $a_i/W \approx 0.3$ arrested and can only be propagated to larger lengths by slightly increasing the external load. For hardened cement paste such an increase of stress during the postpeak plateau (or bump as it has been called in the literature) has been observed in experiments (see, e.g., Van Mier 1991a) for a material like concrete, where the straight crack propagation is violated the plateau in the postpeak regime usually takes the form as shown in Figure A2.1. Thus, for concrete the given LEFM solution is quite approximate and does not cover the observations completely. Nevertheless the approach is simple and gives a first-order insight to what is happening during failure of a tensile specimen. The results show the correct behavior qualitatively; compare to the experimental results shown in Figure 2.11.

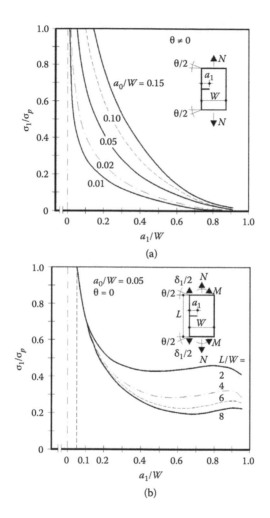

FIGURE A2.2

Relative stress σ_1/σ_p versus relative crack length a_1/W for a single-edge-notched tensile specimen loaded between (a) freely rotating loading platens ($\theta \neq 0$) and (b) nonrotating (or fixed) loading platens ($\theta = 0$). (From Van Mier. 1991a. *Cem. Conc. Res.*, 21(1): 1–15. With permission from Elsevier.)

Finally, it cannot be said too often: it is very important that the boundary conditions are treated with much care. For example, Rots and De Borst (1989) using a softening fracture model based on the crack-band model by Bažant and Oh (1983) conducted various simulations of the effect of the rotational stiffness of the loading platens on the behavior of a tensile specimen, quite similar to the LEFM solution shown here. They concluded that there was no difference between freely rotating and fixed boundaries and in both cases the bump in the softening branch would appear. Unfortunately, an error in the analyses was made, namely the lateral displacement of the loading point was prevented, thereby allowing a bending moment to develop also when the rotational stiffness was assumed to be zero. Such errors are easily made, and may lead to erroneous statements that do not help the debate.

Appendix 3: Stability of Fracture Experiments

In order to measure stable softening curves special experimental equipment is needed. Conventional open-loop systems do not provide feedback that allows capturing the decrease of load with increasing deformation, which is typical for softening. In such a loading device the load increases until something breaks, usually the specimen. Failure is uncontrolled, instantaneous, and a softening curve cannot be measured. By inserting some bars (e.g., aluminum) parallel to the specimen, it is possible to measure part of the softening regime, or, when softening is really gradual, the entire curve. Such systems, using parallel bars in an otherwise conventional load-controlled machine have, for example, been used by Hughes and Chapman (1966) and Evans and Marathe (1968); see Figure A3.1a. Basically the complete specimen-bar system is loaded, and after an experiment the contribution from the bars must be deducted to obtain the curves pertinent to the material under investigation (concrete and rock show a quite similar softening response). In Figure 2.10 some of the results obtained by Evans and Marathe (1968) were shown. These results have been central to the development of the Fictitious Crack Model, which is discussed in some detail in Chapter 2 as well. Modern machines are equipped with electronics that control the deformation and simultaneously can reduce the loading on a specimen. Such so-called servo-hydraulic test machines are fastest when using, as the terminology implies, hydraulic actuators. A double-acting actuator is needed in combination with a servovalve that can react rapidly to signals from the electronic control amplifier. In Figure A3.1b the setup is clarified for a uniaxial tension test (Van Mier and Shi 2002). The test is an illustration of a tensile test between cables. Although the cables are considered a soft element in the entire loading chain that might impair test-control, the obvious advantage is that specimens are loaded at a well defined (centric) point-load.

As shown further on in this appendix, it is still possible to achieve a stable softening response, in spite of the use of cables. The principle of servocontrolled hydraulic testing has been known for a longer time now, and most laboratories are equipped with such loading devices. A review of basic test techniques can be found in Hudson, Crouch, and Fairhurst (1972), Gettu et al. (1996), Van Mier (1997), Van Mier and Shi (2002), and other publications.

Using a fast closed-loop test machine is not the only requirement to measure stable softening behavior. The length over which the control measurement is read is also of great importance. In a simple and straightforward analysis, where the control length is varied from the entire specimen length L to zero, the significance of the measurement length in the test control

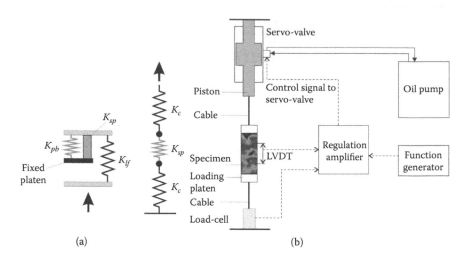

FIGURE A3.1
Schematic overview of the parallel-bar loading device used by Evans and Marathe (1968. *Mater. Struct. (RILEM)*, 1(1): 61–64) (a) and of a modern tensile test in a servocontrolled testing machine (b). *K* denotes the stiffness of the parts indicated; subscripts have the meaning (*pb*) parallel bar, (*sp*) specimen, (*lf*) loading frame, (*c*) cable. (From Van Mier and Shi. 2002. *Int. J. Solids Struct.*, 39: 3359–3372. With permission from Elsevier.)

feedback loop can be shown; see Figure A3.2. The total elongation comprises two parts: the crack opening and the elastic deformation within the control length following,

$$\delta_{tot} = \delta_{crack} + \varepsilon_e \cdot l_{meas} \tag{A3.1}$$

where δ_{crack} = the pure crack opening and $\varepsilon_e.l_{meas}$ is the elastic deformation within the measuring length. Because $\varepsilon_e = \sigma/E$ and in the softening regime σ decreases, the total elastic deformation decreases. When, in the softening regime, the crack opening becomes smaller than the elastic unloading deformation, snap-back behavior will occur, in this case when $l_{meas} = L$, as indicated.

Depending on the speed of the test-control system the snap-back may not be catastrophic, but rather a snap-through, after which the load can be restored and the tail of the softening curve can still be measured as shown in the quite extreme experimental result shown in Figure A3.3. In these experiments 200-mm square plates with a single 25-mm deep notch (as drawn in the inset of Figure A3.3) were loaded between nonrotating (fixed) end-platens. This created a rather critical situation and the servocontrol appeared to be just too slow to prevent the instability. An additional parameter was varied, namely the position of the LVDT used for test control. In the experiments shown in Figure A3.3a test control was over the average deformation measured with four LVDTs placed along the corners of a specimen; in the tests of Figure A3.3b test control was over the average deformation measured with

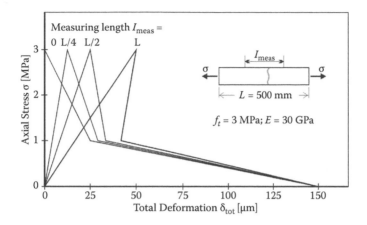

FIGURE A3.2
Effect of measuring length l_{meas} on the postpeak stress–deformation diagram in tension. The measuring length varies from 0 to L as indicated. The results are from a simple series model and show that snap-back may occur when the elastic unloading deformation within the measuring length becomes larger than the actual crack opening. The bilinear relation for $l_{meas} = 0$ is the assumed stress-crack opening behavior of the material.

two centrally placed LVDTs. The measuring length of these LVDTs was 65 mm, which was clearly too long to avoid snap-through. Reducing the measuring length to 35 mm solved the problem. Although one might be tempted to conclude that the results of Figure A3.3 can be used for further analysis, this is not the case. A clear unstable regime appears after the plateau in the softening curve (see Appendix 2 for an explanation of this plateau), and likely the part just after peak stress, that is, the first steep stress-drop in the softening regime was unstable as well. Stability was restored because of the fixed-platen boundary condition. By simply moving the control LVDTs away from the corners, toward the specimen's center, the tests became increasingly more unstable (compare Figures A3.3a and A3.3b). This means that the right choice of the location of the test-control parameter is quite essential with regard to the stability of the experiment. The experiments also showed the importance of the boundary conditions.

In general it is assumed that a stiff loading frame is required, but by using special advanced control parameters, a soft machine is actually quite sufficient. In size-effect tests it always needs to be decided what is scaled, and what is not. The most common scaling is in two dimensions, while keeping the size in the third dimension (thickness) constant. Scaling of the measuring length may be an issue. At best, of course, one would control an experiment over the true crack opening, but because of the construction of LVDTs and other measuring devices it is almost impossible not to include some elastic deformation, at least when unnotched specimens are tested. For notched specimens clip-gauges in the notch might be an appropriate choice. In the experiments shown in Figure 9.5 dog-bone-shaped specimens were loaded in

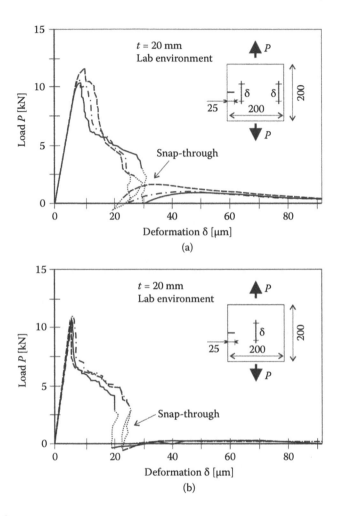

(a)

(b)

FIGURE A3.3

Load-crack opening curves for two times three tests between fixed (nonrotating) loading platens on single-edge notched concrete plates of 20-mm thickness. The snap-through happens
after the bump. At that moment a second crack propagates from the unnotched (right) side of
the specimen. Note that the part just beyond peak, when the crack starts from the notch is quite
steep which hints toward an almost unstable propagation. Test control differs in the tests of (a)
and (b), as indicated in the inset. (After Van Mier and Schlangen. 1989. *Fracture of Concrete and
Rock: Recent Developments.*)

uniaxial tension. A small, but scaled eccentricity was used to ascertain that the main crack leading to softening would always emanate from the same side of a specimen. The curved bays of the dog-bones caused a stress distribution similar as shown in Figures 6.9b and c. Thus, the largest stresses appear over a relatively long part of the specimen; the cross-section of the specimen varies over this length with high stress. Considering the heterogeneity of concrete, the main crack could thus appear anywhere in the high-stressed part of a specimen. A solution was to use an adaptive control system, similar to a method used earlier, among others, by Li, Kulkarni, and Shah (1993). The system comprises a maximum of 16 control LVDTs. The LVDT measuring the largest deformation is the control parameter in the closed-loop servohydraulic loading system. Fast switching between LVDTs is possible through a custom-made switch box. The measuring length of all LVDTs was chosen as short as possible, but also considering that in the largest test specimen a wide enough area was covered. These demands led to a measuring length of 75 mm. In Figure A3.4 an example is shown. The medium-size test specimen has a size $D = 400$ mm, leading to the smallest dimension in the neck of 240 mm. The thickness is 100 mm. In the diagram the curves measured with four LVDTs are shown. Three LVDTs were placed inline along the length of the curved bay at the side of the largest stress as indicated. Along the other side

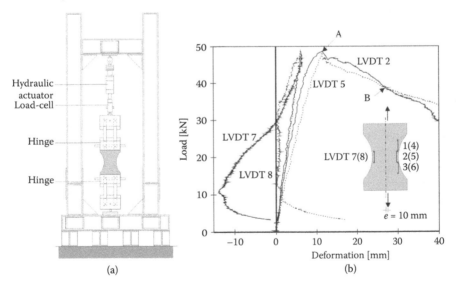

FIGURE A3.4

Example of a specimen loaded in max-control. The LVDT showing the largest deformation at any time during the experiment will be the control LVDT in the closed-loop servohydraulic loading system. Fast switching between LVDTs is possible using a custom-made switch box that selects the critical LVDT. (a) Loading frame; the dog-bone-shaped tensile specimen is loaded between pendulum-bar (which act as hinges) and thus the specimen ends are free to rotate; (b) load-deformation diagrams for four of the LVDTs. (After Van Vliet and Van Mier. 1998. *Proc. IUTAM Symposium on Material Instabilities in Solids.*)

of the specimen the same amount of LVDTs was placed but here just the middle ones on the front and back side were included in the control system. The shown specimen thus had a total of 8 LVDTs connected to the switch box. In the beginning of the test, LVDT-5 (right, back-side) showed the largest deformation, but at peak (point A in Figure A3.4) LVDT-2 (right, front-side) took over. At a deformation of about 27–28 μm (point B), LVDT-5 showed the largest deformation again and control switched to that LVDT. Note that because the specimen was loaded between hinges, the opposite side of the specimen (LVDT-7 and 8) showed compressive deformations. Including the two LVDTs at the left side was important only in the very beginning of an experiment.

The next step is to show that in a "soft" machine stable softening can also be achieved, provided that the electronic and hydraulic parts of the test-control system are fast enough. The logical further development is to cut down the large dog-bone-shaped specimen in Figure A3.4 to a small plate with length in the order of 100 mm, and to replace the outer parts by a steel cable. The cable deformations can then be considered as the elastic deformations of the outer parts of the dog-bone, being actually much "softer" than the dog-bone (mainly due to the dramatically reduced cross-section of the cable). It appears that controlling the deformation rate is needed to achieve stable softening: see Van Mier and Shi (2002). The cable tensile test is shown schematically in Figure A3.1b. The cable lengths can vary, and four different cases were considered: 50-, 100-, 150-, and 200-mm length. The specimen geometry and a photograph of the test setup with a fully instrumented specimen are depicted in Figure A3.5.

In Figure A3.6 it is shown that stable load-deformation response is obtained. The example concerns a test between two 50-mm long cables. The location of the two control LVDTs is shown in the inset of Figure A3.6a. Figure A3.6b shows at which time a certain LVDT is controlling the experiment. When the output value is 0.95 (arbitrary number), LVDT #10 controls the experiment, and when the output value is 0.10 (arbitrary number but deviating from the aforementioned value for the other control LVDT), LVDT #13 is in control. At the end the experiment fails instantaneously when the load on the specimen has decreased to a level equal to the total weight of the lower loading platen and the lower cable. To avoid this sudden rupture, the lower platen should be counterbalanced by the same weight, but this is of interest only when the final part of the tail of the softening curve must be determined. Thus, by using the right combination of hydraulic loading devices and fast electronic control systems stable fracture may be achieved under a variety of loading conditions for specimens of quite large size. It is imperative to have stable softening data in order to draw valid conclusions, therefore it is very important to solve the aforementioned stability issues before a series of experiments is conducted.

Stability of crack propagation can also be considered from an analytical point of view. As mentioned in Chapter 2, with the introduction of the Fictitious Crack Model by Hillerborg and coworkers in 1976 it was proposed

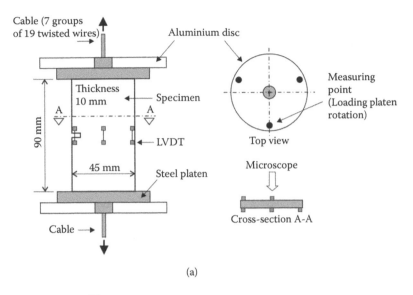

(a)

(b)

FIGURE A3.5

(a) Location of LVDTs on specimen and dimensions of the single-edge-notched specimen used in the cable tensile test, shown in (b). (From Van Mier and Shi. 2002. *Int. J. Solids Struct.*, 39: 3359–3372. With permission from Elsevier.)

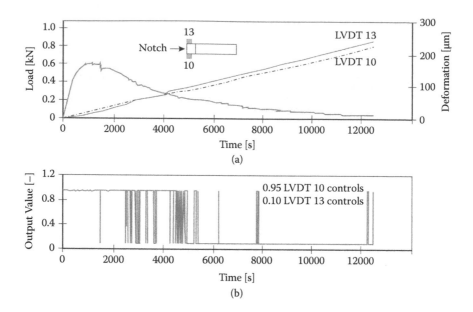

FIGURE A3.6
(a) Load-time and deformation-time diagrams from a cable tensile test controlled in max-rate;
(b) switching between the two control LVDTs: when the output-value is 0.95 the LVDT 10 is
controlling the servovalve, when the output-value is 0.10, LVDT 13 takes over. (From Van Mier
and Shi. 2002. *Int. J. Solids Struct.*, 39: 3359–3372. With permission from Elsevier.)

to measure the softening properties of concrete from a uniaxial tensile
test loaded between fixed (nonrotating) loading platens. Zhou (1988) and
Hillerborg (1989) proposed an equation that can be used to select the dimensions of a tensile specimen in order to obtain a stable result. The inequality
that was derived is:

$$\frac{1}{K_r} < -\frac{6h}{Ebd^3} + \frac{6}{b_c d_c^3}\left(-\frac{dw}{d\sigma}\right)_{min} \tag{A3.2}$$

In this expression b, d, and h are the overall width, depth, and length of the
tensile specimen, b_c and d_c are the width and thickness of the notched section, E is the Young's modulus of the concrete, and $(-dw/d\sigma)_{min}$ is the steepest
part of the softening curve, usually the part just beyond maximum stress.
Nonrotating loading platens have an infinitely large rotational stiffness K_r.
The left part of Equation (A3.2) is always larger than 0, and thus the right-
hand part should also be larger than 0. For a material with steepest slope
in the softening curve $(-dw/d\sigma)_{min}$ the selected specimen dimensions should
then fulfill the following inequality:

$$\frac{hb_c d_c^3}{bd^3} < E\left(-\frac{dw}{d\sigma}\right)_{min} \tag{A3.3}$$

Before deciding which specimen dimensions should be selected it is important to have an idea of the value of the steepest slope. It is therefore debatable if these expressions are really useful. The practical procedure would use the same specimen dimensions for a related material, and check on the test stability. One should be careful in judging whether no local snap-backs occur. Thus it is inevitable to run a few preliminary experiments. Having results such as those presented in Figure A3.3, which are really on the brink of being stable, it is possible to estimate the rotational stiffness of the used loading machine, which is in most cases not known a priori; see Van Mier and Schlangen (1989).

Appendix 4: Crack-Detection Techniques

Before applying fracture mechanics, one should have an idea of the fracture mechanism. This is a fundamental starting point for any fracture mechanics analysis. Fracture mechanisms can be established by trying to measure how cracks develop during loading of a certain test specimen. Basically two stages can be distinguished: visualization and quantification. The first stage is rather demanding in heterogeneous materials: it is not straightforward to distinguish cracks from grain boundaries and interfaces, nor to detect cracks with very small opening, that is, at the submicron scale. In some cases the development of a crack at the [mm]-scale opening is accompanied by many [µm]-scale microcracks. This would, for example, occur in several of the high-performance fiber-reinforced cement-based composites, which are a popular subject of research these days. In such cases it is of importance to combine a large field of view with very high resolution. Quantifying is equally demanding. Cracks in concrete tend to be rather tortuous as a result, again, of its heterogeneity. How are crack length, crack area, crack density, crack roughness, and other parameters defined? How do these parameters relate to the mechanical behavior of concrete? These questions, together with the unavoidable simplifications, are probably the most important debate in this book.

Quite a variety of crack-detection techniques has been developed over the years, applied to different materials, at different size-scales, and under different loading regimes. As mentioned, we have to distinguish between visualization and quantification. Visualizing cracks can be done with the naked eye, or with the aid of optical means (microscopy, such as light- and electron-microscopy and atomic force microscopy, x-ray tomography, etc.). In most cases these are two-dimensional techniques; that is, the surface of a specimen, with or without load, is subjected to close inspection. Getting some insight in the internal crack growth can only be achieved by means of cutting and slicing. In the process the specimen is destroyed, and only the cracking at a given loading stage can be determined. It is interesting that at very high resolution by means of milling with a focused ion-beam cracking at the [nm]- and [µm]-level can be studied. The result usually is stored by means of a (digital) camera, or simply pencil and paper. At some scales and for some techniques it is helpful to enhance the cracks, to make them more visible. With the naked eye it is possible to see cracks with a width of 30–50 µm, provided the light conditions are excellent, and the specimen surface is well prepared, for example, painted white or well polished. Fluorescent epoxy impregnation is a useful enhancement of cracks in concrete, which are presented in Section A4.1. A brief explanation of this technique is accompanied by some results that are meaningful for the debate in Chapters 6 and 10.

In x-ray microscopy/tomography sometimes cracks are enhanced by means of a contrast medium (e.g., Goto 1971 and Otsuka et al. 1998), but this is rather tedious and requires special preparation of the specimen. X-ray tomography is a rapidly developing technique that allows making three-dimensional images of specimens, even under load, showing the internal material structure, but also the development of cracks. In Section A4.2 we discuss x-ray computed tomography in some detail, again highlighting some results that relate to the discussion in Chapters 4 and 11.

By means of digital image correlation it is possible to measure the deformation of a specimen/structure under load. Progressive stages of deformation can be monitored. When discontinuities in the deformation fields arise, cracks have developed, and in this way the technique can be applied in fracture mechanics. In Section A4.3 we briefly discuss the principle of this technique, and show some results that relate to the debate in Chapter 8 on compressive fracture. Most of the examples presented are qualitative in nature: no direct measurements of crack lengths, or crack densities are made. As a matter of fact, this will be a challenge for the coming decades. After having established fracture mechanisms, and after having proposed several possible theories that could well model these mechanisms (see Chapters 4, 10, and 11), the time has come to quantify the results. Even by means of digital computers quantifying crack data is still a challenge, not in least because the cracks are highly tortuous surfaces. In Chapter 10 some attempts are presented to make sense of crack patterns from numerical mesoscale analyses of tensile fracture. Basically this is the same as using crack data from physical experiments. Simplification is at the current state of knowledge unavoidable.

Energy is consumed to create a new crack surface. Measuring parameters indirectly that relate to this energy consumption is quite popular, for example, the use of nondestructive techniques such as acoustic emission (AE) monitoring, measuring changes in the ultrasonic pulse velocity in specimens under load, monitoring temperature changes in a structure due to cracking by means of infrared thermography (e.g., Luong 1990), or, as recently proposed, looking to piezo-nuclear neutron emissions (see Carpinteri and Lacidogna 2010). The first two techniques have been used for many years: in AE monitoring weak elastic waves produced by the nucleation and propagation of cracks and friction in cracks are measured. In ultrasonic pulse velocity experiments, the running time and changes therein of an ultrasonic pulse are determined, and it is attempted to relate the signals to the actual damage in the specimen/structure. These techniques are to some extent quite speculative: it is extremely difficult to relate the acoustic signals to real crack geometries. Nevertheless, the techniques are popular inasmuch as the structure/specimen remains intact during monitoring of internal cracking. In Section A4.4 we discuss AE monitoring and show an example of drying cracking in hardened cement paste. The results from AE monitoring are compared to fluorescent epoxy impregnation on the same specimen at the end of the monitoring period.

In the (near) future destructive techniques where cutting and slicing of a specimen are needed will very likely be replaced by x-ray tomography and AE monitoring, that is, by techniques that can be applied while a specimen is under load, revealing the complete three-dimensional (internal) material structure and changes therein (i.e., crack nucleation and growth).

A4.1 Fluorescent Epoxy Impregnation

The impregnation technique has often been applied over the past decades. The technique developed by Vonk (1992) during his doctoral work has proven to be quite helpful in visualizing the fracture process in concrete subjected to tension and compression. After several small changes, the technique is now applied as follows; see Stähli (2008). A specimen is loaded to the required stress- or deformation-level, after which it is unloaded. To ensure that cracks do not close, one can decide to glue steel platens on the outer surfaces of the specimen. This can be done while the specimen is under load, or after it has been unloaded. In the latter case one must expect that crack widths have reduced. Next, the specimen is placed in a cylindrical container, which can be closed with a heavy (steel) lid. On the lid a valve, a hopper, a vacuum hose, and a manometer are fitted (see the scheme shown in Figure A4.1). Using a vacuum pump, the air pressure in the container can be reduced to 200 mbar (20 kPa). From the hopper fluorescent epoxy can be injected on top of the specimen when the required pressure has been reached. To facilitate the epoxy addition, a hose is fixed to the inside of the lid, ending about 1 cm above the specimen's surface. To prevent spilling epoxy the specimen is

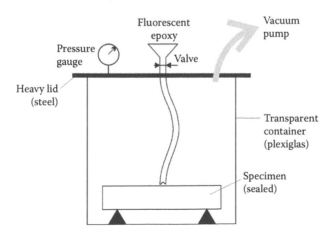

FIGURE A4.1
Setup for fluorescent-epoxy impregnation.

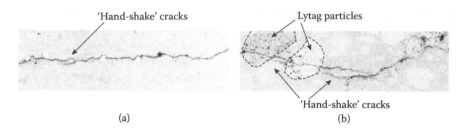

(a)   (b)

FIGURE A4.2
Crack-face bridging in a tensile crack (average crack opening 100 μm) in 2-mm aggregate concrete (a) and 8-mm lytag concrete (b). Note that in Figure (b) the large porous lytag particles are also filled with fluorescent epoxy (cluster of black speckles; two particles have been marked). Bridging in lytag concrete therefore occurs around the largest sand grains; cracks are often found to cross the weak lytag particles. (From Van Mier. 1991a. *Cem. Conc. Res.*, 21(1): 1–15. With permission from Elsevier.)

wrapped in tape before placing it in the container. A small hole just below the hose attached to the hopper allows for filling with epoxy. After the epoxy has been impregnated, atmospheric pressure is restored slowly, taking about 40–50 minutes. Hardening of the epoxy is accelerated by placing the specimen in an oven at 60°C for an entire day. After hardening of the epoxy, the specimen can be cut in slices, and under ultraviolet light the internal crack patterns become visible. Common practice is to take pictures, sometimes at high resolution. Various crack measurements can be done from the images, but first crack mechanisms are studied qualitatively.

In Figure A4.2 two examples of crack-face bridging in concrete subjected to uniaxial tension are shown. The average crack width in the direction of the tensile load was measured with LVDTs around the specimens' circumference, and the specimens were unloaded after reaching 100 μm. In this example the original images were reversed to have a black crack against a white background. Crack-face bridges in concrete develop as crack overlaps (often referred to as "hand-shake cracks") near larger stiff aggregate particles. A small load can be carried by the intact ligament between the two crack-tips; the bridge fails in bending as shown in Van Mier (1991b). The mechanism appears at crack openings larger than 20–25 μm, that is, in the tail of the softening curve; see also Chapter 6 (Figures 6.1 and 6.2) where the mechanism was identified from meso-level analyses of tensile fracture, and Chapter 10 where crack-face bridging forms an integral part of the recently proposed 4-stage fracture model (viz. Section 10.1.4).

Another example of using fluorescent epoxy impregnation is shown in Figure A4.3, demonstrating multiple cracking in fiber-reinforced concrete. The material is actually a fiber-reinforced mortar (sand with d_{max} = 1 mm is used); the fiber content is 3% of 12-mm long, 0.2-mm diameter high-strength steel fibers (1,980–2,180 MPa). The beam is loaded in 4-point bending, in displacement control up to a middle deflection of 1.7 mm. This particular mixture showed the largest prepeak deformations, compared to mixtures containing shorter

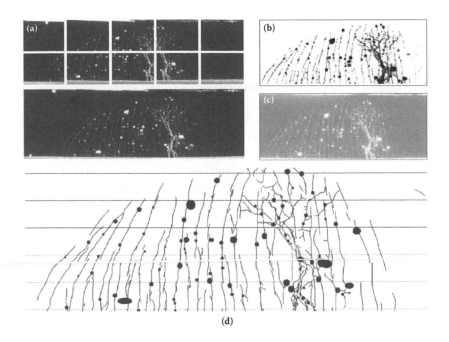

FIGURE A4.3
Crack pattern in a fiber-reinforced concrete beam subjected to 4-point bending. By stitching 10 individual high-resolution UV images together the crack growth is visualized in the entire beam (a). Binary thresholding (b), and grayscale images (c) help to improve the view. After manual tracing the crack pattern of Figure (d) is obtained, which is the basis for various crack measurements. (After Stahli. 2008. *Ultra-Fluid, Oriented Hybrid-Fibre-Concrete*. With kind permission from Dr. Patrick Stähli.)

(6-mm) or longer (30-mm) fibers. At peak (0.9 mm deflection), a diffuse regularly spaced crack pattern is visible; beyond peak in addition a localized crack develops under the right loading point (as shown here). For more information, see Stähli (2008). The image is a mosaic of several images, each measuring 25 × 20 mm^2 (Figure A4.3a), which are stitched together to obtain a single image. Under UV-light the fluorescent epoxy shows in bright yellow, and the concrete in dull blue (here white for epoxy, dark gray for concrete). Figure A4.3b shows the conversion to a binary image; Figure A4.3c to a grayscale image. The latter is used for illustration purposes only (qualitative illustration of the fracture process). After manual tracing of the grayscale image the pattern of Figure A4.3d is obtained. From such figures crack spacing, crack length, crack densities, crack widths, and the like can be measured, which remains, however, a relatively tedious job, see Stähli (2008). The method allows us to see the smallest cracks in the high-resolution subimages, and thus to obtain a very detailed map of all cracks present at a certain loading stage.

Although the impregnated crack patterns reveal much detail of the fracture process, a large disadvantage is that for every loading stage a new specimen must be loaded up to the required deflection. There can be a great deal of scatter in concrete properties, therefore it may take several repetitions of each loading stage before a proper insight in the governing fracture mechanism is obtained. For example, for discovering the crack-face bridging in uniaxial tensile fracture, hundreds of images from 26 specimens of four different concretes were produced. Details can be found in the aforementioned publications. It is not always clear which cracks are "active." that is, show progressive movement of the crack faces during loading. This is hard to show by means of fluorescent epoxy impregnation because the crack pattern is frozen in time. Active cracks can more easily be identified using visualization techniques that work while the specimen/structure is under load, for example, long-distance microscopy (see Van Mier (1991b) and Figure 10.6), or x-ray computed tomography and digital image correlation, which are discussed in the next sections.

Impregnation is quite well suited for demonstrating macroscopic fracture patterns in concrete. Bridging and multiple cracking shown here are just two examples; failure of concrete under compression has been shown using the same techniques by Vonk; see Figure 8.5. Shear-band propagation under triaxial compression was shown by Van Geel (1998); see Figure 8.15. Fluorescent epoxy impregnation is also done on a regular basis on microscopic small samples (for use in light microscopy), for example, in the field of cement and concrete durability. Consequences of various chemical reactions may lead to expansion and thus to internal damage; cracks caused by alkali-aggregate reactions or corrosion-induced crack growth can be shown in quite some detail. The epoxy used in the example of Figure A4.3 (EpoFix, supplied by Struers, mixed with fluorescent dye EpoDye) has a very low viscosity and negligible shrinkage, which are the essential properties, next to a sufficiently long pot-life (1 hour would generally suffice).

A4.2 X-Ray Computed Tomography (CT)

The disadvantage of using multiple specimens for examining the fracture process can be avoided by applying x-ray computed tomography (CT). Of course this technique requires more advanced equipment: an x-ray source and a digital computer for imaging and 3D reconstruction. In recent years table-top x-ray tomography equipment has become standard in many laboratories. The technique can be used at different size-scales: from [μm]-small cement samples to [m]-size concrete prisms. For obtaining some of the results presented in this book (see Chapter 4 and 11) the so-called cone-beam technique (see Persson and Östman 1986) has been applied, which is shown schematically in Figure A4.4. The specimen is rotated in the x-ray beam, and at designated angles an image is captured. In one of the applications, trying to visualize the internal structure of foamed cement and crack growth due

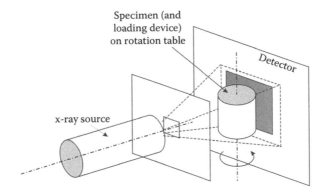

FIGURE A4.4
Computed tomography using the cone-beam technique.

to external loading, typically 4,000 angular steps of 0.09 degrees each were used; see Meyer (2009).

For studying crack growth it is essential that the loading device does not hinder the capturing of images. The parts of the loading frame in the x-ray beam must be translucent to the x-rays, for example, glass can be used, or a carbon–polymer composite used by Meyer (2009), which was selected after consultation with experts from Scanco medical AG, who delivered the table-top x-ray scanner. When a miniature loading device is built into an existing scanner one should expect numerous problems. The samples are very small, and in particular in the case of concrete is it important to check if the material sample is larger than the representative volume (see also Section A4.1 and Chapter 9). Denser materials are less translucent to x-rays than voids (or cracks): on the usual grayscale the densest materials appear white and hollow space is black. In the past contrast media have been used to enhance the visibility of cracks; see for example the work of Goto (1971) and Otsuka et al. (1998). This leads to further complications because methods must be devised to bring the contrast fluid into the cracks. Modern applications function well without additional measures, and as said, function on the basis of density differences only.

In an attempt to obtain detailed insight to crack growth in plain hardened cement, a small loading device was constructed for use in the tomography beamline at the synchrotron of the Paul Scherrer Institute (PSI) at Villigen (see Figure A4.5). Using synchrotron radiation very high resolution images can be obtained, provided the specimens are sufficiently small. The specimens used had a length of 250 μm and diameter of 130 μm. Cylindrical samples were used because this is the ideal geometry when the specimen must be rotated. The field of view is equal to the diameter of the sample, giving the best resolution per pixel. Casting these small specimens was quite difficult. Small holes of 130 μm were made in a sheet of Teflon of 250-μm thickness. The cement was pressed in the holes, and the Teflon could easily be removed

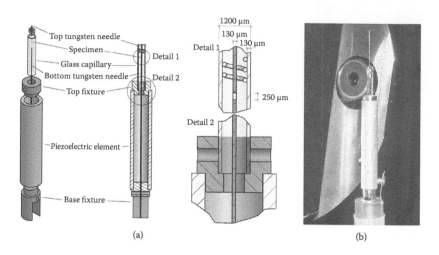

FIGURE A4.5

(a) Small-scale testing device based on piezo-electric actuation, for use in a synchrotron-based microtomography (SRμCT) experiment at the Paul Scherrer Institute in Villigen (CH). (After Trtik et al. 2007. *Proc. 6th Int'l. Conf. on Fracture Mechanics of Concrete and Concrete Structures (FraMCoS-VI).*) (b) Entire loading apparatus in the tomography beamline at PSI. At this scale the specimen is hardly visible; it is located near the top of the glass capillary.

by cutting it away with a sharp razor blade. The specimen was glued between two tungsten needles of the same diameter. The two rods with the specimen in between was fixed in an assembly consisting of a piezo-electric tube and a glass capillary as shown in Figure A4.5a. The glass capillary is translucent to x-rays and was strong enough to function in the loading-rig. The complete device was then placed on top of the rotation table in the tomography beamline at the synchrotron of the Paul Scherrer Institute (see, for instance, Stampanoni et al. 2002). Figure A4.5b shows the loading device in front of the SRμCT-detector. Tensile loading of the cement sample was achieved by varying the voltage on the piezo-electric element. The loading was not controlled: the test was an open-loop arrangement, and only by sheer luck insight to the softening regime would be possible. This was not an obstacle because the primary goal was to obtain information about the fracture process just prior to reaching peak-load.

These experiments are highly demanding. In particular the preparation of the loading device that could be used only once takes an enormous amount of effort. On top of that one simply has to wait until measuring time becomes available at the synchrotron tomography beamline. At most four to five tests could be run each half-year, which is barely sufficient to get a good idea of the statistics. Therefore only a few experiments were conducted, revealing the inner structure of hydrated cement paste, in particular the porosity and unhydrated cement particles that can easily be distinguished because of the large differences in density; see Figure 4.8. Some detail about the cracking process was obtained as well. The cracks were unstable, but nonetheless

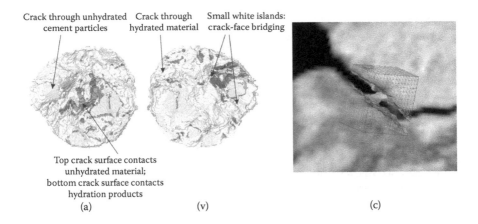

Crack through unhydrated cement particles   Crack through hydrated material   Small white islands: crack-face bridging

Top crack surface contacts unhydrated material; bottom crack surface contacts hydration products

(a)   (v)   (c)

FIGURE A4.6

Crack-face bridging at μm-scale from synchrotron-based microtomography (SRμCT). Two crack-faces of a broken specimen are shown (diameter 130 μm); (a) and (b). The small white islands are the remnants of crack-face bridges. A 3D view of a crack-face bridge connecting the two specimen parts (c). (After Trtik et al. 2007. *Proc. 6th Int'l. Conf. on Fracture Mechanics of Concrete and Concrete Structures (FraMCoS-VI).*)

bridging similar to that shown in Figure A4.2 was observed. In Figure A4.6 the upper and lower crack-face after fracturing a specimen are shown, as well as a 3D view of a crack-face bridge connecting the upper and lower part of the specimen. The main crack runs through the hydrates, but also along the interface between unhydrated and hydrated cement. The light gray in Figures A4.6a and A4.6b indicates the hydrated Portland cement, medium gray is unhydrated cement and dark gray indicates the interface between hydrated and unhydrated cement. The resolution of the whole setup was not quite sufficient to reveal much of the prepeak fracture process, which was the actual goal of the experiment. Other, destructive, techniques are likely more useful, for example, looking to the interior of a (cracked) specimen using a focused ion beam (FIB); see, for instance, Holzer et al. (2004), but then of course it is not possible following cracking throughout the loading sequence in a single specimen.

In this book results from three different applications of x-ray tomography have been used, at three different size-scales, namely the level of the individual aggregate particles and fibers in (fiber-reinforced) concrete ([cm]-scale; see Figure 4.7), pores in foamed cement ([mm]-scale; see Figure 4.6) and the material structure and crack growth in hardened Portland cement ([μm]-scale; see Figures 4.8, 11.9, and A4.6). For each of these applications it was necessary to use different equipment as shown in Table A4.1.

Important crack-data, such as length, tip location, and movement of various parts of a specimen/structure separated by cracks can be quantified from the obtained images using specially developed software; see, for example, Roux et al. (2008), Landis et al. (2007), and many others. In the framework

TABLE A4.1

Overview of X-ray Computed Tomography (CT) Experiments

Size/Scale	Material	Equipment	Theoretical Resolution
[cm]	Concrete	Siemens SOMATON definition 64	0.5–1 mm
[cm]	FRC	Siemens SOMATON definition 64	0.5–1 mm
[mm]	Foamed cement	Scanco Medical μCT40 desk top	6 μm
[μm]	Portland cement	SRμCT, PSI, Villigen	0.7 μm

of this book no further information is given here, and the interested reader is referred to the mentioned literature sources.

A4.3 Digital Image Correlation (DIC)

Digital image correlation is an interesting technique for measuring the deformation of a large part of a whole structure (specimen) under load. Discontinuities in the deformation field are interpreted as being cracks. The predecessor to digital image correlation is stereo-photogrammetry; see, for example, Torrenti, Benaija, and Boulay (1992), which was used in the precomputer era. Making use of digital computers is essential nowadays for storing and analyzing the enormous amount of data collected during a single experiment. Started in two dimensions, fully three-dimensional systems have been developed more recently. The method has even become commercially available, for example, VIC-3D of LIMESS. There is an obvious advantage of working with three-dimensional rather than two-dimensional image correlation: negative influences such as out-of-plane deformations are circumvented, specimen surfaces studied do not need to be perfectly planar, and it is even possible to record more surfaces at the same time. The three-dimensional version of digital image correlation requires stereovision, that is, two digital cameras are needed. In the images shown below, two CCD cameras with a resolution of 2048 × 2048 pixels were used. Three-dimensional digital image correlation has been developed in the last part of the previous century; see Sutton et al. (1983), Helm, McNeill, and Sutton (1996), Kahn-Jetter and Chu (1990), Luo et al. (1993), Synnergen and Sjödahl (1999), Robert et al. (2007), and many others. For calculating the surface displacements two images are needed: a reference image, most often the structure or specimen studied in the unloaded state, and a second image, that is, the same structure/specimen at a given external loading. It is possible to use specific points in the texture of a material as reference points in the analysis. For materials like concrete it may be useful to fill the larger pores on the surface of a specimen, for example, with cement or epoxy, paint the surface white, and finally

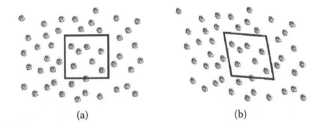

(a)   (b)

FIGURE A4.7
Speckle pattern on unloaded surface of a specimen and definition of subset (a), and situation after deforming the specimen (b).

apply a speckle pattern by spraying black paint by means of an airbrush. The principle is simple. Characteristic parts of a surface of the undeformed specimen can be traced back on the surface of the deformed specimen. The area surrounding the point of interest is called the subset, and should include minimal 3 × 3 speckles, see Figure A4.7.

There are several parameters affecting the results obtained; see, for example, Robert et al. (2007) for details. Image noise cannot be eliminated completely. It is important to keep the conditions (light in particular, and the position of the cameras) the same throughout an experiment in order to reduce the noise. Moreover, the entire setup needs to be free of vibrations. For homogeneous materials under uniform loading very good results can be obtained. For very heterogeneous materials like concrete, in particular when cracks develop, noise will increase and the conclusions may not be as solid as desired. In all applications it is usually better to decrease the image size in order to reveal more details of the fracture process; see Caduff and Van Mier (2010).

As an example, in Figure A4.8 crack patterns determined with 3D-digital image correlation are shown (Figure A4.8a,c), including a comparison with results from vacuum impregnation with fluorescent epoxy (see Section A4.1) of the same specimens after the experiments were terminated (Figure A4.8b,d). The prisms were loaded in uniaxial compression between two different end-conditions: rigid steel platens (Figure A4.8a,c) and steel platens with a special friction-reducing sandwich element inserted between specimen and platen (Figure A4.8b,d). The sandwich elements consist of two sheets of 0.10-mm PFTE (Teflon) with a layer of 0.05-mm grease in between, as proposed by RILEM TC 148-SSC (2000). With this sandwich the friction between loading platen and concrete reduces to approximately 1–2%, and specimens are loaded more uniformly (see also Section 8.2 where the effect of boundary conditions on the stress–strain behavior of concrete in compression is debated, in particular the softening behavior). The results from digital image correlation show that with the method the main crack pattern can be visualized. With the commercial package used it was not possible to look to crack patterns in the prepeak regime, but postpeak the fracture process can

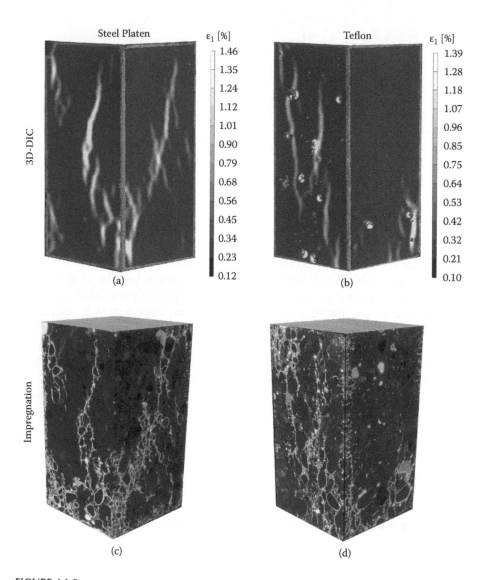

FIGURE A4.8

Crack patterns in 16-mm normal concrete (37-MPa compressive strength), loaded between rigid steel platens (a,c) and steel platens with friction-reducing sandwiches according to RILEM TC 148-SSC (2000) (b,d). The two top images are results obtained from 3D digital image correlation; the two bottom images show the results after vacuum impregnation with fluorescent epoxy. Specimens are prisms $h/d = 2$, $d = 70$ mm. (From Caduff and Van Mier. 2010. *Cem. Conc. Comp.*, 32: 281–290. With permission from Elsevier.)

be captured. Comparison with the impregnated specimens indicates that with the latter method all fine details in the crack pattern are revealed; yet it is not obvious which cracks are active at a certain moment, and here digital image correlation is a great help. In spite of its limitation in capturing all fine cracks, knowing the main mechanisms is helpful for constructing average crack models such as the 4-stage fracture model presented in Chapter 10.

A4.4 Acoustic Emission (AE) Monitoring

Nondestructive techniques for monitoring crack growth include acoustic emission monitoring. The method is based on the small elastic waves produced in the material when cracks nucleate or propagate, or when there is friction between the crack-faces. By mounting several sensors on a specimen/structure where cracks appear, for example, caused by external loading or restraint, the approximate location of the events can be determined. Location analysis is based on assumptions of the speed at which signals travel through the material, which, for a material like concrete, can only be a very rough average. The signals can be very small indeed, and one must be careful not to perform measurements in the background noise. Signals occurring close to sensors may also turn out to give erroneous results, and location analysis is hampered when more cracks develop at the same time. In spite of such difficulties there is great interest in developing sound and robust methods. Research in the field is on-going. As a general reference the reader is referred to the recent book by Grosse and Ohtsu (2008).

It is still difficult to correlate acoustic events with the geometry of the real cracks.

Again, we show the method through an example. Due to drying, hardened cement paste shrinks, and when the deformations are restrained cracks may develop. The ensuing cracks may affect the strength of a cement or concrete specimen; see, for example, Van Mier (1991a) and Van Mier and Schlangen (1989). Combining experience from two research teams drying shrinkage cracking was studied using both AE source location and fluorescent epoxy impregnation. Several circular concrete slabs with a thickness of 42 mm and a diameter of 235 mm, made of two different materials, were subjected to a controlled drying regime; see Bisschop (2002) and Shiotani, Bisschop, and Van Mier, (2003). Some slabs were made of pure hardened cement paste (hcp) with a w/c-ratio of 0.45; the others were also made of hcp but contained 35% (volume) of 6-mm glass spheres. Glass, with a Young's modulus of 77 GPa is an excellent substitute for aggregates, with the great advantage that the shape of the grains is better controlled and that a single size fraction can be used. The disadvantage is that the bond of the glass spheres to the surrounding cement matrix is very low, which can be improved, however, by sand-blasting the surface of the glass spheres. In the experiment shown here, smooth spheres were used. There was no danger for the glass to react with the cement because the total duration of the (short-term) experiments was 7 days for controlled hardening (1 day in the mould; 6 days in CH-saturated tap water at room temperature), and an additional 16-h or 282-h controlled drying in an environmental cabin at 25% ± 5% RH and 31°C ± 0.5°C. During 16-h drying the hcp specimen lost 22.1 g of moisture, the composite with glass spheres 18.1 g. At the end of the experiments the specimens were impregnated by means of fluorescent epoxy, that is, using the technique explained in Section A4.1. Crack growth during the drying period was monitored by means of 6 AE sensors (diameter 6 mm × height 6 mm; resonant freq. 500 kHz) that were attached to the top surface of a specimen, which was the drying surface. All other surfaces of the specimens were sealed with three layers of adhesive tape, which proved to be a sufficient barrier against drying. A Mistras AE System (Physical Acoustic Corporation) was used for acoustic monitoring. Because six sensors were used location analysis was possible, which was limited to two dimensions because all sensors were attached to the top (drying) surface of the slabs. Next to the location analysis various AE parameters were recorded, such as the peak amplitude, the rise time, the ring-down count, duration time, energy/absolute energy, the signal strength, and initiation frequency. In this appendix we only include the cumulative AE events, the cumulative absolute energy and the location analysis of the largest AE events and compare these results to the crack patterns from the fluorescent-epoxy impregnation.

Due to drying shrinkage, cell-like crack patterns develop, quite similar to the crack patterns observed in drying clay or adobe. In Figure A4.9 the crack patterns obtained from the vacuum fluorescent-epoxy impregnation are shown, both for the plain hcp (Figure A4.9b) and for the glass sphere–cement composite (Figure A4.9c). Figure A4.9a shows the complete specimen, indicating the location of the six AE sensors (black circles).

Both for the hcp and the composite the shrinkage cracks develop in cell with varying shape. In analyzing the geometry of these crack patterns in more detail Bisschop (2002) found that the cell size decreased toward the edges of the specimens. Explanations are rather speculative at this moment. The formation of the cells can be explained from overcoming capillary forces that are present between the cement particles, as elucidated in Section 11.2 (see also Van Mier 2007). A comparison between the AE events (small crosses) and the optical cracks is difficult because source-location was carried out in two dimensions only. Some of the sources appear to overlap with the cracks, in other cases there is a mismatch. Note that a 30-dB threshold was adopted, implying that only the strongest events are recorded. More interesting was the temporal evolution of the acoustic events, which is the major advantage of this technique.

In Figure A4.10a the results of the hcp-drying test and the composite are compared. Not only the cumulative AE events are shown, but also the

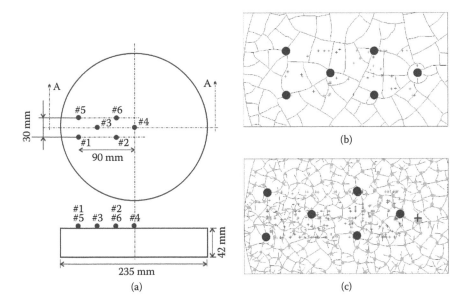

FIGURE A4.9

Specimen used in short-term drying experiments (a); crack pattern (top view) in hardened cement paste after 16-h drying (b), and in glass sphere–cement composite (c). Note that in the composite the top surface has been slightly ground to show the top of the glass spheres close to the surface. The black circles indicate the location of the AE sensors; the small crosses in (b) and (c) show the locations of the most energetic AE events. (From Shiotani, Bisschop, and Van Mier, 2003. *Engng. Fract. Mech.*, 70(12): 1509–1525. With permission from Elsevier.)

cumulative absolute energy. At a later stage a number of additional tests were performed with a longer drying time of 282 hours. The results from these long-term tests have been included in Figure A4.10b. In the short-time experiment the hcp appears to develop a complete crack pattern within the first hour; after that hardly any additional AE events are recorded. In contrast the composite shows not only a larger number of AE events (factor 3.5 larger than the hcp specimen), but also the cumulative absolute energy is much larger (factor 4.5 compared to the hcp specimen). Obviously, with the progressive drying front over the thickness of the slab, shrinkage of the cement past around the stiff glass particles, which give additional restraint (see Bisschop 2002), leads to a continuous development of interface cracks along the glass particles. The shrinkage restraint caused by the glass particles is commonly referred to as "aggregate-restraint," and will continue as long as the drying front goes deeper and drying shrinkage of the cement matrix proceeds into the interior of the specimen. The longer drying experiments depicted in Figure A4.10b show a continuation of acoustic activity in the composite, whereas in the hcp specimen the activity only slightly increases. The immediate cracking of the hcp specimen, which occurred at 6 minutes after the start of the drying sequence, is caused by the steep moisture gradients that

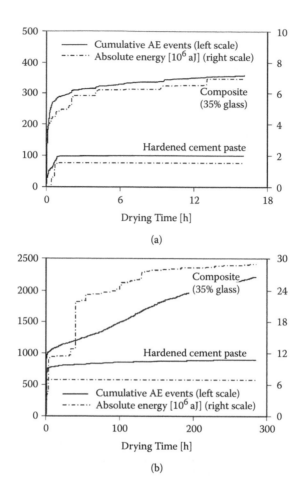

(a)

(b)

FIGURE A4.10
Acoustic activity in hcp and composite specimens up till 16 hours of drying (a). At the end of the test the specimens were impregnated. The optical crack patterns are shown in Figure A4.9. (b) Acoustic activity in two additional experiments in which the drying time was extended to 282 hours. (After Bisschop. 2002. *Drying Shrinkage Microcracking in Cement-based Materials.* Reprinted with kind permission of Dr. Jan Bisschop.)

develop in the specimen; see Bisschop (2002). The phenomenon is usually called "self-restraint." The two types of restraint, aggregate-restraint and self-restraint are explained in Figure A4.11. An in-depth analysis may give further details about the shape of the acoustic emission events. For those results the interested reader is referred to the relevant literature. For the purpose of this book no further analysis is needed here.

One final remark relates to the geometry of the drying shrinkage cracks in the third dimension. In Figure A4.12 two cross-sections of the hcp and composite slabs are shown, straight through the middle of the specimens as indicated in Figure A4.9a. After an initial straight crack growth perpendicular to

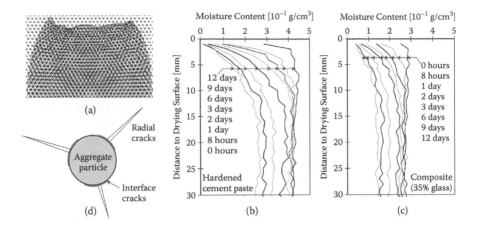

FIGURE A4.11

Self-restraint is caused by a moisture gradient in the structure/specimen under consideration. Where the humidity has decreased most, drying shrinkage deformations are largest, that is, close to the drying surface. A simple numerical analysis based on a transient flow analysis leads to the deformed structure of (a); drying occurs only along the top surface. (After Sadouki and Van Mier 1996. *HERON*, 41(4): 267–286.) The moisture profiles from NMR experiments on both the hardened cement paste (b) and the composite containing 35% glass particles (c) show a gradual drying in time. (After Bisschop, Pel, and van Mier. 2001. *Creep, Shrinkage and Durability Mechanics of Concrete and Other Quasi-Brittle Materials, Proc. CONCREEP-6.*) The drying starts near the surface, almost immediately, and then progressively moves deeper into the specimen. Note that after 12 days of drying a moisture gradient still persists, which is the reason for the differential deformations in (a). When the figures are presented relative to the absolute amount of cement in the samples, (b) and (c) are identical (see Bisschop 2002 for details). Aggregate restraint is caused by the presence of stiff aggregate particles: radial cracks grow from the interface into the surrounding cement matrix. The mechanism was shown by Golterman (1995). The aggregate particle is subjected to hydrostatic compression.

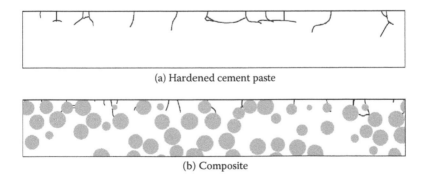

FIGURE A4.12

Crack profiles over Section A-A (see Figure A4.9) in the hcp specimen (a) and the composite specimen (b). The drying surface is the top surface. Gray circles in Figure (b) indicate the glass spheres. (After Bisschop. 2002. *Drying Shrinkage Microcracking in Cement-Based Materials.* Reprinted with kind permission of Dr. Jan Bisschop.)

the drying surface, the cracks in the hcp specimen (Figure A4.12a) reorient in a horizontal plane a few mm below the drying surface. This is not quite visible in the composite, likely because the additional aggregate restraint prevents the development of the horizontal cracks. There are different explanations for the observed behavior. Van Mier (2007) offers an explanation based on capillary forces between the cement particles. When the bonds between the cement particles are broken, due to shrinkage deformations around kernels in neighboring areas, the cell patterns of cracks develop. When the cracks penetrate into the specimens' interior, additional drying occurs along the crack-faces. In addition capillary suction may assist in the development of the horizontal crack branches. We do not further discuss these mechanisms here because it is part of the debate in Section 11.2; see also Van Mier (2007) for further information. Important, also for the discussion in Chapter 9, is the notion that in hcp a full shrinkage crack pattern develops within the first hour after start of drying, characterized by vertical cracks growing per-pendicular to the drying surface to a limited depth, followed by horizontal branching. In composite materials the cracks proceed for much longer inter-vals as a direct consequence of aggregate restraint; also the cracks appear to develop at larger depths from the drying surface.

In conclusion, combining various crack detection techniques may help to unravel details of the fracture process. Where one technique cannot be applied, another may be helpful. As shown in this last section, fluorescent-epoxy impregnation gives great detail of the crack geometry, but only at one selected point of the experiment. The temporal development of the fracture process can be monitored conveniently using the AE source location, but detailed information on crack geometry cannot be obtained.

Appendix 5: Active and Passive Confinement

Confinement is a good way to improve the properties of a material. There are two ways: the first possibility, "active confinement," is a more structural measure, whereas the second option, "passive confinement," must be considered as a definite change of the material structure. In both cases the density and orientation of microcracks can be affected, which may lead to enhancement of the mechanical properties of the material specimen, albeit sometimes only in a specific direction.

A5.1 Active Confinement—Multiaxial Compression

Multiaxial compression is the most common form of active confinement. Not only (compressive) stresses are applied in the first principal direction, but also along the second or third principal axis. In Chapter 8 we showed a variety of different stress–deformation curves from multiaxial compression tests. Tests are commonly done in a standard triaxial device (Hoek-cell), in which two principal stresses are always equal; triaxial compression with $\sigma_1 < \sigma_2 = \sigma_3 < 0$ (compressive stresses are negative), for instance tests by Richardt, Brandtzaeg, and Brown (1929), Kotsovos and Newman (1977), Jamet, Millard, and Nahas (1984); and triaxial extension $\sigma_1 > \sigma_2 = \sigma_3$ where σ_1 can be either tensile or compressive, and σ_2 and σ_3 are both compressive, for example, experiments by Visser and Van Mier (1994) and Visser (1998; Figure A5.1a) or in a "true-multi-axial testing machine," which allows for free variation of the three principal stresses, for instance, Schickert and Winkler (1977) and Van Mier (1984; Figure A5.1b).

Tests in a standard triaxial test will always have two equal lateral stresses, $\sigma_2 = \sigma_3$, whereas the axial stress can be varied independently. All failure points lie on two intersecting lines with the failure surface in three dimensions, namely the compressive meridian and the tensile meridian, both of which are indicated in Figure A5.2c. More possible stress combinations can be explored in "true" multiaxial experiments using cubes, prisms, or plates, which can, in principle, cover the entire stress space. In Figure A5.2a,b, a summary is given of 2D failure contours from experiments by Kupfer, which are considered the best available data at present. These experiments were carried out with steel brushes to transmit stresses to the platelike specimens. Likewise, 3D failure contours for normal concrete are shown in Figure

(a)

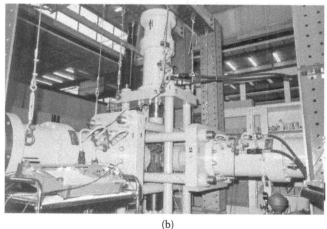

(b)

FIGURE A5.1

Examples of a triaxial cell and a "true" multiaxial machine: (a) interior of a triaxial cell suitable for extensile tests on concrete and rock using 100-mm diameter cylindrical specimens. The entire specimen is fitted in a steel cylinder; by means of fluid pressure on the sides of the cylinder lateral confinement is applied; in thraxial direction tensile loading is applied by means of a servocontrolled hydraulic actuator. (After Visser and Van Mier. 1994. *Computer Methods and Advances in Geomechanics*.) (b) Machine suitable for testing 100-mm cubes allowing for free variation of all three principal stresses; in each direction compressive stresses up to a maximum of −200 MPa (or tensile stresses up to 140 MPa) can be applied by means of a servocontrolled hydraulic actuator, allowing to measure the complete stress–strain behavior including the softening regime. (After Van Mier. 1984. *Strain-Softening of Concrete under Multiaxial Loading Conditions*.)

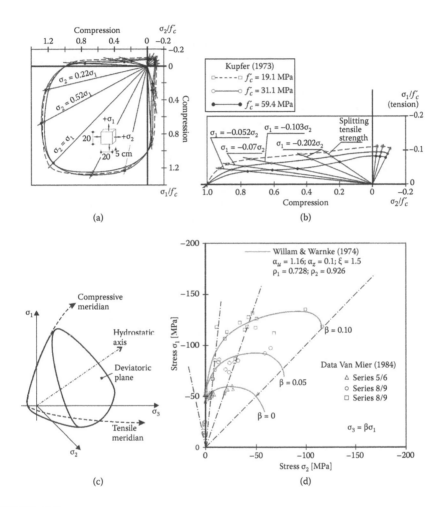

FIGURE A5.2

Failure contours for normal concrete: (a) 2D biaxial envelopes measured by Kupfer (1973); (b) enlarged portion of the biaxial tensile/compressive regime and the tensile/tensile regime; (c) 3D sketch of the failure envelope, indicating the important surfaces and contours; and (d) results from 3D proportional stress-path and proportional deformation path tests (From Van Mier. 1984. *Strain-Softening of Concrete under Multiaxial Loading Conditions*). Note that the axes in (a) and (b) are dimensionless with respect to the uniaxial compression strength, whereas in (c) absolute values are shown. (Parts (a) and (b) are from Kupfer. 1973. *Behaviour of Concrete under Short Term Multiaxial Loading, with Emphasis on Biaxial Loading*. With permission of Deutscher Ausschuss für Stahlbeton.)

A5.2d. Here the apparatus shown in Figure A5.1b was used; loading was also applied by means of steel brushes. Only part of the stress-space was investigated using proportional stress-paths $\sigma_2 = \alpha\sigma_1$ and displacement-paths $u_2 = \alpha u_1$ (α = +0.2, 0, –0.1, and –0.33). As a matter of fact the individual curves of Figure A5.2d lie on tilted planes $\sigma_3 = \beta\sigma_1$ (with β = 0, 0.05, and 0.10). Going from 2D- to 3D-confined tests results in a rapidly expanding stress contour, and it will be obvious that soon the loading equipment will fail to register the ultimate stress if the minor confining stress keeps increasing. Therefore it is assumed that the failure contour for concrete, as shown schematically in Figure A5.2c, is an open-ended cone in the 3D-compression regime, with the top lying in the triaxial tensile regime. The shape of the cone in deviatoric sections (which are oriented perpendicular to the hydrostatic axis) varies from rounded-triangular (at low σ_0) to circular (for high σ_0). The data in Figure A5.2d are compared to the five-parameter model developed by Willam and Warnke (1974). The parameter fit is shown in the inset. The match between data and model is quite agreeable. In line with the approach followed in this book, the better option is to calculate the failure strength (and the complete stress deformation behavior) for a variety of stress combinations from ab initio analyses. In that case no separate equations for the failure envelope would be needed. In other words, the failure envelope is an integral part of the constitutive equations.

In Figure 8.14 in Chapter 8 the two distinct failure modes that are found under multiaxial compression were mentioned. The first mode is found under uniaxial compression, and leads to a more-or-less symmetrical pattern of short inclined cracks. This so-called "cylindrical mode" is found along the entire compressive meridian, at least below the brittle-to-ductile transition. Above this transition it is not entirely clear how the material fails, nor if it is possible to separate a test specimen in several distinct parts. The other type of failure mode is found along the tensile meridian, and is called the "planar mode" or extensile failure. It is characterized by a horizontal splitting-type or shear crack in the plane of the two largest compressive stresses. If an extension test is carried out in a conventional triaxial cell, the failure plane is oriented perpendicular to the axis of the cylindrical specimen. We do not further discuss extensile failure in the context of this book. Instead, we take a closer look at the "cylindrical" failure mode. The orientation of the microcracks in an experiment can be measured using stereological methods (see, e.g., Stroeven 1979 who applied the method, which is described in detail in Underwood 1970). By means of fluorescent petroleum the microcracks on the surface of a specimen under load can easily be visualized (see also Section A4.1). The petroleum will be sucked into the concrete, in particular where cracks are present. The fluorescent particles will concentrate along the edges of cracks. In Figure A5.3 two examples are shown of crack-orientation distributions using the so-called "roses-of-intersections" from stereology. Simply counting the numbers of intersecting cracks along radial lines from the center of the specimen, and normalizing the count over the length of the

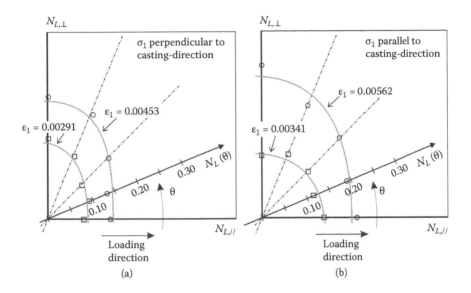

FIGURE A5.3
Two examples of roses-of-numbers-of-intersections per unit testline length for 100-mm concrete cubes loaded perpendicular to the casting direction (a), and parallel to the direction of casting (b). The diagrams are a measure of the damage-anisotropy of the concrete specimens at different stages of loading. (After Van Mier. 1984. *Strain-Softening of Concrete under Multiaxial Loading Conditions.*)

measuring line, is a good indication for the number of cracks and their orientation. Figure A5.3 shows the orientations for two 100-mm cubes loaded in uniaxial compression at two stages in the postpeak regime, at approximately −40 MPa (corresponding to axial strains $\varepsilon_1 = -0.00291$ and -0.00341 as indicated in the diagrams) and at −20 MPa (which relates to axial strains $\varepsilon_1 = -0.00453$ and -0.00562, that is, well in the softening regime).

The specimen of Figure A5.3a was loaded perpendicular to the casting direction; the one of Figure A5.3b parallel to casting, which results in a different orientation of initial damage from casting. Just beyond peak, at −0.00291 and −0.00341 strain the rose-of-number-of-intersections is elongated in Figure A5.3a, and more circular in Figure A5.3b. This means that the microcracks are better aligned to the loading direction in the first specimen, which was loaded perpendicular to the direction of casting. This simple result, which confirms our intuitive idea about the importance of the casting direction, shows that the method can be useful. When the specimens are further strained in the postpeak regime the elongated shape of the perpendicular-loaded specimen becomes more pronounced, but the parallel-loaded specimen now also shows a more elongated orientation distribution. Obviously, as a result of the applied uniaxial compressive load new microcracks develop, which are oriented in the loading direction. In three dimensions the circular and elliptical roses translate to oblate- and prolate-ellipsoids, assuming symmetric cracking behavior in these uniaxial-loaded specimens. The degree of crack-anisotropy can be derived from the relative length of the three axes of the ellipsoids. This is exactly what drives the behavior under multiaxial compression: the state of stress determines where cracks can or cannot develop, and as a result the crack orientation and crack density distributions are affected.

It is interesting to extend this idea about crack orientation distributions in experiments where the specimens are subjected to a second multiaxial loading regime after being damaged during a first loading history with completely different orientation. Without having to handle the specimens between the two loading regimes, 90° rotations are the best (only) possible choice when a "true" multiaxial machine is available (note: for cylindrical specimens applying torsion in addition to a triaxial loading may also provide rotation of the loading axes, but this presents a rather severe complication to otherwise rather simple conventional triaxial cells). In Figure A5.4 two results are shown from experiments where a second loading is applied, which, in both cases, is 90° rotated with respect to the first loading. In the test of Figure A5.4a the first loading is almost similar to uniaxial compression, except that a small symmetric confinement is present; that is, $\sigma_2 = \sigma_3 = -1$ MPa. The specimen is loaded approximately to $\varepsilon_1 = -0.0045$, well after the peak. Next, after unloading the situation is changed: σ_3 becomes the main loading direction, whereas $\sigma_1 = \sigma_2 = -1$ MPa. This specimen thus represents an example of the (cylindrical) failure mode observed along the compressive meridian (see Figure 8.14b). The second specimen, Figure A5.4b, is initially loaded with $\sigma_2 = 0.5\sigma_1$ and $\sigma_3 = -1$ MPa up to approximately $\varepsilon_1 = -0.005$; subsequently, after unloading the major and minor stresses are exchanged and σ_2 is coupled to σ_3, namely following $\sigma_2 = 0.5\sigma_3$ and $\sigma_1 = -1$ MPa. Displacement-control was used in all these experiments in the direction of the major principal stress. Thus the second test is representative for the second, planar failure mode (Figure 8.14a).

The specimen in Figure A5.4a fails symmetrically due to the first loading, and the expected damage orientation distribution takes the shape of an oblate ellipsoid; the second loading is also symmetric, but now in the σ_3-direction and the resulting damage orientation distribution will change to a more spherical shape as indicated. Interestingly the strength measured after the second loading does not reach the initial level of −53.46 MPa, but reaches just −42.90 MPa, that is, about 20% lower. The reason will be obvious: because damage has already been inflicted from the first loading, and, more importantly, oriented in the right direction, the second loading will simply cause damage to increase from the already substantial level. Therefore a critical state will be reached much earlier, and the second peak load (σ_{3p}) decreases. The σ_3–ε_3-curve appears to join the first curve (σ_1–ε_1) almost exactly at the right spot in the softening regime, as if just a cyclic loading was applied without rotating the loading direction.

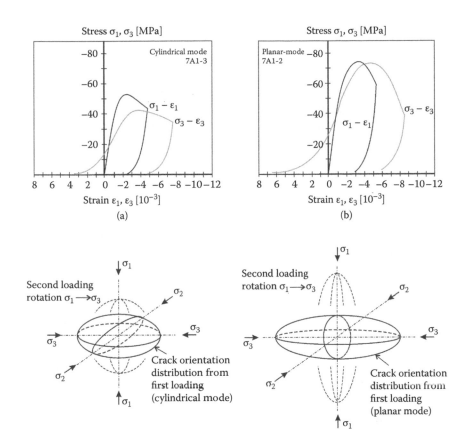

FIGURE A5.4
Stress–strain diagrams for two specimens that were loaded two times: (a) cylindrical failure mode, and (b) planar failure mode. The loading situation is explained in the main text. The damage-orientation distributions due to the first and second loadings are indicated below each diagram. (After Van Mier. 1984. *Strain-Softening of Concrete under Multiaxial Loading Conditions.*)

The planar-mode test of Figure A5.4b behaves quite differently: now the second loading has about the same failure strength as measured during the first loading (viz. −74.29 MPa and −72.62 MPa before and after rotation), and except for the long shallow part in the beginning of the σ_3–ε_3-diagram the curves look identical. The damage orientation distribution from the first loading is a prolate ellipsoid, and simply insufficient damage has been done along the σ_3-axis to affect the second diagram. Thus, a new damage orientation distribution, again with a prolate ellipsoidal shape has to develop: a new, second crack system is needed to fail the specimen after rotation. The shallow part of the σ_3–ε_3-curves in Figures A5.4a and A5.4b is caused by the damage from the first loading history and is (in part) oriented perpendicular to σ_3. These cracks must be closed before significant stresses can develop and quite obviously, larger deformations are required. This is a contact effect caused by the geometrical mismatch between the two opposite crack-faces.

These examples show very nicely how anisotropy in the damage orientation distribution affects the mechanical behavior of a specimen when it is loaded again. It is not unthinkable that such situations also develop in real practice. The amounts of damage (microcracks) as well as the damage orientations are very important parameters on which a useful constitutive equation for concrete should be based. Before this can be achieved, however, much experimental work will be needed, for instance, measuring the crack topography for different stress combinations at varying loading levels by means of x-ray computed tomography (see Section A4.2).

A5.2 Passive Confinement—Fiber-Reinforced Concrete

Mixing fibers in concrete can have the same effect as applying external confinement. It was first recognized by Yin et al. (1990) in biaxial experiments on fiber-reinforced concrete. The biaxial failure envelopes, more specifically, the biaxial compression regime, expanded when fibers were added to the concrete, indicating that the fibers restrain crack growth in the free (unconfined) direction. Thus, like adding a small confining stress in the third direction, fibers will help to restrain crack growth. The two mechanisms thus appear to be similar; see Figure A5.5.

Another form of passive confinement is the use of spiral reinforcement, or steel cladding in columns to ensure that the concrete is fully contained in steel reinforcement. When axial load is applied, the concrete will expand

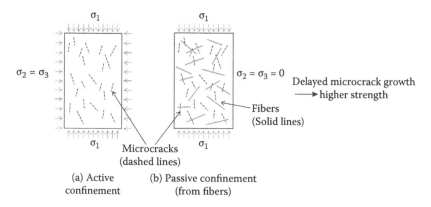

(a) Active confinement

(b) Passive confinement (from fibers)

FIGURE A5.5
Delayed microcrack growth in general leads to a higher strength. Either active (external) confinement (a) or adding fibers to the concrete (b) is an effective measure.

in the lateral direction, which is restrained by the spiral reinforcement, or steel cylinders. Much work has been done to optimize the steel content to obtain the best results against the lowest costs. We do not dwell further on this form of passive confinement, but have a little closer look at the effect of (steel) fiber reinforcement.

In a conference in 1992 in Sheffield (see Swamy 1992) my impression was that delegates were rather disappointed because for 40 years no appreciable progress had been made. In spite of all the research done no real applications of FRC had emerged at that time. This has changed a lot in the last decade because mechanisms are better understood, and fibers are not just added to a concrete mixture. They are incorporated in the mixture design from the very first considerations. Fibers work best when no large aggregates are present in the concrete, because in that way the fiber distribution is not affected. Moreover, the bonding of fibers to finer-grained mixtures is much better, as the fibers are better contained in the particle structure. At best the grain-size distribution is adjusted to the size of the fibers, most importantly the diameter, as shown by Rieger and Van Mier (2010). In addition one should take care that the bonding length is sufficiently long to assure pull-out of the fibers from the concrete, rather than breaking them. Smaller fibers may help to bridge and arrest the smallest cracks, whereas longer and thicker fibers are needed to guarantee an appreciable effect when large fractures develop. In 1987 Rossi, Acker, and Mallier made this hypothesis and suggested adding fibers of two different sizes to concrete: the smaller fibers would restrain microcracking in the prepeak regime (Rossi et al. called this the "material level") and larger fibers would arrest cracks in the softening regime (to which Rossi et al. referred to as the "structural level"). High-modulus fibers, such as steel, are usually considered the best choice inasmuch as they will be better capable of redistributing stresses after cracks develop in the concrete matrix. The concept of 1987 was tried and shown to work by Rossi and Renwez (1996). These ideas are spreading, not least due to the many conferences organized in the last two decades; see, for instance, the series of HPFRCC-Conferences, the last one in 2007 in Mainz (Reinhardt and Naaman 2007).

Much can be written about fiber concrete, but here we limit ourselves to a few significant results that are also of interest for the development of the universal 4-stage fracture model in Chapter 10. The idea of adding fibers of different size, that is, developing hybrid fiber concrete may lead to a substantial increase of flexural strength as shown in Figure A5.6. The results shown all relate to hybrid fiber concrete containing steel fibers of two different sizes (Markovic, Walraven, and Van Mier 2003 and Van Gunsteren 2003) or three different sizes (Stähli and van Mier 2004). The fiber-factor is a parameter that weighs the contributions of the different types of fibers, via the aspect ratio l/d (where l is the fiber length and d the fiber diameter) and volume fraction V of each fiber type in a single (dimensionless) number, following:

$$V_{fib} = \sum \left(V[\%] \cdot \frac{l}{d} \right) \qquad\qquad (A5.1)$$

Figure A5.6 clearly shows that adding fibers of one, two or three different sizes, and at increasingly higher volumes, leads to a substantial increase of the flexural strength (in comparison: plain concrete has a flexural strength around 8–10 MPa). An important condition is that the concrete remains workable when cast and that the fibers do not segregate; the addition of new and improved (super-) plasticizers is a key factor in these developments. What is of more interest here is not so much the strength increase, but rather the improvement in the load deformation behavior, which changes from quasi-brittle behavior for plain concrete to a highly ductile response for hybrid fiber concrete. In Figure A5.7 some results from Markovic (2006) are shown. In the lower left corner, the behavior of plain concrete is shown. Adding 2% of 13-mm long fibers has a large effect on the flexural strength; in the prepeak behavior not so much deformation is registered, but postpeak the deformations (midpoint deflection) are huge. When 1% of 60-mm long fibers are present in the mixture, a slightly lower strength is found compared to the mixture with 13-mm fibers, but the deformations up to peak are larger, as is the postpeak ductility. Obviously crack-face bridging from the long fibers is more effective than from the shorter fibers (note: an example of fiber bridging is shown in Figure 10.13 in Section 10.4). Now, if the two curves are added, the dotted line is obtained. This result would be reached if no synergy occurred between the two types of fibers in hybrid fiber concrete. Some synergy occurs, however, as experiments on hybrid fiber concrete containing

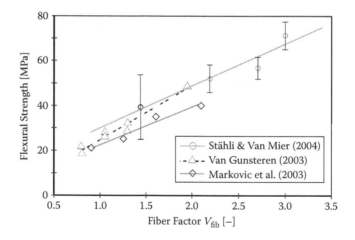

FIGURE A5.6

Increase of flexural strength from 4-point bending tests with increasing fiber factor V_{fib}. (After Stähli and Van Mier. 2004. *Proc. 5th Int'l Conf. on Fracture of Concrete and Concrete Structures (FraMCoS-V)*.)

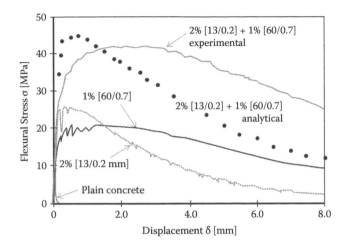

FIGURE A5.7
Behavior of different concretes subjected to 4-point bending: single fiber concrete containing either 2% 13-mm long fibers (0.2 mm diameter) or 1% 60-mm long fibers (0.7 mm diameter) behaves significantly better than plain concrete. The sum of the behavior of these single-fiber mixtures leads to the dotted line, which comes close to the response of hybrid fiber concrete containing a cocktail of both fibers as indicated. (After Markovic. 2006. *High-Performance Hybrid-Fibre Concrete.* With kind permission from Dr. Ivan Markovic.)

2% 13-mm fibers and 1% 60-mm fibers show: although the strength is almost equal to the analytical result, the deformations before and after peak increase significantly. Thus, mixing different types of fibers seems like a good solution to optimize the properties of concrete, see for example in Markovic (2006) and Stahli (2008). Microcrack growth can be delayed considerably, that is, the fibers work perfectly as internal confinement, and overall ductility, especially in the postpeak regime is significantly improved due to the additional effect from fiber bridging (see Figure 10.13).

Finally, the directionality of the fibers may further enhance the properties of these concretes in some directions. Aligning the fibers is possible using a variety of methods, for example, sprinkling fibers through narrow slits in molds for producing Slurry Infiltrated Fiber CONcrete (SIFCON; Van Mier and Timmers 1992) or by means of magnetic positioning as proposed by Linsel (2005). Here we used the fluidity of the hybrid fiber concrete as a tool to align fibers by controlling the flow in the longitudinal direction of the beams. In that way fibers are well aligned parallel to the main tensile direction, and better performance is expected. Note that due to the alignment the material will become anisotropic, and properties are likely improved in just one direction, and may even get worse in other directions; see also Van Mier and Timmers (1992). In Figure A5.8 some of the results from the flow tests are shown.

The results in Figure A5.8 suggest that a relation between the number of fibers in a cross-section and the nominal bending strength exists: the higher

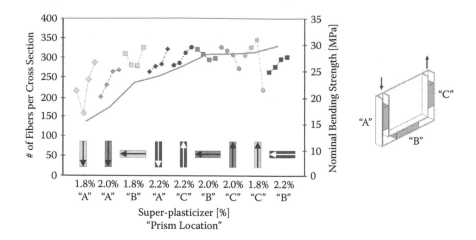

FIGURE A5.8

Relation between numbers of fibers per cross-section (symbols, left axis), nominal bending strength (full line, right axis) and the flow direction of fiber concrete containing 3% 30-mm long steel fibers (0.6-mm diameter). U-shaped molds are used; the concrete is always filled at one side, flows along a horizontal part of the U-mold and rises in the second leg, as indicated in the inset. The capitals "A," "B," and "C" refer to three different prism positions as indicated. Three different mixtures were tested with varying content of plasticizer: 1.8%, 2.0%, and 2.2%. (From Stähli et al. 2008. *Ultra-Fluid, Oriented Hybrid-Fibre-Concrete.* With permission from Elsevier.)

the amount of fibers is, the higher the flexural strength. Higher fiber counts and higher flexural strengths are observed in the "B" and "C" prisms, that is, the horizontal and rising branches of the U-shaped mold. Clearly, fibers are better aligned in those parts; in the "A"-specimens the concrete simply is cast from the top and does not really flow. Thus flowing of concrete appears to be a key factor in fiber alignment and may help to obtain better strength. Using dog-bone-shaped specimens, and pumping the concrete in a vertical standing mold will also result in highly aligned fibers in the neck of the dog-bone. Stähli (2008) showed that the tensile strength also depends on the casting method, quite similar to the results shown here for bending. Thus, as in actively confined concrete, directionality in the properties may be achieved in fiber concrete as well. The results may in the future find an application in the precast concrete industry where local properties of a building element may be enhanced by casting it in the right way.

References

Akita, H., Koide, H., Tomon, M., and Sohn, D. (2003). A practical method for uniaxial tension test of concrete. *Mater. Struct. (RILEM)*, 36: 365–371.

Akono, A.-T. Reis, P.M., and Ulm, F.-J. (2011). Scratching as a fracture process: From butter to steel. *Phys. Rev. Lett.*, 106: 204302–1–4.

Alava, M.J., Nukala, Ph.K.V., and Zapperi, S. (2006). Statistical models of fracture. *Adv. Phys.*, 55(3–4): 349–476.

Alexander, K.M. and Wardlaw, J. (1960). Dependence of cement-aggregate bond-strength on size of aggregate. *Nature*, 197(4733): 230–231.

Alexander, K.M., Wardlaw, J., and Gilbert, D.J. (1965). Aggregate-cement bond, cement paste strength and the strength of concrete. In *Proc. Int'l. Conf. on the Structure of Concrete*. London: Cement & Concrete Association, 59–81.

Akazawa, H. (1943). *J. Jpn. Civ. Eng. Inst.*

Anson, M. and Newman, K. (1966). The effect of mix proportions and method of testing on the Poisson's ratio for mortar and concretes, *Mag. Conc. Res.*, 18(1), 115–130.

Ashby, M.F. and Hallam, S.D. (1986). The failure of brittle solids containing small cracks under compressive stress states. *Acta Metall.*, 34(3): 497–510.

Attard, M.M. and Setunge, S. (1996). Stress-strain relationship of confined and unconfined concrete. *ACI Mater. J.*, 93: 432–442.

Bagi, K. and Kuhn, M.R. (2004). Different rolling measures for granular assemblies. In P.A. Vermeer, W. Ehlers, H.J. Herrmann, and E. Ramm (Eds.), *Modelling of Cohesive-Frictional Materials*, London: Taylor & Francis, 3–12.

Banville, J. (2009). *The Infinities*. UK and New York: Picador.

Barenblatt, G.I. (1962). The mathematical theory of equilibrium of cracks in brittle fracture. *Adv. Appl. Mech.*, 7: 55–129.

Bazant, M.Z. and Kaxiras, E. (1996). Modeling of covalent bonding in solids by inversion of cohesive energy curves. *Phys. Rev. Lett.*, 77(21): 4370–4373.

Bazant, M.Z., Kaxiras, E., and Justo, J.F. (1997). Environment-dependent inter-atomic potential for bulk silicon. *Phys. Rev. B*, 56(14): 8542–8552.

Bažant, Z.P. (1984). Size effect in blunt fracture: Concrete, rock, metal. *J. Eng. Mech.*, 110: 518–535.

Bažant, Z.P. (1989). Identification of strain-softening constitutive relation from uniaxial tests by series coupling model for localization. *Cem. Conc. Res.*, 19: 973–977.

Bažant, Z.P. (2004). Scaling theory for quasibrittle structural failure. *Proc. Nat. Acad. Sci.*, 101(37): 13400–13407.

Bažant, Z.P. and Oh, B.-H. (1983). Crack band theory for fracture of concrete. *Mater. Struct. (RILEM)*, 16: 155–177.

Bažant, Z.P. and Pfeiffer, P.A. (1986), Shear fracture tests of concrete, *Mater. Struct. (RILEM)*, 19: 111–121.

Bažant, Z.P., Prat, P.C., and Tabbara, M.R. (1990). Anti-plane shear fracture tests (Mode III). *ACI Mater. J.*, 87: 12–19.

Bažant, Z.P., Xi, Y., and Reid, S.G. (1991). Statistical size effect in quasi-brittle structures: Part I: Is Weibull theory applicable? *J. Eng. Mech.*, 117(11): 2609–2622.

Bažant, Z.P. and Yavari, A. (2005). Is the cause of size effect on structural strength fractal or energetic statistical? *Eng. Fract. Mech.*, 72: 1–31.

Bentz, D.P. (1997). Three dimensional computer simulation of Portland cement hydration and microstructure development. *J. Amer. Ceramic Soc.*, 80: 3-21.

Bentz, D.P., Coveney, P.V., Garboczi, E.J., Kleyn, M.F., and Stutzman, P.E. (1994). Cellular automaton simulations of cement hydration and microstructure development. *Model. Simul. Mater. Sci. Eng.*, 2: 783–808.

Bentz, D.P., Quenard, D.A., Kunzel, H.M., Baruchel, J., Peyrin, F., Martys, N.S., and Garboczi, E.J. (2000). Microstructure and transport properties of porous building materials. II: Three-dimensional x-ray tomographic studies. *Mater. Struct. (RILEM)*, 33: 147–153.

Beranek, W.J. and Hobbelman, G.J. (1992). Handboek voor het mechanisch gedrag van metselwerk (Glossary of the mechanical behaviour of masonry). *Report C-77*, CUR Foundation, Gouda (in Dutch).

Beranek, W.J. and Hobbelman, G.J. (1994). Constitutive modelling of structural concrete as an assemblage of spheres. In H. Mang, N. Bicanic, and R. de Borst (Eds.), *Computational Models of Concrete Structures, Proceedings EURO-C*, Swansea, UK: Pineridge Press, 535–542.

Berlage, A.C.J. (1987). *Strength Development of Hardening Concrete*. PhD thesis, Delft University of Technology (in Dutch).

Birchall, J.D., Howard, A.J., and Kendall, K. (1981). Flexural strength and porosity of cements. *Nature*, 289: 388–390.

Bishoi, A. and Scrivener, K. (2009). μic: A new platform for modeling the hydration of cements. *Cem. Conc. Res.*, 39: 266–274.

Bisschop, J. (2002). *Drying Shrinkage Microcracking in Cement-Based Materials*. PhD thesis, Delft University of Technology.

Bisschop, J., Pel, L., and van Mier, J.G.M. (2001). Effect of aggregate size and paste volume on drying shrinkage microcracking in cement-based composites. In F.-J. Ulm, Z.P. Bažant, and F.H. Wittmann (Eds.), *Creep, Shrinkage and Durability Mechanics of Concrete and Other Quasi-Brittle Materials, Proc. CONCREEP-6*, Amsterdam: Elsevier Science, 75–80.

Bluhm, J.I. and Morrissey, R.J. (1965). Fracture in a tensile specimen. In T. Yokobori, T. Kawasake, and J.L. Swedlow (Eds.), *Proc. ICF-1*, Sendai, 1739–1780.

Bobet, A. and Einstein, H.H. (1998). Fracture coalescence in rock-type materials under uniaxial and biaxial compression. *Int. J. Rock Mech. Mining Sci.*, 35(5): 863–888.

Bolander, J. and Kobayashi, Y. (1995). Size effect mechanisms in numerical concrete fracture. In F.H. Wittmann (Ed.), *Proc. 2nd Int'l. Conf. on Fracture Mechanics of Concrete and Concrete Structures (FraMCoS-2)*, Freiburg: AEDIFICATIO, 535–542.

Broek, D. (1983). *Elementary Engineering Fracture Mechanics*. The Hague: Martinus Nijhoff.

Burgoyne, C. and Scantlebury, R. (2006). Why did Palau bridge collapse? *Struct. Eng.*, 6: 30–37.

Caduff, D. and Van Mier, J.G.M. (2010). Analysis of compressive fracture of three different concretes by means of 3D-digital image correlation and vacuum impregnation. *Cem. Conc. Comp.*, 32: 281–290.

Carneiro, F. and Barcellos. (1944). Resistance a la traction des Bétons. Brochure des Instituto Nacional de Technologica, Rio de Janeiro.

Carpinteri, A. (1994). Scaling laws and renormalization groups for strength and toughness of disordered materials. *Int. J. Solids Struct.*, 31: 291–302.

Carpinteri, A., Chiaia, B., and Cornetti, P. (2003). On the mechanics of quasi-brittle materials with a fractal microstructure. *Eng. Fract. Mech.,* 70: 2321–2349.

Carpinteri, A. Chiaia, B., and Ferro, G. (1995). Size effects on nominal strength of concrete structures: Multi-fractality of material ligaments and dimensional transition from order to disorder. *Mater. Struct. (RILEM)*, 28: 311–317.

Carpinteri, A., Chiaia, B., and Invernizzi, S. (2004). Numerical analysis of indentation fracture in quasi-brittle materials. *Eng. Fract. Mech.,* 71: 567–577.

Carpinteri, A. and Ferro, G. (1994). Size effects on tensile fracture properties: A unified explanation based on disorder and fractality of concrete microstructure. *Mater. Struct. (RILEM)*, 27:563–571.

Carpinteri, A. and Lacidogna, G. (2010). Energy emissions from fracture of concrete: acoustic, electromagnetic, piezo-nuclear. In B.-H. Oh, O.-C. Choi, and L. Chung (Eds.), *Fracture Mechanics of Concrete and Concrete Structures, Proc. FraMCoS-7,* Seoul: Korea Concrete Institute, 21–28.

CEB-FIP model code (1990)., 1993 ed., London: Thomas Telford.

Célarié, F., Prades, S., Bonamy, D., Ferrero, L., Bouchaud, E., Guillot, C., and Marlière, C. (2003). Glass breaks like metal, but at the nanometer scale. *Phys. Review Lett.,* 90(7): 21, February.

Chen, W.F. (1970). Double-punch test for tensile strength of concrete. *ACI Mater.J.,* 67(2): 993–995.

Chiaia, B. (2001). Fracture mechanisms induced in a brittle material by a hard cutting indenter. *Int. J. Solids Struct.,* 38: 7747–7768.

Chiaia, B., Vervuurt, A., and van Mier, J.G.M. (1997).Lattice model evaluation of progressive failure in disordered particle composites. *Eng. Fract. Mech.,* 57(2/3): 301–318.

Constantinides, G. and Ulm, F.J. (2004). The effect of two types of C-S-H on the elasticity of cement-based materials: Results from nano-indentation and micromechanical modeling. *Cem. Conc. Res.,* 34(1): 67–70.

Cornelissen, H.A.W., Hordijk, D.A., and Reinhardt, H.W. (1986a). Experimental determination of crack softening characteristics of normalweight and lightweight concrete. *HERON,* 31(2): 45–56.

Cornelissen, H.A.W., Hordijk, D.A., and Reinhardt, H.W. (1986b). Experiments and theory for the application of fracture mechanics to normal and lightweight concrete. In F.H. Wittmann (Ed.), *Fracture Toughness and Fracture Energy.* Amsterdam: Elsevier, 565–575.

Cundall, P.A. and Strack, O.D.L. (1979). A discrete numerical model for granular assemblies. *Géotechnique* 29(1):47–65.

CUR. (2003). Advanced analysis of civil engineering structures. Report 2003-3, CUR-Committee A36 Concrete Mechanics, Gouda, the Netherlands: CUR Foundation.

Danzer, R. (2006). Some notes on the correlation between fracture and defect statistics: Are Weibull statistics valid for very small specimens? *J. Eur. Ceram. Soc.,* 26: 3043–3049.

Danzer, R., Lube, T., and Supanic, P. (2001). Monte Carlo simulations of strength distributions of brittle materials: Type of distribution, specimen and sample size. *Z. Metallkd.,* 92: 773–783.

Danzer, R., Supancic, P., Pascual, J., and Lube, T. (2007). Fracture statistics of ceramics: Weibull statistics and deviations from Weibull statistics. *Eng. Fract. Mech.,* 74: 2919–2932.

De Borst, R. (1986). *Non-linear Analysis of Frictional Materials*. PhD thesis, Delft University of Technology.

Dempsey, J.P., Adamson, R.M., and Mulmule, S.V. (1999b). Scale effects on the in-situ tensile strength and fracture of ice: Part II First-year sea ice at Resolute N.W.T. *Int. J. Fract.*, 95: 346–378.

Dempsey, J.P., DeFranco, S.J., Adamson, R.M., and Mulmule, S.V. (1999a). Scale effects on the in-situ tensile strength and fracture of ice: Part I Large grained freshwater ice at Spray Lakes Reservoir, Alberta. *Int. J. Fract.*, 95: 325–345.

Desrues, J. (1998). Localisation patterns in ductile and brittle geomaterials. In E. van der Giessen and R. de Borst (Eds.), *Material Instabilities in Solids*. Chichester: John Wiley & Sons, 137–158.

Desrues, J. and Viggiani, G. (2004). Strain localization in sand: An overview of the experimental results obtained in Grenoble using stereophotogrammetry. *Numer. Anal. Meth. Geomech.*, 28(4): 279–321.

Diamond, S. (2000). Mercury porosimetry: An inappropriate method for the measurement of pore size distributions in cement-based materials. *Cem. Conc. Res.*, 30: 1517–1525.

Donev, A., Cisse, I., Sachs, D., Variano, E.A., Stillinger, F.H., Connelly, R., Torquato, S., and Chaikin, P.M. (2004). Improving the density of jammed disordered packings using ellipsoids. *Science*, 303: 990–993.

Dugdale, D.S. (1960). Yielding of sheets containing slits. *J. Mech. Phys. Sol.* 8: 100–108.

Dyskin, A.V., Germanovich, L.N., and Ustinov, K.B. (1999). A 3-D model of wing crack growth and interaction. *Eng. Fract. Mech.*, 63: 81–110.

Elfgren, L. (Ed.) (1989). *Fracture Mechanics of Concrete Structures: From Theory to Applications*. London: Chapman & Hall.

Elfgren, L. (1992). Round robin analysis and tests of anchor bolts. In Z.P. Bažant (Ed.), *Proc. FraMCoS-1 Fracture Mechanics of Concrete Structures*. London/New York: Elsevier Applied Science, 865–869.

Elices, M., Guinea, G.V., and Planas, J. (1992). Measurement of the fracture energy using three-point bend tests: Part 3: Influence of cutting the P-δ tail. *Mater. Struct. (RILEM)*, 25(150): 327–334.

Evans, R.H. and Marathe, M.S. (1968). Microcracking and stress-strain curves for concrete in tension. *Mater. Struct. (RILEM)*, 1(1): 61–64.

Foote, R.M.L., May, Y.-W., and Cotterell, B. (1986). Crack growth resistance curves in strain-softening materials. *J. Mech. Phys. Solids*, 34(6): 593–607.

Fortt, A.L. and Schulson, E.M. (2007). The resistance to sliding along Coulombic shear faults in ice. *Acta Mater.*, 55: 2253–2264.

Freudenthal, A.M. (1968). Statistical approach to brittle fracture. In H. Liebowitz (Ed.), *Fracture*, Volume 2, New York/London: Academic Press, 591–619.

Garboczi, J. and Bentz, D.P. (1991). Digital simulation of the aggregate-cement paste interfacial zone in concrete. *J. Mat. Res.*, 6: 196–201.

Gatty, L., Bonnamy, S., Feylessoufi, A., Clinard, C., Richard, P., and Van Damme, H. (2001). A transmission electron microscopy study of interfaces and matrix homogeneity in ultra-high-performance cement-based materials. *J. Mater. Sci.*, 36: 4013–4026.

Gettu, R., Mobasher, B., Carmona, S., and Jansen, D.C. (1996). Testing of concrete under closed-loop control. *Adv. Cem. Bas. Mat.*, 3: 54–71.

Golterman, P. (1995). Mechanical predictions of concrete deterioration: Part 2: Classification of crack patterns. *ACI Mater. J.*, 92(1): 58–63.

Goto, Y. (1971). Cracks formed in concrete around deformed tension bars. *J. Amer. Conc. Inst.* 68(4): 244–251.

Griffith, A.A. (1921). The phenomena of rupture and flow in solids. *Phil. Trans. Roy. Soc. London, Ser. A* 221: 16–198.

Griffith, A.A. (1924). The theory of rupture. In B. Biezeno and J.M. Burgers (Eds.), *Proc. 1st Int'l. Congress for Applied Mechanics.* Delft: Waltman, 55–63.

Grosse, C.U. and Ohtsu, M. (2008). *Acoustic Emission Testing.* Berlin, Heidelberg: Springer-Verlag.

Guinea, G.V., Planas, J., and Elices, M. (1992). Measurement of the fracture energy using three-point bend tests: Part 1: Influence of experimental procedures. *Mater. Struct. (RILEM)*, 25(148): 212–218.

Gupta, P.K. (2002). Strength of glass fibres. In M. Elices and J. Llorca (Eds.), *Fibre Fracture.* Oxford, UK: Elsevier Science, 127–153.

Hallbauer, D.K. Wagner, H., and Cook, N.G.W. (1973). Some observations concerning the microscopic and mechanical behaviour of quartzite specimens in stiff triaxial compression tests. *Int.J. Rock Mech. Min. Sci. Geomech. Abstr.*, 10: 713–726.

Hashin, Z. (1965). On elastic behaviour of fibre reinforced materials of arbitrary transverse phase geometry. *J. Mech. Phys. Solids*, 13: 119–134.

Hashin, Z. (1983). Analysis of composite materials: A survey. *J. Appl. Mech.*, 50: 481–505.

Hashin, Z. and Shtrikman, S. (1963). A variational approach to the theory of the elastic behaviour of multiphase materials. *J. Mech. Phys. Solids*, 11:127–140.

Hassanzadeh M. (1992). *Behaviour of Fracture Process Zones in Concrete Influenced by Simultaneously Applied Normal and Shear Displacements.* PhD thesis, Lund University of Technology.

Hassanzadeh, M. (1995). Fracture mechanical properties of rocks and mortar/rock interfaces. In S. Diamond, S. Mindess, F.P. Glasser, L.W. Roberts, J.P. Skalny, and L.D. Wakeley (Eds.), *Microstructure of Cement-Based Systems/Bonding and Interfaces in Cementitious Materials*, Pittsburgh, PA: MRS, 370: 377–382.

Helbing, A., Alvaredo, A., and Wittmann, F.H. (1991). Round robin tests on anchor bolts. In L. Elfgren (Ed.), *Round-Robin Analysis of Anchor Bolts: RILEM TC-90 FMA Fracture Mechanics of Concrete: Applications*, Lulea University of Technology, Sweden, 8:1–8:22.

Helm, J., McNeill, S., and Sutton, M. (1996). Improved three-dimensional image correlation for surface displacement measurement. *Optical Engng.*, 35(7): 1911–1920.

Herrmann, H.J., Hansen, A., and Roux, S. (1989). Fracture of disordered elastic lattices in two dimensions. *Phys Rev B.*, 39(1): 637–648.

Hettema, M. (1996). *The Thermo-Mechanical Behaviour of Sedimentary Rock: An Experimental Study.* PhD thesis, Delft University of Technology.

Hillerborg, A. (1985). The theoretical basis of a method to determine the fracture energy G_f of concrete. *Mater. Struct. (RILEM)*, 18: 291–296.

Hillerborg, A. (1989). Stability problems in fracture mechanics testing. In B. Barr and S.E. Swartz, (Eds.). *Fracture of Concrete and Rock: Recent Developments*, London: Elsevier Applied Science, 369–378.

Hillerborg, A. (1990). Fracture mechanics concepts applied to moment capacity and rotational capacity of reinforced concrete beams. *Eng. Fract. Mech.*, 35: 233-240.

Hillerborg, A. (1992). Effect of fibre modified fracture properties on shear resistance of reinforced mortar and concrete beams. In A. Carpinteri (Ed.), *Applications of Fracture Mechanics to Reinforced Concrete*, London/New York: Elsevier Applied Science, 487–501.

Hillerborg, A., Modeér, M., and Peterson, P.-E. (1976). Analysis of crack formation and crack growth in concrete by means of fracture mechanics and finite elements. *Cem. Conc. Res.*, 6: 773–782.

Hilsdorf, H. (1965). Determination of the biaxial strength of concrete (Die Bestimmung der Zweiachsiger Festigkeit des Betons). *Deutscher Ausschuss für Stahlbeton*, Heft 173 (in German).

Holland, D. and Marder, M. (1999). Cracks and atoms. *Adv. Mater.*, 11(10): 793–806.

Holzer, L., Indutyi, F., Gasser, P.H., Münch, B., and Wegmann, M. (2004). Three-dimensional analysis of porous BaTiO3 ceramics using FIB nanotomography. *J. Microscopy*, 216: 84–95.

Hordijk, D.A. (1991). *Local Approach to Fatigue of Concrete*. PhD thesis, Delft University of Technology.

Horii, H. and Nemat-Nasser, S. (1986). Brittle failure in compression: Splitting, faulting and brittle-ductile transition. *Phil. Trans. Roy. Soc. London: Ser. A. Math. Phys. Sci.*, 319(1549): 337–374.

Hrennikoff, A. (1941). Solution of problems of elasticity by the framework method, *J. Appl. Mech.* A169-A175.

Hsu, T.T.C., Slate, F.O., Sturman, G.M., and Winter, G. (1963). Microcracking of plain concrete and the shape of the stress-strain curve. *ACI-J.*, 60(14): 209–224.

Hudson, J.A., Crouch, S.L., and Fairhurst, C. (1972). Soft, stiff and servo-controlled testing machines: A review with reference to rock failure. *Engng. Geol.*, 6(3): 155–189.

Hughes, B.P. and Chapman, G.P. (1966). The complete stress-strain curve for concrete in direct tension. *RILEM Bull.*, 30: 95–97.

Hull, D. (1993). Tilting cracks: The evolution of fracture surface topology in brittle solids, *Int. J. Fract.*, 62: 119–138.

Hurlbut, B. (1985). *Experimental and Computational Strain-Softening Investigation of Concrete*. MSc thesis, Civil Engineering Department, University of Colorado, Boulder.

Iacono, C. (2007). *Procedures for Parameter Estimates of Computational Models for Localized Failure*. PhD thesis, Delft University of Technology.

Ince, R. Arslan, A. and Karihaloo, B.L. (2003). Lattice modelling of size effect in concrete strength. *Engng. Fract. Mech.*, 70: 2307–2320.

Ingraffea, A.R. and Panthaki, M.J. (1986). Analysis of shear fracture tests of concrete beams. In *Finite Element Analysis of Concrete Structures*, Tokyo: ASCE, 151–173.

Ingraffea, A.R. and Saouma, V. (1984). Numerical modeling of discrete crack propagation in reinforced and plain concrete. In G.C. Sih and A. di Tommaso (Eds.), *Application of Fracture Mechanics to Concrete Structures*. Dordrecht: Martinus Nijhoff, Ch. 4.

Irwin, G.R. (1958). Fracture. In S. Flugge (Ed.), *Handbuch der Physik*, Berlin: Springer-Verlag, vol. 6, 551–590.

Iwashita, K. and Oda, M (2000). Micro-deformation mechanism of shear banding process based on modified distinct element methods. *Powder Techn.*, 109: 192–205.

Jamet, P., Millard, A., and Nahas, G. (1984). Triaxial behaviour of a micro-concrete complete stress-strain curves for confining pressures ranging from 0 to 100 MPa. In *Proc. Int'l. Conf. on Concrete under Multiaxial Conditions*. Toulouse: Presses de l'Université Paul Sabatier, 133–140.

Jansen, D.C. and Shah, S.P. (1997). Effect of length on compressive strain-softening of concrete, *J. Eng. Mech. (ASCE)*, 124(1): 25–35.

Jayatilaka, A.,De, S., and Trustrum, K. (1977). Statistical approach to brittle fracture. *J. Mater. Sci.*, 12: 1426–1430.

Jennings, H.M. (2000). A model for the microstructure of calcium silicate hydrate in cement paste. *Cem. Conc. Res.*, 30: 101–116.

Kahn-Jetter, Z. and Chu, T. (1990). Three-dimensional displacement measurements using digital image correlation method and photogrammetric analysis. *Exp. Mech.*, 30(1): 10–16.

Kaplan, M.F. (1961). Crack propagation and the fracture of concrete, *J. Amer. Conc. Inst.*, 58(5): 591–609.

Karihaloo, B.L. and Nallathambi, P. (1989). An improved effective crack model for the determination of fracture toughness of concrete, *Cem. Conc. Res.*, 19: 603–610.

Kelly, A. and MacMillan, N.H. (1986). *Strong Solids*. Oxford, UK: Oxford University Press.

Kemeny, J.M. and Cook, N.G.W. (1991). Micromechanics of deformation in rocks. In S.P. Shah (Ed.), *Toughening Mechanisms in Quasi-Brittle Materials*. Dordrecht: Kluwer Academic, 155–188.

Koenders, E.A.B. (1997). *Simulation of Volume Changes in Hardening Cement-Based Materials*. PhD-thesis, Delft University of Technology.

Kotsovos, M.D. (1983). Effect of testing techniques on the post-ultimate behaviour of concrete in compression. *Mater. Struct. (RILEM)*, 16(1): 3–12.

Kotsovos, M.D. and Newman, J.B. (1977). Behaviour of concrete under multiaxial stress. *J. Amer. Conc. Inst.*, 74: 443–446.

Kupfer, H. (1973). Das Verhalten des Betons unter Mehrachsiger Kurzzeitbelastung unter Besonderer Berücksichtigung des Zweiachsiger Beanspruchung (Behaviour of concrete under short term multiaxial loading, with emphasis on biaxial loading). *Deutscher Ausschuss für Stahlbeton*, Berlin, Heft 229 (in German).

Labuz, J.F., Shah, S.P., and Dowding, C.H. (1985). Experimental analysis of crack propagation in granite. *Int. J. Rock Mech. Min. Sci. Geomech. Abstr.*, 22(2): 85–98.

Landis, E.N., Zhang, T., Nagy, E.N., Nagy, G., and Franklin, W.R. (2007). Cracking, damage and fracture in four dimensions. *Mater. Struct. (RILEM)*, 40: 357–364.

Lawn, B. (1993). *Fracture of Brittle Solids*, 2nd ed. Cambridge, UK: Cambridge University Press.

Leicester, R.H. (1973). Effect of size on the strength of structures. *CSIRO, Report #71*, Australian Forest Production Laboratory, Division of Building Research Technology.

Leung, K.-T. and Néda, Z. (2000). Pattern formation and selection in quasistatic fracture. *Phys. Rev. Lett.*, 85(3): 662–665.

Li, V.C. (2010). Driving infrastructure sustainability with strain hardening cementitious composites (SHCC). In G.P.A.G. van Zijl and W.P. Boshoff (Eds.), *Advances in Cement-Based Materials*. London: Taylor & Francis Group, 181–191.

Li, Z., Kulkarni, S.M., and Shah, S.P. (1993). New test-method for obtaining softening response of un-notched concrete specimen under uniaxial tension. *Exp. Mech.*, 9: 181–188.

Lilliu, G. (2007). *3D Analysis of Fracture Processes in Concrete*. PhD thesis, Delft University of Technology.

Lilliu, G. and Van Mier, J.G.M. (1999). Analysis of crack growth in the Brazilian test. In R. Eligehausen (Ed.), *Construction Materials: Theory and Application*. Stuttgart: IBIDEM Verlag, 123–138.

Lilliu, G. and Van Mier, J.G.M. (2003). 3D lattice type fracture model for concrete. *Engng. Fract. Mech.*, 70(7/8): 927–942.

Lilliu, G. and Van Mier, J.G.M. (2007). On the relative use of micro-mechanical lattice analysis of 3-phase particle composites. *Engng. Fract. Mech.*, 74(7): 1174–1189.

Lingen, F.J. (2000). *Design of an Object-oriented Finite Element Package for Parallel Computers*. PhD thesis, Delft University of Technology.

Linsbauer, H.N. and Tschegg, E.K. (1986). Fracture energy determination of concrete with cube-shaped specimens. *Zement und Beton*, 31: 38–42.

Linsel, S. (2005). *Magnetic Positioning of Steel Fibres in Cementitious Materials*. PhD thesis, Technical University Berlin (in German).

Lockner, D.A. and Byerlee, J.D. (1992). Fault growth and acoustic emission in confined granite. *Appl. Mech. Rev.*, 45(3): 165–173.

Luding, S. (2004), Micro-macro transition for anisotropic frictional granular packing. *Int. J. Solids Struct.* 41: 5821–5836.

Luo, P., Chao, Y., Sutton, M., and Peters, W. (1993). Accurate measurement of three-dimensional deformations in deformable and rigid bodies using computer vision. *Exp. Mech.*, 33(1): 123–132.

Luong, M.P. (1990). Infrared thermo-vision of damage processes in concrete and rock. *Eng. Fract. Mech.*, 35: 291–301.

Maekawa, K., Chaube, R., and Kishi, T. (1999). *Modelling of Concrete Performance Hydration, Microstructure Formation and Mass Transport*. London: E&FN Spon.

Man, H.-K. (2009). *Analysis of 3D Scale and Size Effects in Numerical Concrete*. PhD thesis #18776, ETH Zurich, Switzerland.

Man, H.-K. and Van Mier, J.G.M. (2008a). Influence of particle density on 3D size effects in the fracture of (numerical) concrete. *Mech. Mater.*, 40(6): 470–486.

Man, H.-K. and Van Mier, J.G.M. (2008b). Size effect on strength and fracture energy for numerical concrete with realistic aggregate shapes. *Int. J. Fract.*, 154(1–2): 61–72.

Man, H.-K. andVan Mier, J.G.M. (2011). Damage distribution and size effect in numerical concrete from lattice analyses. *Cem. Conc. Comp.*, 33: 867–880.

Marchand, N., Parks, D.M., and Pelloux, R.M. (1986). K_I-solutions for single edge notched specimens under fixed end displacements. *Int. J. Fract.*, 32: 53–65.

Margoldová, J. and Van Mier, J.G.M. (1994). Simulation of compressive fracture in concrete. In K.-H. Schwalbe and C. Berger (Eds.), *Proc. ECF10 Structural Integrity: Experiments, Models, Applications*, Warley, UK: EMAS,1399–1408.

Markeset, G. (1993). *Failure of Concrete under Compressive Strain Gradients*. PhD thesis, Norges Tekniske Hogskole.

Markovic, I. (2006). *High-Performance Hybrid-Fibre Concrete*. PhD-thesis, Delft University of Technology.

Markovic, I., Walraven, J.C., and Van Mier, J.G.M. (2003). Development of high performance hybrid fibre concrete. In A.E. Naaman and H.W. Reinhardt (Eds.), *Proc. 4th Int'l. RILEM Workshop on High Performance Hybrid Fibre Concrete (HPFRCC-4)*, Bagneux: RILEM S.A.R.L., 277–300.

Meakin, P. (1991). Simple models for material failure and deformation. In J.G.M. van Mier, J.G. Rots, and A. Bakker (Eds.), *Fracture Processes in Concrete, Rocks and Ceramics*, London/New York: E&FN Spon, 213–229.

Mechtcherine, V. (2007). Testing behaviour of strain-hardening cement-based composites in tension: Summary of recent research. In H.W. Reinhardt and A.E. Naaman (Eds.), *High Performance Fibre-Reinforced Composites (HPFRCC-5) Proceedings 53*, Bagneux, France: RILEM S.A.R.L., 13-22.

Meda, A. (2003). Tensile behaviour in natural building stone: Serena sandstone. *Mater. Struct. (RILEM)*, 36: 553–559.

Meyer, D. (2009). *Foamed Cement-based Materials*, PhD-thesis #18701, Zurich: ETH.

Meyer, D., Man, H.-K., and Van Mier, J.G.M. (2009). Fracture of foamed cementitious materials: A combined experimental and numerical study. In H. Zhao and N.A. Fleck (Eds.), *Mechanical Properties of Cellular Materials*. New York: IUTAM Bookseries 12, Springer Science+ Business Media, 115–123.

Mindess, S. (1991). Fracture process zone detection. In S.P. Shah and A. Carpinteri (Eds.), *Fracture Mechanics Test Methods for Concrete*. London/New York: Chapman & Hall, 231–261.

Mokni, M. (1992). *Relations entre Déformations en Masse et Déformations Localisées dans les Matériaux Granulaires*. PhD-thesis, Université Joseph Fourier, Grenoble.

Moore, D.E. and Lockner, D.A. (1995). The role of microcracking in shear-fracture propagation in granite. *Struct. Geol.* 17(1): 95–114.

Mourkazel, C, and Herrmann, H.J. (1992). A vectorizable random lattice. *J. Stat. Phys.*, 68: 911–923.

Muguruma, Y., Tanaka, T. and Tsuji, Y. (2000). Numerical simulation of particle flow with liquid bridge between particles. *Powder Techn.*, 109: 49–57.

Nádai, A. (1924). Beobachtungen der Gleitflächenbildung an Plastischen Stoffen (Observations of shear-band development in plastic materials). In C.B. Bienzeno and J.M. Burgers (Eds.), *Proc. 1st Int. Conf. Appl. Mech.*, Delft: Waltman, 318–325.

Newman, J.B. (1979). Concrete under complex stress. In F.D. Lydon (Ed.), *Developments in Concrete Technology -1* 151–220. London, UK: Applied Science Publishers.

Newman, K. (1968). The structure and properties of concrete: An introductory review. In *Proc. Int'l. Conf on The Structure of Concrete*. London: Cement & Concrete Association, xiii–xxiii.

Nijenhuis, W. (1973). *De Verplaatsingsmethode: Toegepast voor de Berekening van (staaf) Constructies*, Agon-Elsevier, Amsterdam/Brussel (in Dutch).

Nilsson, S. (1961). The tensile strength of concrete determined by splitting tests on cubes. *RILEM Bull.*, 2(11): 63–67.

Nooru-Mohamed, M.B. (1992). *Mixed-Mode Fracture of Concrete: An Experimental Approach*. PhD thesis, Delft University of Technology.

Nooru-Mohamed, M.B., Schlangen, E., and Van Mier, J.G.M. (1993). Experimental and numerical study on the behavior of concrete subjected to biaxial tension and shear. *Adv. Cem. Based Mater.*, 1(1): 22–37.

Olesen, J.F., Ostergaard, L., and Stang, H. (2006). Nonlinear fracture mechanics and plasticity of the split cylinder test. *Mater. Struct. (RILEM)*, 39: 421–432.

Otsuka, K., Date, H., and Kurita, T. (1998). Fracture process zone in concrete tension specimens by X-ray and AE techniques. In H. Mihashi and K. Rokugo (Eds.)*Fracture Mechanics of Concrete Structures, Proceedings FraMCoS-3*. Freiburg: Aedificatio, 3–16.

Ottosen, N.S. (1984). Evaluation of concrete cylinder tests using finite elements. *J. Engng. Mech.*, 110(3): 465–481.

Paterson, M.S. (1978). *Experimental Rock Deformation: The Brittle Field*. New York: Springer Verlag.

Pellenq, R.J.-M. and Van Damme, H. (2004). Why does concrete set? The nature of cohesive forces in hardened cement-based materials. *MRS Bulletin*, May: 319–323.

Persson, S. and Östman, E. (1986). The use of computed tomography in non-destructive testing of polymeric materials, aluminium and concrete. Part I: Basic principles. *Polymer Test*. 6(6): 407–414.

Petersson, P.-E. (1980). Fracture energy of concrete: Method of determination, *Cem. Conc. Res.*, 10: 78–89.

Pignat, C., Navi, P., and Scrivener, K. (2005). Simulation of cement paste microstructure hydration, pore space characterization and permeability determination. *Mater. Struct. (RILEM)*, 38: 459–466.

Planas, J., Elices, M., and Guinea, G.V. (1992). Measurement of the fracture energy using three-point bend tests: Part 2: Influence of bulk energy dissipation. *Mater. Struct. (RILEM)*, 25(150): 305–312.

Planas, J., Guinea, G.V., Gálvez, J.C., Sanz, B., and Fathy, A.M. (2007). Indirect tests for stress-crack-opening curve. In J. Planas (Ed.), *RILEM TC 187-SOC Report Experimental Determination of the Stress-Crack Opening Curve for Concrete in Tension*, Report 39. Bagneux, France: RILEM S.A.R.L., Ch. 3, 13–27.

Prado, E.P. and Van Mier, J.G.M. (2003). Effect of particle structure on mode I fracture process in concrete. *Engng. Fract. Mech.*, 70(14): 1793–1807.

Pruijssers, A.F. (1988). *Aggregate Interlock and Dowel Action under Monotonic and Cyclic Loading*. PhD thesis, Delft University of Technology.

Rabinovich, Y.I., Esayanur, M.S., and Moudgil, B.M. (2005). Capillary forces between two spheres with a fixed volume liquid bridge: Theory and experiment. *Langmuir*, 21(24): 10992–10997.

Raiss, M.E., Dougill, J.W., and Newman, J.B. (1989). Observation of the development of fracture process zones in concrete. In B. Barr and S.E. Swartz (Eds.), *Fracture of Concrete and Rock: Recent Developments*. London: Elsevier Applied Science, 243–253.

Reinhardt, H.W. (1984), Fracture mechanics of an elastic softening material like concrete. *HERON*, 29(2): 1–37.

Reinhardt, H.W. Cornelissen, H.A.W. and Hordijk, D.A. (1989). Mixed-mode fracture tests on concrete. In S.P. Shah and S.E. Swartz (Eds.), *Fracture of Concrete and Rock*. New York: Springer Verlag, 117–130.

Reinhardt, H.W. and Naaman, A.E. (Eds.) (2007). *Proc. 5th Int'l. Conf. on High Performance Fibre Reinforced Cement Composites (HPFRCC-V)*. Bagneux, France: RILEM, 518 p.

Richardt, F.E., Brandtzaeg, A., and Brown, R.L. (1929). The behaviour of plain and spirally reinforced concrete in compression. *University of Illinois Engng. Exp. Station, Bulletin No. 190*.

Rieger, C. (2010). *Micro-Fiber Cement: Pullout Tests, Uniaxial Tensile Tests and Material Scaling*. PhD thesis #19396, ETH Zurich.

Rieger, C. and Van Mier, J.G.M. (2009). Pullout of microfibers from hardened cement paste. In G.P.A.G. van Zijl and W.P. Boshoff (Eds.), *Advances in Cement-Based Materials*. London: Taylor & Francis Group, 67–73.

Rieger, C. and Van Mier, J.G.M. (2010). Scaling of fracture properties of fibre reinforced cement. In B.-H. Oh, O.C. Choi, and L. Chung (Eds.) *Proc. 7th Int'l. Conf. on Fracture Mechanics of Concrete and Concrete Structures (FraMCoS-VII)*, May 23–28, Jeju, Korea, 216–223.

RILEM TC 50-FMC (1985). Determination of the fracture energy of mortar and concrete by means of three-point bend tests on notched beams (draft RILEM Recommendation). *Mater. Struct. (RILEM)*, 18(4):287–290.

RILEM TC 148-SSC (2000). Test method for measurement of the strain-softening behaviour of concrete under uniaxial compression: Test recommendation. *Mater. Struct. (RILEM)*, 33(230): 347–351.

Robert, L., Nazaret, F., Cutard, T., and Orteu, J.J. (2007). Use of 3-D digital image correlation to characterize the mechanical behavior of a fibre reinforced refractory castable. *Exp. Mech.*, 47: 761–773.

Rocco, C., Guinea, G.V., Planas, J., and Elices, M. (1999a). Mechanisms of rupture in the splitting test. *ACI Mater. J.*, 96(1): 52–60.

Rocco, C., Guinea, G.V., Planas, J., and Elices, M. (1999b). Size effect and boundary conditions in the Brazilian test: Experimental verification. *Mater. Struct. (RILEM)*, 32: 210–217.

Rocco, C., Guinea, G.V., Planas, J., and Elices, M. (1999c). Size effect and boundary conditions in the Brazilian test: Theoretical analysis. *Mater. Struct. (RILEM)*, 32: 437–444.

Roelfstra, P.E., Sadouki, H., and Wittmann, F.H. (1985). Le béton numérique. *Mater. Struct. (RILEM)*. 18(107): 327–335.

Rossi, P., Acker, P., and Mallier, Y. (1987). Effect of steel fibres at two stages: The material and the structure. *Mater. Struct (RILEM)*, 20: 436–439.

Rossi, P. and Renwez, S (1996). High performance multi-modal fibre reinforced cement composites (HPMFRCC). In *Proc. 4th In'l. Symp. on High Strength/High Performance Concrete*. Paris: RILEM, 687–694.

Rots, J.G. and De Borst, R. (1989). Analysis of concrete fracture in "direct" tension. *Int. J. Solids Struct.*, 25(12): 1381–1394.

Roux, S. and Guyon, E. (1985). Mechanical percolation: A small beam lattice study. *J. Physique Lett.*, 46: L999–L1004.

Roux, S., Hild, F., Voit, P., and Bernard, D. (2008). Three-dimensional image correlation from X-ray computed tomography of solid foam. *Comp. Part A: Appl. Sci. Manufact.*, 39(8): 1253–1265.

Sadouki, H. and Van Mier, J.G.M. (1996). Analysis of hygral-induced crack growth in multi-phase materials, *HERON*, 41(4): 267–286.

Samani, A.K. and Attard, M.M. (2010). A stress-strain model for uniaxial compression and triaxially confined concrete incorporating size effect. *University of New South Wales Report #R-457*, 46 p.

Sammis, C.G. and Ashby, M.F. (1986). The failure of brittle porous solids under compressive stress states. *Acta Metall.*, 34(3): 511–526.

Schärer, R. (2005), *Micromechanical Properties of Portland Cements*, MSc thesis, ETH Zurich, Department of Civil Engineering.

Schickert, G. (1980). Threshold values in compression experiments (Schwellenwerte beim Betondruckversuch). *Deutscher Auschuss für Stahlbeton*, Vol. 312, Berlin (in German).

Schickert, G. and Danssmann, J. (1984). Behaviour of concrete stressed at high hydrostatic compression. In *Proc. Int'l. Conf. on Concrete under Multiaxial Conditions*, Volume II. Toulouse: Presses de l'Université Paul Sabatier, 69–84.

Schickert, G. and Winkler, H. (1977). Strength and deformation of concrete under multiaxial compressive stress. *Deutscher Ausschuss für Stahlbeton*, Berlin, Vol. 277 (in German).

Schlangen, E. (1993). *Experimental and Numerical Analysis of Fracture Processes in Concrete*. PhD thesis, Delft University of Technology.

Schlangen, E. and Garboczi, E.J. (1996). New method for simulating fracture using an elastically uniform random geometry lattice. *Int. J. Engng. Sci.*, 34(10): 1131–1144.

Schlangen, E. and Garboczi, E.J. (1997). Fracture simulations of concrete using lattice models: Computational aspects. *Engng. Fract. Mech.* 57(2/3): 319–332.

Schlangen, E. and Van Mier, J.G.M. (1991). Lattice model for numerical simulation of concrete fracture. In V.E. Saouma, R. Dungar, and D. Morris (Eds.), *Proc. Int'l EPRI Conference on Dam Fracture*, Boulder, CO, September 11-13. Palo Alto, CA: Electric Power Research Institute, 511–527.

Schlangen, E. and Van Mier, J.G.M. (1992a). Experimental and numerical analysis of the micro-mechanisms of fracture of cement-based composites. *Cem. Conc. Comp.*, 14(2): 105-118.

Schlangen, E. and Van Mier, J.G.M. (1992b). Shear fracture in cementitious composites. Part II: Numerical simulations. In Z.P. Bažant (Ed.), *Proc. 1st Int. Conf. on Fracture Mechanics of Concrete Structures (FraMCoS-1)*, Breckenridge, CO, June 1–5. London/New York: Elsevier Applied Science, 671–676.

Schlangen, E. and Van Mier, J.G.M. (1992c). Lattice model for simulating fracture of concrete. In F.H. Wittmann (Ed.), *Proc. 1st Bolomey Workshop on Numerical Models in Fracture Mechanics of Concrete*. Rotterdam: Balkema, 195–205.

Schlangen, E. and Van Mier, J.G.M. (1994). Fracture simulations in concrete and rock using a random lattice. In H. Siriwardane and M.M. Zaman (Eds.), *Computer Methods and Advances in Geomechanics*. Rotterdam: Balkema, 1641–1646.

Schlangen, E. and Van Mier, J.G.M. (1995). Crack propagation in sandstone: A combined experimental and numerical approach. *Rock Mech. Rock Engng.*, 28(2): 93–110.

Schulson, E.M. and Gratz, E.T. (1999). The brittle compressive failure of orthotropic ice under triaxial loading. *Acta Mater.*, 47(3): 745–755.

Scrivener, K. (1989). The microstructure of concrete. In J. Skalny (Ed.), *Materials Science of Concrete: I*. Columbus, OH: American Ceramic Society, 127–161.

Sempere, J.-C. and Macdonald, K.C. (1986). Overlapping spreading centers: Implications from crack growth simulation by the displacement discontinuity method. *Tectonics*, 5: 151–163.

Shah, S.P. (1990). Experimental methods for determining fracture process zone and fracture parameters. *Engng. Fract. Mech.*, 35(1/2/3): 3–14.

Shi, C., Van Dam, A.G., Van Mier, J.G.M., and Sluys, L.J. (2000). Crack interaction in concrete. In F.H. Wittmann (Ed.), *Materials for Buildings and Structures: EUROMAT 99*, Volume 6. Weinheim, Germany: Wiley-VCH, 125–131.

Shi, C. and Van Mier, J.G.M. (2000). Mode I fracture in concrete specimens with off-centred notches. In G.C. Sih, (Ed.), *Proc. Meso-Mechanics 2000*, Beijing: Tsinghua University Press, 1111–1120.

Shiotani, T., Bisschop, J., and Van Mier, J.G.M. (2003). Temporal and spatial development of drying shrinkage cracking in cement-based materials. *Engng. Fract. Mech.*, 70(12): 1509–1525.

Shioya, T., Iguro, M., Nojiri, Y., Akiyama, H., and Okada, T. (1989). Shear strength of large reinforced concrete beams. *ACI Convention Seattle 1987*, SP-118: 259–279.

Stähli, P. (2008). *Ultra-Fluid, Oriented Hybrid-Fibre-Concrete*. PhD thesis #17996, ETH Zurich, 202p.

Stähli, P., Custer, R., and Van Mier, J.G.M. (2008). On flow properties, fibre distribution, fibre orientation and flexural behaviour of FRC, *Mater. Struct. (RILEM)*, 41: 189-196.

Stähli, P. and Van Mier, J.G.M. (2004). Three-fibre-type hybrid fibre concrete. In V.C. Li, C.K.Y. Leung, K.J. Willam, and S.L. Billington (Eds.), *Proc. 5th Int'l Conf. on Fracture of Concrete and Concrete Structures (FraMCoS-V)*; Vail, CO, April 12–16. Evanston, IL: IA-FraMCoS, 1105–1112.

Stampanoni, M., Borchert, G., Wyss, P., Abela, R., Patterson, B., Hunt, S., Vermeulen, D., and Ruegsegger, P. (2002). High resolution X-ray detector for synchrotron-based microtomography. *Nucl. Inst. Meth.* A 491(1–2): 291–301.

Stillinger, F.H. and Weber, T.A. (1985). Computer simulation of local order in condensed phases of silicon. *Phys. Rev. B*, 31(8): 5262–5271.

Stroeven, M. (1999). *Discrete Numerical Modelling of Composite Materials: Applications to Cementitious Materials.* PhD thesis, Delft University of Technology.

Stroeven, P. (1973). *Some Aspects of the Micromechanics of Concrete.* PhD thesis, Delft University of Technology.

Stroeven, P. (1979). Geometric probability approach to the examination of microcracking in plain concrete. *J. Mater. Sci.*, 14: 1141–1151.

Suresh, S. (1991). *Fatigue of Materials.* Cambridge, UK: Cambridge University Press.

Sutton, M.A., Wolters, W.J., Peters, W.J., Ranson, W.F., and McNeill, S.R. (1983). Determination of displacements using an improved digital correlation method, *Image Vision Comp.*, 1: 133–129.

Swamy, R.N. (Ed.) (1992). Fibre reinforced cement and concrete. *Proc. 4th Int'l. RILEM Symposium, Sheffield, July 20–23.* London/New York: E&FN Spon.

Swartz, S.E. and Taha, N.M. (1990). Mixed-mode crack propagation and fracture in concrete. *Engng. Fract. Mech.*, 35(1/2/3): 137–144.

Synnergren, P. and Sjödahl, M. (1999). A stereoscopic digital speckle photography system for 3-D displacement field measurements. *Optical Lasers Engng*, 31: 425–443.

Tada, H., Paris, P.C., and Irwin, G.R. (1973). *The Stress Analysis of Cracks Handbook.* Hellertown, PA: Del Research Corporation.

Tennis, P.D. and Jennings, H.M. (2000). A model for two types of calcium silicate hydrate in the microstructure of Portland cement pastes. *Cem. Conc. Res.*, 30: 855–863.

Termonia, Y. and Meakin, P. (1986). Formation of fractal cracks in a kinetic fracture model. *Nature*, 320: 429–431.

Thornton, C. and Antony, S.J. (2000). Quasi-static shear deformation of a soft particle system. *Powder Techn.* 109: 179–191.

Tijssens, M.G.A. (2001). *On the Cohesive Surface Methodology for Fracture of Brittle Heterogeneous Solids.* PhD thesis, Delft University of Technology.

Timoshenko, S.P. (1983). *History of Strength of Materials* (unabridged and unaltered republication of the 1953 edition). New York: Dover.

Timoshenko, S.P. and Goodier, J.N. (1970). *Theory of Elasticity*, 3rd ed. Tokyo: McGraw-Hill, Kogakusha, .

Topin, V., Delenne, J.-Y., Radjai, F., Brendel, L., and Mabille, F. (2007). Strength and failure of cemented granular matter. *Eur. Phys. J. E*, 23: 413–429.

Torrenti, J.M., Benaija, E.H., and Boulay, C. (1992). Strain localization in concrete in compression: The influence of boundary conditions. In Z.P. Bažant (Ed.), *Fracture Mechanics of Concrete Structures, Proc. FraMCoS-1.* London/New York, Elsevier Applied Science, 281–286.

Trtik, P., Stähli, P., Landis, E.N., Stampanoni, M., and Van Mier, J.G.M. (2007). Micro-tensile testing and 3D imaging of hydrated Portland cement. In A. Carpinteri, P. Gambarova, G. Ferro, and G. Plizzari (Eds.), *Proc. 6th Int'l. Conf. on Fracture Mechanics of Concrete and Concrete Structures (FraMCoS-VI)*. London: Taylor & Francis, 1277–1282.

Trtik, P., Van Mier, J.G.M., and Stampanoni, M. (2005). Three dimensional crack detection in hardened cement pastes using synchrotron-based computer micro-tomography (SRμCT). In: *Proc. 11th Int'l. Conf. of Fracture (ICF-11), Symposium 34 Physics and Scaling in Fracture*, Torino, Italy, March 20–25 (CD-ROM).

Tschegg, E.K. and Linsbauer, H.N. (1986). Prüfeinrichtung zur Ermittlung von Bruchmechanische Kennwerten (Testing procedure for determination of fracture mechanics parameters). *Patentschrift No. A-233/86*, Ostereichisches Patentamt (in German).

Underwood, E.E. (1970). *Quantitative Stereology*. Reading, MA: Addison Wesley.

Van Geel, H.J.G.M. (1998). *Concrete Behaviour in Multiaxial Compression*, PhD thesis, Eindhoven University of Technology.

Van Gunsteren, E. (2003). *Development of a High-Performance Hybrid-Fibre Concrete with Reference to Fracture under Tension and Shear*. MSc thesis, ETH Zurich.

Van Mier, J.G.M. (1984). *Strain-Softening of Concrete under Multiaxial Loading Conditions*, PhD thesis, Eindhoven University of Technology.

Van Mier, J.G.M. (1985). Influence of damage orientation distribution on the multi-axial stress-strain behaviour of concrete. *Cem. Conc. Res.*, 15(5): 849–862.

Van Mier, J.G.M. (1986a). Multiaxial strain-softening of concrete. *Mater. Struct. (RILEM)*, 19(3): 179–200.

Van Mier, J.G.M. (1986b). Fracture of concrete under complex stress. *HERON*, 31(3): 1–90.

Van Mier, J.G.M. (Ed.) (1987). Examples of non-linear analysis of reinforced concrete structures with DIANA. *HERON*, 32(3):1–147.

Van Mier, J.G.M. (1991a). Mode I fracture of concrete: Discontinuous crack growth and crack interface grain bridging. *Cem. Conc. Res.*, 21(1): 1–15.

Van Mier, J.G.M. (1991b). Crack face bridging in normal, high strength and lytag con-crete. In J.G.M. van Mier, J.G. Rots, and A. Bakker (Eds.), *Fracture Processes in Concrete, Rock and Ceramics*, London/New York: Chapman & Hall, 27–40.

Van Mier, J.G.M. (1997). *Fracture Processes of Concrete: Assessment of Material Parameters for Fracture Models*. Boca Raton, FL: CRC Press.

Van Mier, J.G.M. (1998). Failure of concrete under uniaxial compression: An overview. In H. Mihashi and K. Rokugo (Eds.), *Fracture Mechanics of Concrete Structures: Proceedings FraMCoS-3*. Freiburg: AEDIFICATIO, 1169–1182.

Van Mier, J.G.M. (2000). *De Kunst van Breken and Scheuren (The Art of Fracturing)*. Inaugural Lecture, Delft University of Technology (in Dutch).

Van Mier, J.G.M. (2004a). Reality behind fictitious cracks? In V.C. Li, C.K.Y. Leung, K.J. Willam and S.L. Billington (Eds.), *Proc. 5th International Conference on Fracture of Concrete and Concrete Structures (FraMCoS-V)*, Vail, CO, April 12–16. Evanston, IL: IA-FraMCoS,11–30.

Van Mier, J.G.M. (2004b). Lattice modelling of size effect in concrete strength. Discussion on a paper by R. Ince, A. Arslan, and B.L. Karihaloo, *Engng. Fract. Mech.*, 71(11): 1625–1628.

Van Mier, J.G.M. (2004c). Cementitious composites with high tensile strength and ductility through hybrid fibres. In M. Di Prisco, R. Felicetti, and G. Plizzari (Eds.), *Proc. 6th RILEM Symposium on Fibre Reinforced Concrete (BEFIB 2004)*. Bagneux, France: RILEM, 219–236.

Van Mier, J.G.M. (2007). Multi-scale interaction potentials (*F-r*) for describing fracture of brittle disordered materials like cement and concrete. *Int. J. Fract.*, 143(1): 41–78.

Van Mier, J.G.M. (2008). Framework for a generalized four-stage fracture model of cement-based materials. *Engng. Fract. Mech.*, 75: 5072–5086.

Van Mier, J.G.M. (2009). Mode II fracture localization in concrete loaded in compression. *J. Eng. Mech. (ASCE)*, 135(1): 1–8.

Van Mier, J.G.M. and Lilliu, G. (2001). Three-dimensional lattice model for fracture analysis of particle composites. In N. Bicanic (Ed.), *Proc. 4th Int'l. Conference on Analysis of Discontinuous Deformation, ICADD-4*, University of Glasgow, June 6–8, 121–133.

Van Mier, J.G.M. and Lilliu, G. (2002). Shear fracture in concrete? In A.V. Dyskin, X. Hu, and E. Sahouryeh (Eds.), *Structural Integrity and Fracture*. Lisse: Swets & Zeitlinger, 333–342.

Van Mier, J.G.M. and Man, H.-K. (2009). Some notes on microcracking, softening, localization and size effects. *Int. J. Damage Mech.*, 18: 283–309.

Van Mier, J.G.M. and Nooru-Mohamed, M.B. (1989). Fracture of concrete under tensile and shearlike loadings. In H. Mihashi, H. Takahashi, and F.H. Wittmann (Eds.), *Fracture Toughness and Fracture Energy: Test Methods for Concrete and Rock*, Rotterdam: AA Balkema, 549–563.

Van Mier, J.G.M., and Nooru-Mohamed, M.B. (1990). Geometrical and structural aspects of concrete fracture. *Engng. Fract. Mech.*, 35(4/5): 617–628.

Van Mier, J.G.M. and Schlangen, E. (1989). On the stability of softening systems. In B. Barr and S.E. Swartz (Eds.), *Fracture of Concrete and Rock: Recent Developments*. London: Elsevier Applied Science, 387–396.

Van Mier, J.G.M., Schlangen, E., and Nooru-Mohamed, M.B. (1992). Shear fracture in cementitious composites. Part I: Experimental observations. In Z.P. Bažant (Ed.), *Proc.1st Int. Conf. on Fracture Mechanics of Concrete Structures (FraMCoS-1)*, Breckenridge, CO, June 1–5. London/New York: Elsevier Applied Science, 659–670.

Van Mier, J.G.M., Schlangen, E., and Vervuurt, A. (1995). Lattice type fracture models for concrete. In H.-B. Mühlhaus (Ed.), *Continuum Models for Materials with Microstructure*, Chichester, UK: John Wiley & Sons, 341–377.

Van Mier, J.G.M., Shah, S.P. et al. (1997), Strain-softening of concrete under uniaxial compression, *Mater. Struct. (RILEM)*, 30(4): 195–209.

Van Mier, J.G.M. and Shi, C. (2002). Stability issues in uniaxial tensile tests on brittle disordered materials. *Int. J. Solids Struct.*, 39: 3359–3372.

Van Mier, J.G.M. and Timmers, G. (1992). Shear fracture in slurry infiltrated fibre concrete (SIFCON). In H.W. Reinhardt and A.E. Naaman (Eds.), *Proc. RILEM/ACI Workshop on High Performance Fibre Reinforced Cement Composites* Mainz, June 24–26, 1991. London/New York: E&FN Spon/Chapman & Hall, 348–360.

Van Mier, J.G.M., Van Vliet, M.R.A., and Wang, T.K. (2002). Fracture mechanisms in particle composites: Statistical aspects in lattice type analysis. *Mech. Mater.*, 34: 705–724.

Van Mier, J.G.M. and Van Vliet, M.R.A. (2003). Influence of microstructure of concrete on size/scale effects in tensile fracture. *Eng. Fract. Mech.*, 70(16): 2281–2306.

Van Mier, J.G.M. and Vonk, R.A. (1991). Fracture of concrete under multiaxial stress: Recent developments. *Mater. Struct. (RILEM)*, 24(1): 61–65.

Van Vliet, M.R.A. (2000). *Size Effect in Tensile Fracture of Concrete and Rock*, PhD thesis, Delft University of Technology.

Van Vliet, M.R.A. and Van Mier, J.G.M. (1996). Experimental investigation of concrete fracture under uniaxial compression. *Mech. Coh.-frict. Mater.*, 1(1): 115–127.

Van Vliet, M.R.A. and Van Mier, J.G.M. (1998). Experimental analysis of size/scale effects in brittle disordered materials: Concrete. In R. de Borst and E. van der Giessen (Eds.), *Proc. IUTAM Symposium on Material Instabilities in Solids*. Delft, June 9–13, 1997. Chichester: John Wiley & Sons, 185–206.

Van Vliet, M.R.A. and Van Mier, J.G.M. (1999). Effect of strain gradients on the size effect of concrete in uniaxial tension. *Int. J. Fract.*, 95(1/4): 195–219.

Van Vliet, M.R.A. and Van Mier, J.G.M. (2000). Experimental investigation of size effect in concrete and sandstone under uniaxial tension. *Eng. Fract. Mech.*, 65(2/3): 165–188.

Vervuurt, A. (1997), *Interface Fracture in Concrete*, PhD thesis, Delft University of Technology.

Vervuurt, A., Schlangen, E., and Van Mier, J.G.M. (1993a). Numerical study of pull-out of anchors embedded in concrete. In H.P. Rossmanith (Ed.), *Fracture and Damage of Concrete and Rock (FDCR-2)*, London/New York: Chapman & Hall/E&FN Spon, 569–578.

Vervuurt, A., Schlangen, E., and Van Mier, J.G.M. (1993b). A numerical and experimental analysis of the pull-out behaviour of steel anchors embedded in concrete. *TU-Delft Report no. 25.5-93-1*.

Vervuurt, A. and Van Mier, J.G.M. (1995). Interface fracture in cement-based materials. In F.H. Wittmann (Ed.), *Proc. 2nd Int'l. Conf. on Fracture Mechanics of Concrete and Concrete Structures (FraMCoS-2)*. Freiburg: AEDIFICATIO, 295–304.

Vervuurt, A., Van Mier, J.G.M., and Schlangen, E. (1994). Analysis of anchor pull-out in concrete. *Mater. Struct. (RILEM)*, 27(169): 251–259.

Vervuurt, A., Van Vliet, M.R.A., Van Mier, J.G.M., and Schlangen, E. (1995). Simulations of tensile fracture in concrete. In F.H. Wittmann (Ed.), *Proc. 2nd Int'l. Conf. on Fracture Mechanics of Concrete and Concrete Structures (FraMCoS-2)*. Freiburg: AEDIFICATIO, 353–362.

Vile, G.W.D. (1968). The strength of concrete under short-term static biaxial stress. In *Proc. Int'l. Conf. on The Structure of Concrete*. London: Cement & Concrete Association, 275–288.

Visser, J.H.M. (1998). *Hydraulically Driven Fracture of Concrete and Rock*. PhD thesis, Delft University of Technology.

Visser, J.H.M. and Van Mier, J.G.M. (1994). Hydraulic fracture in the tensile regime. In H.J. Siriwardane and M.M. Zaman (Eds.), *Computer Methods and Advances in Geomechanics*. Rotterdam: Balkema, 1647–1652.

Vonk, R.A. (1992). *Softening of Concrete Loaded in Compression*. PhD thesis, Eindhoven University of Technology.

Vonk, R.A., Rutten, H.S., Van Mier, J.G.M., and Fijneman, H.J. (1989). Influence of boundary conditions on softening of concrete loaded in compression. In S.P. Shah, S.E. Swartz, and B. Barr (Eds.), *Fracture of Concrete and Rock: Recent Developments*. Elsevier Applied Science, London/New York, 711–720.

Vořechovský, M. (2007). Interplay of size effects in concrete specimens under tension studied via computational stochastic fracture mechanics. *Int. J. Solids Struct.*, 44: 2715–2731.

Walsh, P.F. (1972). Fracture of plain concrete. *Indian Conc. J.*, 46(11): 469–470.

Wang, J. (1994). *Development and Application of a Micromechanics-Based Numerical Approach for the Study of Crack Propagation in Concrete*, PhD thesis, EPFL Lausanne.

Weibull, W. (1939). A statistical theory of strength of materials. *Roy. Swedish Inst. Engrg. Res.*, 151:1–45.

Weibull, W. (1951). A statistical distribution function of wide applicability. *J. Appl. Mech.*, 18: 293–298.

Willam, K.J. and Warnke, E.P. (1974). Constitutive model for the triaxial behaviour of concrete. In *Proc. Coll. On Concrete Structures Subjected to Triaxial Stresses, IABSE Report 19*. Zurich: IABSE.

Wischers, G. (1978). Aufnahme und Auswirkungen von Druckbeanspruchungen auf Beton. *Betontechnische Berichte*, 19:31–56 (in German).

Wissing, B. (1988). *Acoustic Emission of Concrete*, MSc thesis, Delft University of Technology, Dept. of Civil Eng. (in Dutch).

Wittmann, F.H. (1978). The cause and technological importance of capillary shrinkage of concrete. *Betonwerk+Fertigteil Techn.*, 5: 272–276 (in German).

Wittmann, F.H., Roelfstra, P.E., Mihashi, H., Huang, Y.-Y., Zhang, X.-H., and Nomura, N. (1987). Influence of age of loading, water-cement ratio and rate of loading on fracture energy of concrete, *Mater. Struct. (RILEM)*, 20(116): 103–110.

Wittmann, F.H., Sadouki, H., and Steiger, T. (1993). Experimental and numerical study of effective properties of composite materials. In C. Huet (Ed.), *Micromechanics of Concrete and Cementitious Composites*. Lausanne: Presses Polytechniques et Universitaires Romandes, 59–80.

Xu, D. and Reinhardt, H.W. (1989). Softening of concrete under torsional loading. In B. Barr, S.E. Swartz, and S.P. Shah (Eds.), *Fracture of Concrete and Rock: Recent Developments*. London/New York: Elsevier Applied Science, 39–50.

Yacoub-Tokatly, Z., Barr, B., and Norris, P. (1989). Mode II fracture: A tentative geometry. In B. Barr, S.E. Swartz, and S.P. Shah (Eds.), *Fracture of Concrete and Rock: Recent Developments*. London/New York: Elsevier Applied Science, 596–604.

Ye, G. (2003). *The Microstructure and Permeability of Cementitious Materials*. PhD thesis, Delft University of Technology.

Yin, W.S., Su, E.C.M., Mansur, M.A., and Hsu, T.T.C. (1990). Fibre-reinforced concrete under biaxial compression. *Engng. Fract. Mech.*, 35(1/2/3): 261–268.

Zech, B. and Wittmann, F.H. (1978). A complex study on the reliability assessment of the containment of a PWR. Part II Probabilistic approach to describe the behaviour of materials. *Nucl. Engng. Des.*, 48: 575–584.

Zhang, M.H. and Gjørv, O.E. (1990). Microstructure of the interfacial zone between lightweight aggregate and cement paste. *Cem. Conc. Res.*, 20: 610–618.

Zhou, F.P. (1988). Some aspects of tensile fracture behaviour and structural response of cementitious materials. *Report TVBM 1008*, Lund Institute of Technology.

Zimbelmann, R. (1985). A contribution to the problem of cement-aggregate bond. *Cem. Conc. Res.*, 15(5): 801–808.

Index

Milton Keynes UK
Ingram Content Group UK Ltd.
UKHW031141141024
449569UK00024B/1159